Climate Risks, Regional Integration, and Sustainability in the Mekong Region

The **Strategic Information and Research Development Centre (SIRD)** is an independent publishing house founded in January 2000 in Petaling Jaya, Malaysia. The SIRD list focuses on Malaysian and Southeast Asian studies, economics, gender studies, social sciences, politics and international relations. Our books address the scholarly community, students, the NGO and development communities, policymakers, activists and the wider public. SIRD also distributes titles (via its sister organization, **GB Gerakbudaya Enterprise Sdn Bhd**) published by scholarly and institutional presses, NGOs and other independent publishers. We also organize seminars, forums and group discussions. All this, we believe, is conducive to the development and consolidation of the notions of civil liberty and democracy.

CLIMATE RISKS, REGIONAL INTEGRATION, AND SUSTAINABILITY IN THE MEKONG REGION

Edited by

Louis Lebel (USER),
Chu Thai Hoanh (IWMI),
Chayanis Krittasudthacheewa (SEI),
Rajesh Daniel (SEI)

SUMERNET

SEI STOCKHOLM ENVIRONMENT INSTITUTE

SIRD

Strategic Information and Research Development Centre
Petaling Jaya, Malaysia

First published in 2014 by
Strategic Information and Research Development Centre
No. 11, Lorong 11/4E, 46200 Petaling Jaya, Selangor, Malaysia.
Email: gerak@gerakbudaya.com Website: www.gerakbudaya.com
and
SUMERNET Secretariat
Stockholm Environment Institute (SEI) – Asia
15th Floor, Witthyakit Building, 254 Chulalongkorn University
Chulalongkorn Soi 64, Phyathai Road, Pathumwan, Bangkok 10330, Thailand.
Email: secretariat@sumernet.org Website: www.sumernet.org

Supported by

SWEDEN

Climate & Development Knowledge Network

CDKN is funded by

UKaid from the British people

Ministry of Foreign Affairs of the Netherlands

Perpustakaan Negara Malaysia / Cataloguing-in-Publication Data

Climate Risks, Regional Integration and Sustainability in the Mekong Region /
Edited by Louis Lebel, Chu Thai Hoanh, Chayanis Krittasudthacheewa, and Rajesh Daniel
Includes index
Bibliography: page 356
ISBN 978-967-0630-25-0 (pbk)
ISBN 978-967-0630-33-5 (hbk)
1. Sustainable agriculture – Mekong River. 2. Climatic changes – Mekong River.
3. Agriculture – Environmental aspects – Mekong River. I. Lebel, Louis.
II. Chu, Thai Hoanh. III. Krittasudthacheewa, Chayanis. IV. Daniel, Rajesh.
338.10959

Cover image: Bridge across the Mekong River, Lao PDR (Roengchai Kongmuang for SEI – Asia)
Project editor: Dayaneetha De Silva, Saga Seed Books
Cover design and layout: Janice Cheong

Printed by Vinlin Press Sdn Bhd
2, Jalan Meranti Permai 1, Jalan Puchong, 47100 Puchong, Selangor, Malaysia.

Contents

Abbreviations

ADB	Asian Development Bank
ADEME	Agence de l'Environnement et de la Maîtrise de l'Energie
APAEC	ASEAN Plan of Action for Energy Cooperation
ASEAN	Association of Southeast Asian Nations
CAFTA	China–ASEAN Free Trade Area
CDKN	Climate and Development Knowledge Network
CGCM3	Coupled Global Climate Model
CFSC	Committee for Flood and Storm Control
DAFO	District Agriculture and Forestry Office
DFID	Department for International Development
FAO	United Nations Food and Agriculture Organization
FCPF	Forest Carbon Partnership Facility
FDI	foreign direct investment
FSIV	Forest Science Institute of Vietnam
FTA	free trade agreement
GCM	global circulation model
GHG	greenhouse gases
GMS	Greater Mekong Subregion
ICRAF	World Agroforestry Centre
IFReDI	Inland Fisheries Research and Development Institute
IPCC	Intergovernmental Panel on Climate Change
IUCN	World Conservation Union
LARReC	Living Aquatic Resources Research Centre
LMB	Lower Mekong Basin
LVI	Livelihood Vulnerability Index
MDG	Millennium Development Goals
MA	Millennium Ecosystem Assessment
MRA	multiple regression analysis
MRC	Mekong River Commission

MRV	Measurement, Reporting and Verification
NMT	non-motorized transport
NUOL	National University of Laos
NTFP	non-timber forest products
OCSB	Office of the Cane and Sugar Board
OECD	Organization for Economic Cooperation and Development
PAFO	Provincial Agriculture and Forestry Office
PFES	Payment for Forest Environmental Services
PSROI	Participatory Social Returns On Investment
RCWG	risk communication working group
REDD	Reducing Emissions from Deforestation and Forest Degradation in Developing Countries
SLF	Sustainable Livelihood Framework
SUMERNET	Sustainable Mekong Research Network
TEV	Total Economic Value
TSP	tourism service provider
UNEP	United Nations Environment Program
VFU	Vietnam Forestry University
WRI	World Resources Institute

Measurements

cm	centimeter
GW	gigawatt
kg	kilogram
km	kilometer
ha	hectare
m^3	cubic meter
mm	millimeter
mt	metric tonne
tCO_2-eq	tonnes of carbon dioxide equivalent
$GtCO_2$-eq	gigatonnes of carbon dioxide equivalent

Acknowledgements

It would not have been possible to share new knowledge and insights for the policies highlighted in this book with the wider public without the great collective efforts of SUMERNET partners who have contributed as chapter authors, research mentors, and internal and external reviewers and provided valuable suggestions for improving the chapters. We thank the SUMERNET Secretariat team who provided various kinds of support to the implementation of the collaborative studies and preparation of this book.

This book was primarily funded by the Swedish Government through the Swedish International Development Cooperation Agency (SIDA), by the UK Department for International Development (DFID), and the Netherlands Directorate for International Cooperation (DGIS) through the Climate and Development Knowledge Network (CDKN), and delivered through the SUMERNET program. Their continuing support on bridging science and policies related to sustainable development for the benefit of developing countries is gratefully acknowledged.

We are grateful to the reviewers of the chapters in this book for their comments and advice: Tim Boyle, Jake Brunner, Shabbir Gheewala, Eric Kemp-Benedict, Kanokwan Manorom, Carl Middleton, Beatrice Mosello, Oraphan Nabangchang, Babette P. Resurrección, Albert Salamanca, and Kien Tran Mai.

Foreword

It is a pleasure to introduce this book, the major output of the second phase of the Sustainable Mekong Research Network (SUMERNET). Since its start in 2005, the SUMERNET program has aimed to strengthen the knowledge-base for policy processes in pursuit of a sustainable Mekong region. With support from the Government of Sweden and other donors, the program provides grants and practical support to regional research teams who carry out original research and engage directly as well as through boundary partners with policy and planning at multiple levels. In keeping with SUMERNET's regional focus, grantees form consortia that include institutions from at least two and often three or four countries.

As Asia Centre Director of the Stockholm Environment Institute (SEI), I have particular interest in the program and this book because SEI's Asia Centre has hosted the SUMERNET Secretariat since its inception, and will continue to host it in Phase 3, which has just begun. The SUMERNET program's strategy and goals are directly aligned with SEI's mission to bridge science and policy for sustainable development. SEI, like SUMERNET, aims to inform policy processes through its research and assessment activities. Given its importance to SEI's mission, the SUMERNET program is a keystone of SEI's work in the Mekong region.

As explained well in the introduction, and as demonstrated throughout the book's in-depth case studies, the Mekong region is complex and dynamic, but its people, ecosystems, and livelihoods are also particularly vulnerable to climate change. As cross-border connections strengthen in the course of regional integration, some vulnerabilities are lessened, while others increase. The rising volume of cross-border flows of people, goods, money, and information also puts pressure on ecosystems and natural resources, creating challenges for long-term sustainability. The intersection of climate risk, regional integration, and sustainability is taken up throughout the book. Section I contains four "think pieces" that explore central policy issues: energy and climate; ecosystem services; transboundary flows; and urbanization. Section II then goes into detail by documenting research carried out under the SUMERNET program. Each chapter within Section II presents one or more case studies that put the general comments from Section I into local context, a set of "place-based lessons" that are summarized in Section III, the synthesis.

Rising climate risk, increasing connections between countries, and rising pressures on ecosystems are not just issues for the Mekong region; they are defining trends for the world of today. Recent reports from the IPCC and the World Bank suggest that global temperatures could rise by four degrees centigrade over pre-industrial levels by the end of the century.

This book, one of the SUMERNET Book Series, is therefore relevant beyond the Mekong itself. It offers insights and methods that apply to other parts of the world as we work to understand – and respond to – the challenges and opportunities for long-term sustainability arising from climate change and international integration.

Eric Kemp-Benedict
Centre Director
SEI – Asia

1

Introduction: Pursuing Sustainability in the Mekong Region

Chayanis Krittasudthacheewa, Louis Lebel, and Chu Thai Hoanh

The Mekong and its changes

The Mekong region is not only rich in natural biodiversity and culturally diverse but also has one of the fastest growing regional economies in the world. Returns from economic growth have raised incomes and improved people's well-being, but many social and economic challenges remain. It has proven difficult to effectively integrate social, economic, and environmental objectives in pursuing sustainability in the region. Rapid change and the regional interdependence across the Mekong countries and China's Yunnan province has diverse consequences (SEI 2009). Looking back, there is both credit and blame; looking forward, both opportunities and threats.

A major challenge in the region is how to improve the livelihoods of those who depend heavily on natural resources. A combination of rapid economic development, demographic shifts, and rising living standards is posing a new set of challenges to meeting increased food and energy demands. Even as economies industrialize and service sectors expand, agriculture and fisheries remain fundamental to human well-being across the region (Be et al. 2008). Unfortunately, the expansion and intensification of crop production have been accompanied by land degradation, which now affects 10–40 percent of the land area in each Mekong country (IWMI-

SEA 2009). Deforestation and a decline in water quantity and quality are also important environmental issues. The rich biodiversity in the region has already been greatly affected by land-use changes and remains vulnerable to climate change.

Demand for energy is projected to grow between 7 percent and 16 percent per annum (ADB 2007). This increase is faster than the expected rate of economic growth, placing great stress on existing energy systems and also on the region's ecosystems. At the same time, these energy demands mask great disparities in the use of energy. Fast urbanization and motorization is a driver; at the same time 50 of the 260 million people in the Mekong region still have no electricity and must rely on traditional energy sources—fuelwood, charcoal, and farm residues (ADB 2007). Lack of access to sustainable and clean energy remains a significant cause of poverty in the region.

A range of socioeconomic mechanisms (trade, migration, demand for goods and services, unemployment) as well as natural ones (river flow, wind) transmit pressure from one country to another. Pressure on forests, fisheries, land, energy, and mineral resources has come, in part, from cross-border demand. Such demand has also driven foreign investment and both legal and illegal trade. International migration has also contributed significantly to the integration of economies. The movement of goods has been supported by connecting infrastructure. These linkages and complexities redefine the policymaking challenges for pursuing regional sustainable development (Kaosa-ard and Dore 2003). Despite signs of ever-expanding regional economic integration, some knowledge gaps still remain about the real extent of intra-region linkages, the consequences of integration, and the causes of inequitable development and losses of environmental services (Kummu et al. 2008).

Embedded within many of these integration efforts are conflicts over the use and management of natural resources. For instance, conflicts over water—both within and between countries—are intensifying from escalating industrial and agricultural demands for water and energy, interference with natural river flows from large hydropower dams, river-linking and diversion schemes, and altered sediment and nutrient loads affecting river ecosystems (Molle et al. 2009). Likewise, land for growing food and making a living is increasingly contested—especially where large investors are able to obtain vast land concessions at the expense of residents (Schönweger et al. 2012). In the Mekong region, rapid urbanization is another critical process of change, especially, in the dynamic peri-urban areas where opportunities and challenges for sustainability are often in sharpest relief.

Why this book?

The literature on the Mekong region is growing rapidly. This volume brings together a new set of multi-country empirical case studies that contribute to this growing understanding of the complexities of resource management and governance in the Mekong region. In addition it identifies several significant lessons for engaging with policy and planning processes in the region in ways that support the pursuit of regional sustainability through the research efforts of a regional network, the Sustainable Mekong Research Network (SUMERNET).

Addressing sustainable development challenges

The Sustainable Mekong Research Network was set up to inform and influence policies relevant to the sustainable development of the Mekong region through collaborative research and assessment. Launched in 2005 with support from the Swedish government, SUMERNET grew from 14 founding member institutes at the beginning to include 47 member institutes in 2013 with additional funding support from the Climate and Development Knowledge Network (CDKN). The SUMERNET Secretariat has been hosted at the Asia Centre of the Stockholm Environment Institute (SEI – Asia) in Bangkok since it begun.

Building on past research collaboration the network constructively engages with policy processes at multiple levels—local through regional. Fostering, supporting, and sustaining high quality and policy-relevant research is at the heart of SUMERNET.

From the wide vista of sustainable development challenges in the region, SUMERNET identified the following research themes as its focus for its second phase (2010–2013) that have contributed to the studies in this volume:

Ecosystem services: How are ecosystem services being used, conserved, and governed?

Transboundary issues: What are the long-term impacts of regional economic integration on transboundary issues?

Poverty and livelihoods: How are livelihoods of the poor impacted by changes in ecosystem services and transboundary issues?

Urbanization: What have been the consequences of urbanization for society and the environment and how can regional and urban planning be redirected to support sustainability?

Energy and climate change: How can equitable and sustainable energy be pursued in the context of a changing climate?

This volume documents some of the work from the 10 collaborative studies granted from a wider set of more than 50 concept notes submitted through three competitive calls for demand-driven, policy-relevant research that included a multi-country and participatory design (see Fig. 1.1 SUMERNET study sites).

The recipient of grants in SUMERNET Phase 2 (2010–2013) were provided with mentoring throughout the study period and participated in several capacity-building workshops to help with writing, communication, and policy engagement. Another important feature is that all studies explicitly identified 'boundary partners' in their proposals which then became very important actors for policy engagement. Boundary partners, as originally identified by IDRC (2001), are those individuals, groups, or organizations with whom the program interacts directly and with whom the program can anticipate opportunities for influence.

These studies involved more than 100 researchers and about 30 national teams. Some of the insights and knowledge generated from the implementation of these projects in the past two years are shared in ten chapters of Part II of this volume.

Organization of this book

This book contains 17 chapters, including this Introduction, which are organized in three sections.

Part I contains four chapters produced from a series of 'policy think pieces' on key sustainable development issues in the Mekong region identified by a wider group from the SUMERNET network, i.e. Energy, economy, and climate change in the Mekong region (Chapter 2); Valuing ecosystem services in the Mekong region (Chapter 3); Transboundary flows of resources, people, goods, and services in the Mekong region (Chapter 4); and Urbanization and sustainable development in the Mekong region (Chapter 5).

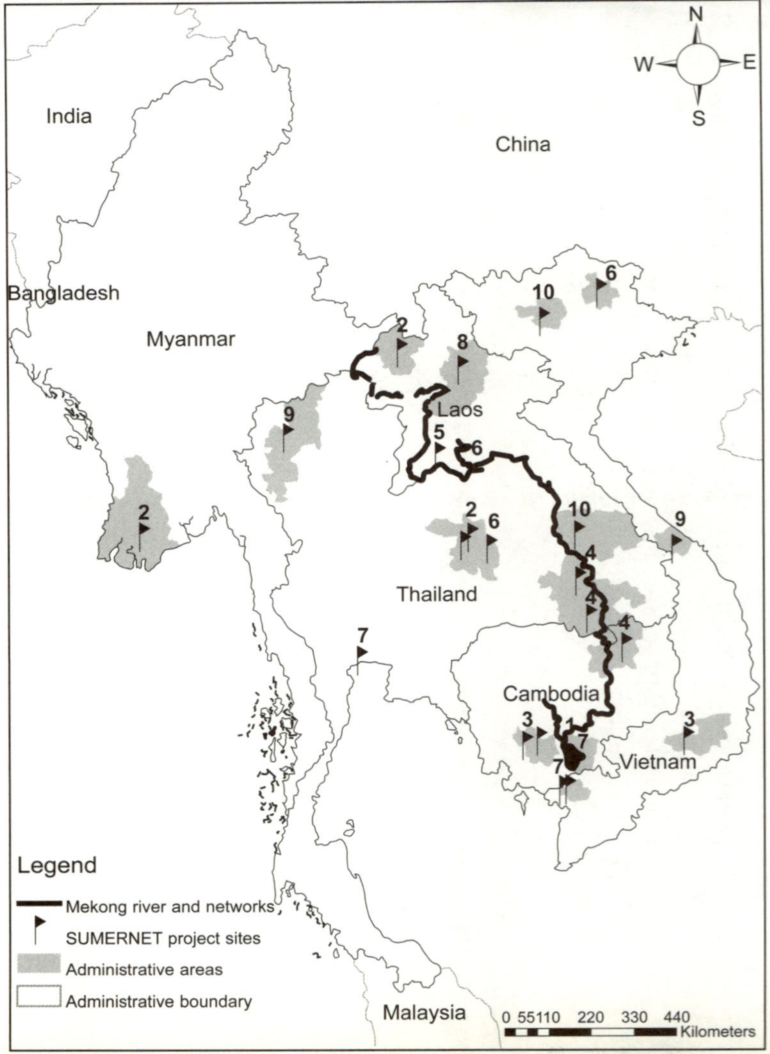

Fig. 1.1 SUMERNET study sites

1 Vulnerability Assessment of Livelihoods in the Lower Mekong Basin (LMB): Adaptation Options for Enhancing Capacity of People Living in the Most Vulnerable Flood-prone Areas in Cambodia and Vietnam
2 Making Economic Integration Work for the Rural Poor through Contract Farming Practices
3 Evaluation of Pilot Program on Payment for Forest Environmental Services in Vietnam and Policy Implications for Laos and Cambodia
4 Transboundary Fish Trade in the LMB: Impacts on Fisheries and Rural Employment in Cambodia, Lao PDR and Thailand
5 Impact of Urban Expansion on the Hinterland and Local Responses in the Mekong Region: A Study in Khon Kaen, Thailand, and Vang Vieng, Lao PDR
6 Research on Integrating Community-based Participatory Carbon Measurement and Monitoring with Satellite Remote Sensing and GIS in a Measurement, Reporting and Verification (MRV) System for Reducing Emissions from Deforestation and Forest Degradation-Plus (REDD+) and Agroforestry Carbon Sequestration Activities
7 Communicating Water-related Climate Change Risks to Improve Local Adaptation in the Deltas of the Mekong Region
8 Climate Change Implications to Food Security and Livelihoods of Small-scale Farmers
9 Sustainable Urban Tourism through Low Carbon Initiatives: Experiences from Hue and Chiang Mai
10 Participatory Social Returns on Investment (PSROI): Greater Mekong Basin Climate Change Adaptation Planning and Costing Project

Part II includes eleven chapters resulting from the cross-border collaborative studies mentioned above, which were guided by the four think pieces in the previous section and implemented by almost 30 national teams from Cambodia, China, Lao PDR, Myanmar, Thailand, Vietnam, and the Philippines.

The final section, Part III: Synthesis, is a concluding chapter (Chapter 17), emphasizing that the pursuit of sustainability takes place at multiple levels. It synthesizes the findings from all chapters to derive three key messages for the book.

Key messages

1. To understand drivers, opportunities, and constraints on sustainability at multiple levels in the Mekong region, it is important to take into account the linkages to higher and lower levels.

2. Insufficient progress on regional sustainability could undermine the climate resilience of societies in the Mekong region.

3. The impacts of economic integration, urbanization, and climate change vary greatly among places and communities. Development gains and environmental impacts are uneven among places and social groups. All of these underline the urgent need for policies and plans that put sustainability as a core objective and measurable target.

I

Policy Think Pieces

2

Energy, Economy, and Climate Change in the Mekong Region

Lailai Li and Tatirose Vijitpan

This chapter aims to analyze the main challenges to sustainable development in the Mekong region, with a focus on the linkages between energy and climate change. Using a systems approach, we review the strategies and policies adopted by the countries in the region in response to these changes, identify policy gaps, and discuss the opportunities and policy options needed to achieve desired goals and objectives.

The Mekong region—Cambodia, Lao PDR (Laos), Myanmar, Thailand, Yunnan Province (China), and Vietnam—is beset by self-reinforcing poverty cycles, characterized by several pairs of paradoxical conditions: low income vs fast economic growth; poor access to energy vs rapidly increasing energy demands; extremely low carbon dioxide (CO_2) emission levels vs extremely fast growth of emissions; and areas/populations with the least responsibility for climate change being the most affected and vulnerable to its effects. In a systems approach, these challenges can be seen to be forming positive and negative feedback loops, which in the absence of proper intervention, could drive systemic collapse as positive feedback loops drive growth, explosion, erosion, and collapse in systems. A system with an unchecked positive loop ultimately will destroy itself (Meadows et al. 2004).

Poverty and fast economic growth

Economic poverty

In monetary terms, poverty is measured by the proportion of a population whose income or consumption falls below an objectively defined level considered necessary to meet per capita minimum nutritional requirements, i.e. the poverty line. By commonly used World Bank standards, US$1–$2 a day constitutes the poverty line; by this reckoning 70 percent of the population of the Mekong region, about 159 million people, live in poverty (UNDP 2006). In the first decade of the twenty-first century, the Mekong countries have defined their respective national poverty lines, however (see Box 2.1). Using these national standards, a much smaller percentage of the population, 45 million, in the region live below the poverty line (Fig. 2.1) (ibid.).

Box 2.1 GMS: Nationally defined poverty lines

- Vietnam: VND1,906,950, and the food poverty line VND1,372,774 per person per annum (Ministry of Labor, Invalids and Social Affairs, 2002).
- Thailand: THB473 per capita per month (1988), and THB922 per capita per month (2002) (about US$0.96 a day in 2002).
- Laos: LAK85,000 per capita per month, where LAK100,000 per capita per month (urban poverty) and LAK82,000 per capita per month (rural poverty) ($1~ LAK8,035), (Participatory Poverty Assessment and Lao Expenditure and Consumption Survey, National Statistics Center/Committee for Planning and Investment, 2001).
- China: CNY1,196 (US$176) per capita per annum; US$0.48 per day (2009)
- Cambodia: KHR1,826 (US$0.45) per person per day or KHR9,130 (US$2.25) per day for a family of five (2004). About 80 percent of this is for food and 20 percent for non-food basic needs (clothing, housing, etc.)

Fig. 2.1 GMS: Share of population living below the poverty line

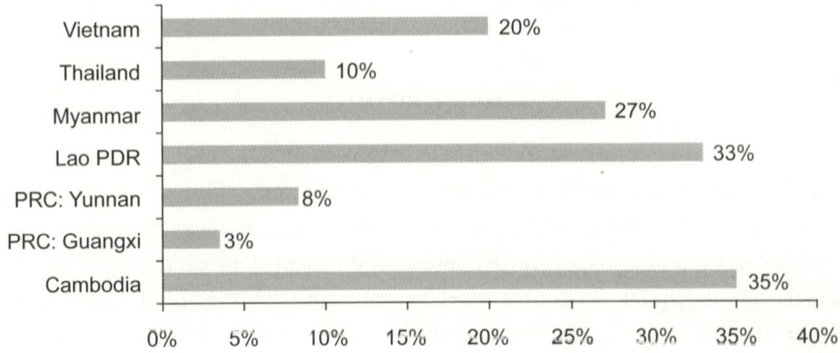

Poverty is also indicated by access to basic needs—food, clean drinking water, housing, health care, and education. To take some examples, at the regional level, 21 percent of the population does not have access to clean water, and some 30 percent lacks access to closed sanitation systems (U.S. Central Intelligence Agency 2010). In Cambodia, 30 percent of the total population consumes less than the minimum dietary needs, and 50 percent of children under the age of 5 were underweight in 2005; access to water resources was 60 percent in the urban areas and 40 percent in the countryside, but even this was only an average of 3 m^3 per year, the minimum volume of water required for a human being (Yamakushi and Promphakping 2007). In Laos, 40 percent of children under the age of 5 are underweight, and 30 percent of the population consumes less than the minimum dietary needs; there is 70 percent and 40 percent access to improved water sources in the urban and rural areas respectively.

Thailand's Human Development Index value, as measured by the UNDP, is ranked medium, 0.668 for 2006 (UNDP 2006), which is above all the other countries in the region except for China. However, there are huge disparities across income groups and regions. A large proportion of people living in Northeast Thailand, which has the largest area of non-irrigated agricultural land and one-third of the country's population, still depends on rivers and other natural resources for their livelihoods.

In Vietnam, poverty is largely defined geographically and ethnically. Two-thirds of those living in poverty are in the northern uplands, the Mekong Delta, and the North Central Coast; the ethnic minority groups that make up 15 percent of Vietnam's total population in remote rural areas represent 30 percent of the poor. In these regions, nearly one-third of children under the age of 5 are underweight; maternal mortality rates are ten times higher in isolated rural areas than in the cities and towns.

Energy poverty

Closely associated with economic poverty is inaccessibility to energy with which to meet basic livelihood and economic needs. This is described as energy poverty, "the absence of sufficient choice in accessing adequate, affordable, reliable, high quality, safe and environmentally benign energy services to support economic and human development" (Reddy 2000: 57). In the Mekong region, more than 20 percent of the population has no access to electricity. Energy usage in the region is lower than the world average and far below that of World Bank data on low and middle-income countries in Europe and Central Asia (Euro area)[1] as indicated in Fig. 2.2 (World Bank 2001–2010).

Fig. 2.2 Energy use per capita (kg oil equivalent)

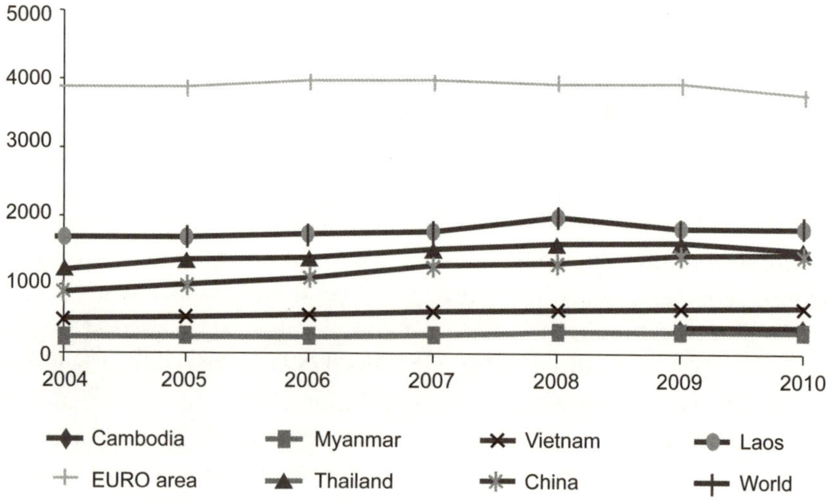

Source: Based on World Bank 2001–2010.

At the national level, energy poverty is also reflected in energy structures featuring poor access to cleaner energy. In addition to charcoal, traditional biomass (paddy husks, bagasse, woodchips, palm waste, rice straw, etc.) makes up a high share of the energy mix for rural households: 80 percent in Laos; 83 percent in Cambodia; more than 50 percent in

Vietnam; and 40 percent in Yunnan province. In Myanmar, with its rural population of 38 million, 64 percent of the primary energy comes in the form of fuelwood, charcoal, and biomass (ADB 2009a). On average, the share of energy from biomass in the region is far above the average for the world, including the Euro area (Fig. 2.3a).

Fig. 2.3 Energy mix 2006–2010

a. Energy from biomass

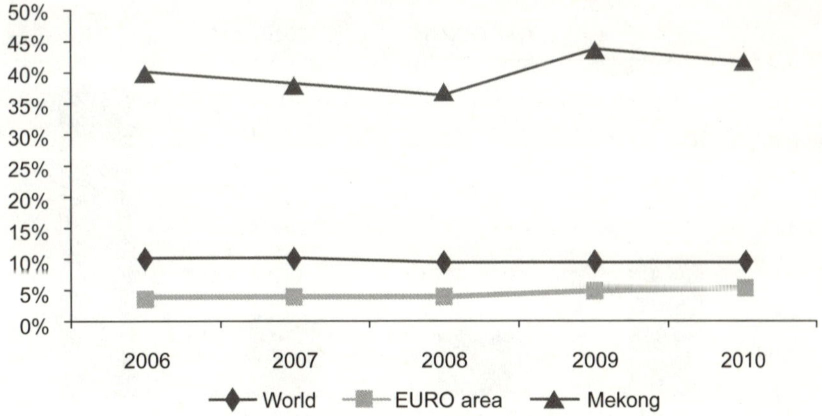

b. Primary energy demand, 2006

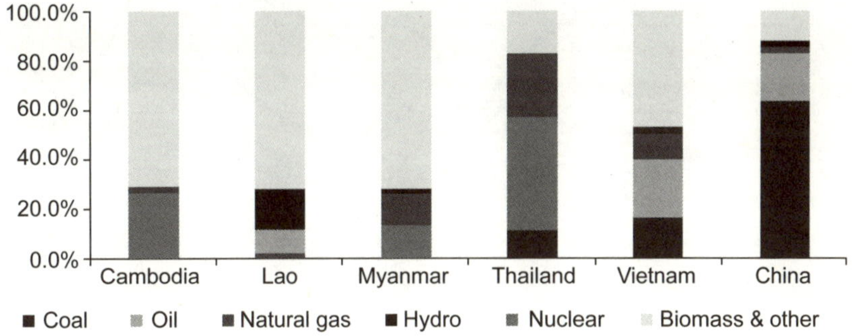

Source: Based on ADB (2009a).

The heavy dependence on traditional biomass energy is a serious health hazard. Each day 4,000 deaths are reported from indoor air pollution, most of the victims being women and children—more than half of the latter being below the age of five (ADB 2009a). Identifying poverty in economic terms as well as in terms of energy access enables decision-makers to better evaluate more sustainable development policies for poverty reduction.

At 7.4 percent the region also exceeded the world's average growth rates despite the Asian financial crisis in 1997 (Fig. 2.4). This is twice the world average, over three times that of the major advanced economies (G7), and over four times that of the lower and middle-income economies in the Euro area.

Fig. 2.4 Comparison of GDP growth, 1990–2010 (constant prices)

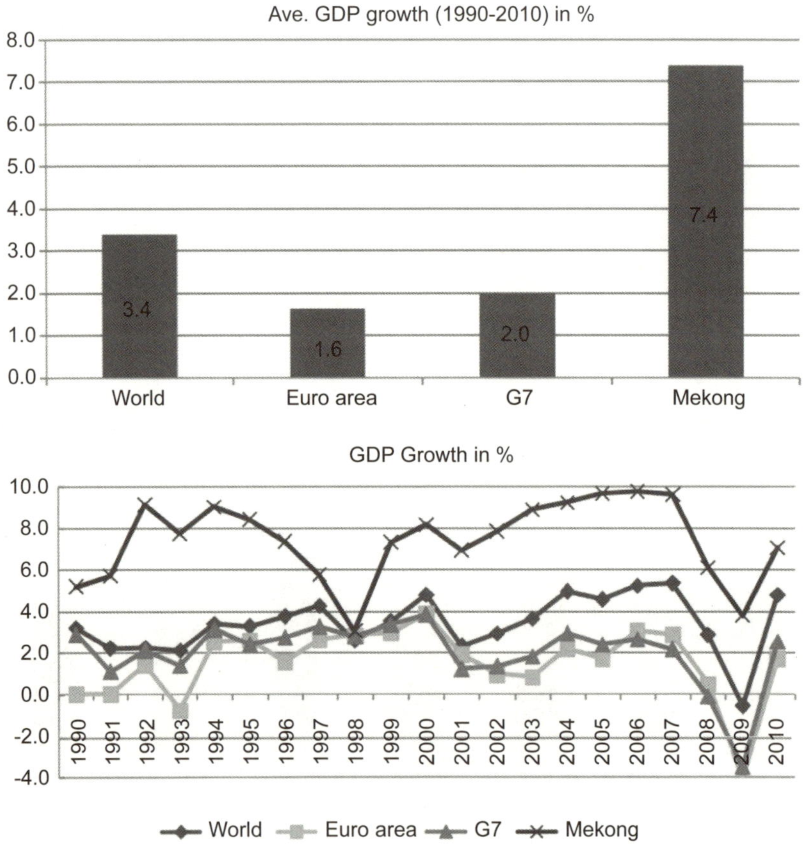

Source: International Monetary Fund, *World Economic Outlook Database*, October 2010.

The current traditional pattern of growth and poverty allevation via rapid economic development in the Mekong region is driving the demand for energy and therefore intensifying the challenge of energy security. Heavy dependence on oil imports is a common phenomenon.

For example, in Laos, all the petroleum products imported are largely used in the transportation sector; 62 percent of domestic energy demand is met by imports from Thailand. On the other hand, Laos is an exporter of hydropower to Thailand through mega dam projects.

Low emissions, fast growth, and low energy efficiency

Low energy access and consumption levels result in low CO_2 emissions. However, as noted, rapid economic growth in the region is driving up CO_2 emissions because of increasing energy demands, as indicated by Fig. 2.5.

Fig. 2.5 CO_2 emissions per capita (mt) and emissions growth

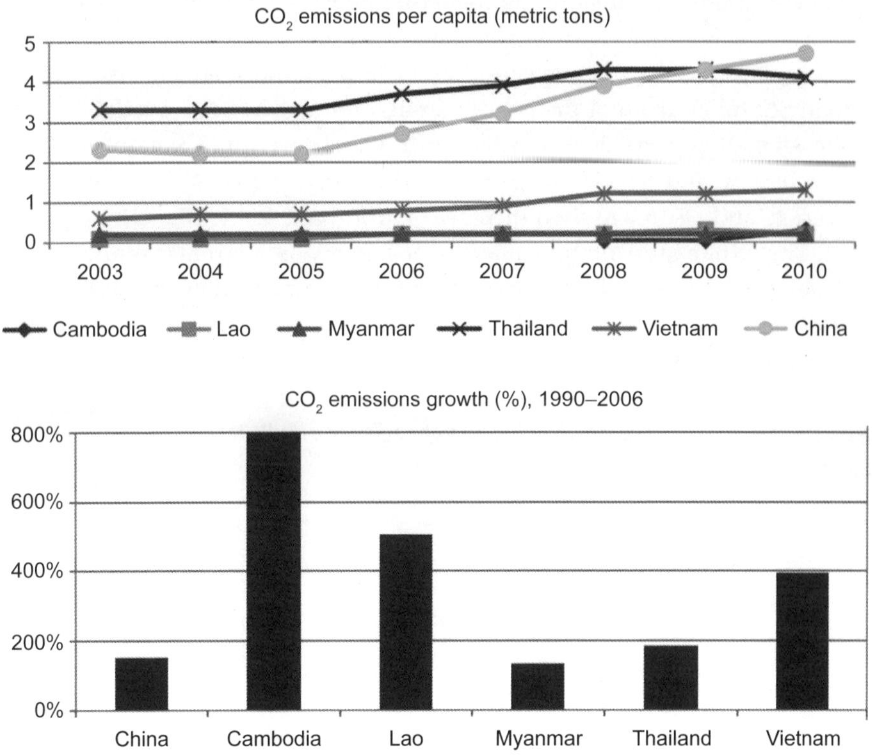

Note: Data for Cambodia is not available for the period 2003–7.
Sources: World Bank (2001–10).

In 2010, the average CO_2 emissions across the Mekong region were 1.8 metric tons (mt) per capita, as against 8.4 mt in the Euro area, and the world average of 4.4 mt. CO_2 emissions have increased rapidly in the last decade. In just five years China's emissions per capita went up from a little over 2 mt to 4.8 mt, surpassing the world average. Fig. 2.5a shows the growth of emissions over 15 years. In this period, emissions from the six Mekong countries as a whole increased 363 percent, the growth for the lower income group was 136 percent; the world average growth was 34 percent, and in the Euro area this figure was 4 percent (World Bank 2001–10).

Two factors in particular account for the rapid rise in CO_2 emissions. One is related to historical worldwide patterns, where economic development drives energy demand and consumption and in turn increases CO_2 emissions. Areas with the fastest economic growth show an accompanying outstanding increase in energy consumption and thus emissions. Another factor is associated with the low energy efficiency in the region. This can be measured by GDP per unit of energy use (PPP$/ tonne of oil equivalent or toe), illustrated by Fig. 2.6. The available data shows that during 2004–10, this figure was US$4.8 in Thailand, US$4.0 in Vietnam, and US$3.9 in China, in comparison to US$4.8 as the world average and US$6.9 in the Euro area (World Bank 2001–10).

The other side of low energy efficiency is a great potential for developing a low-carbon economy, featuring improved energy efficiency and the substitution of fossil fuels with renewable and cleaner energy, which will be discussed later.

Fig. 2.6 GDP per unit of energy use (PPP$/toe)

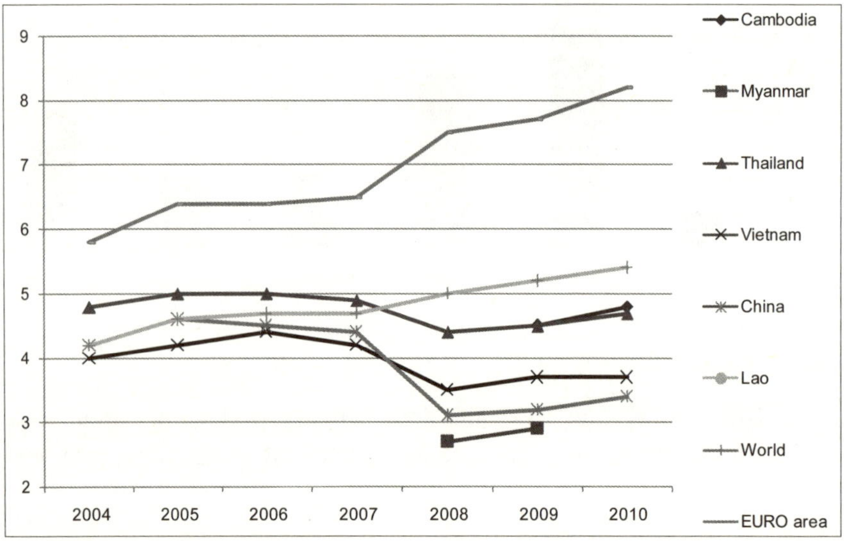

Sources: World Bank (2001–10).

Impacts of climate change

The increase in fossil fuel consumption for economic development directly leads to an increase of CO_2 emissions as the major man-made (anthropogenic) cause of climate change. Compared to the climate in 1961–90, the Intergovernmental Panel on Climate Change (IPCC) projected that average global temperatures will increase by +1°C in 2010–39 and +3°C to +4°C in 2070–99, average rainfall will decrease by –20 mm in 2010–39, but increase by +60 mm in 2070–99. Globally, water has been projected to be a key indicator of these changes (IPCC 2007).

Across the Mekong region, temperatures are rising and have risen by 0.5 to 1.5°C in the past 50 years (WWF 2009). According to Eastham et al. (2008), by 2030 the Mekong basin's mean temperature is likely to increase by 0.79°C, with greater increases for the colder catchments in the north of the basin. In this projection, the Mekong Delta and other low-lying coastal areas will suffer the most significant negative consequences. The projections indicate that the Mekong region is already getting hotter. Thailand's temperatures have reportedly increased from 1.0 to 1.8°C in the past 50 years; average daytime temperatures in April have been particularly high at 40°C; and Vietnam's temperatures increased by 0.7°C during this same period (ADB 2009a). Daily maximum and minimum temperatures are also increasing (Helsinki University of Technology and SEA START RC 2009: 51–3).

The Mekong region is expected to be one of the most vulnerable to climate change, which will amplify the existing threats to the region's terrestrial, freshwater, estuarine, and marine ecosystems (WWF 2009). The projected impacts of climate change by 2050 range from low (e.g., reduced water availability), to moderate (e.g., increasing temperatures), to potentially high (e.g., decreasing food production and sea level rises in the Mekong Delta) (Grumbine and Xu 2011).

The Himalayan glaciers which feed the headwaters of the region's rivers are melting at a rapid rate, threatening the flows of the Mekong, Irrawaddy, Salween, and Red rivers, upon which millions of people rely on for their livelihoods. In the past decade or two, there have been shifts in the rainy seasons coupled with more frequent extreme weather events, floods, and storms in the region. For instance, at Kratie, on the banks of the Mekong in eastern Cambodia, the frequency of 'extreme wet' flood events is likely to increase from an annual probability of 5 percent under historic conditions to a 76 percent probability under climate projections.

As it is, the Mekong Delta, which is the most productive and densely populated part of the Basin, is prone to severe flooding in the wet season.

During the dry months the Delta's agricultural areas are prone to seawater intrusions, and salination. Grumbine and Xu's study (2011) summarized that the key impacts under future projections for climate and population in 2030 include increasing flood risk, episodic food scarcity, and likely changes in the productivity of fisheries through hydrological impacts on the ecology of rivers, water bodies, and flood plains. Another study pointed out a similar result—an increase in mean temperatures by approximately 0.8°C, which will likely increase droughts and floods, decrease crop yields, exacerbate threats to biodiversity, and enhance endemic morbidity and mortality from disease throughout the region (Cruz et al. 2007).

Under one realistic scenario, the sea level will be about 40 cm higher than it is today by the end of the twenty-first century. Changes or impacts are already being felt. The city of Bangkok is sinking by 5–10 mm each year. Land subsidence and groundwater extraction, combined with rising sea levels, could leave Bangkok under 50–100 cm of water by 2025 (UNEP 2009).

The Mekong region is one of the most biologically diverse places on Earth. It is home to a diversity of landscapes such as the Greater Annamites, the Lower Mekong Dry Forests, and the Kayeh Karen Tenasserim Ecoregions; all three areas offer a high diversity of plant and animal species and are important because they harbor many rare, endemic, and endangered species (WWF 2009). However, in the complex and connected structures and processes of ecosystems and ecosystem service provision, the homes of biodiversity will likely shrink under predicted climate change (Ohlemuller et al. 2008), because the habitats, e.g. forests or wetlands, that support this diversity are shrinking or disappearing. At the same time other drivers e.g., land use change, invasive species, unsustainable harvesting practices, and hunting, have reduced the buffering capacity of these habitats from climate change. Although some species will be able to adapt without dispersing (Bradshaw and Holzapfel 2006), many will not, potentially resulting in extinctions (Stork et al. 2007).

The poor will suffer the most from the impacts of climate change; they are the least prepared for, but most exposed to, the changes.

Summary of challenges

In systems thinking, the dilemma or the challenges faced by the Mekong region can be depicted as an unchecked positive feedback loop that will drive the system to collapse if no proper action is taken to intervene.

Fig. 2.7 Energy, climate change, and poverty

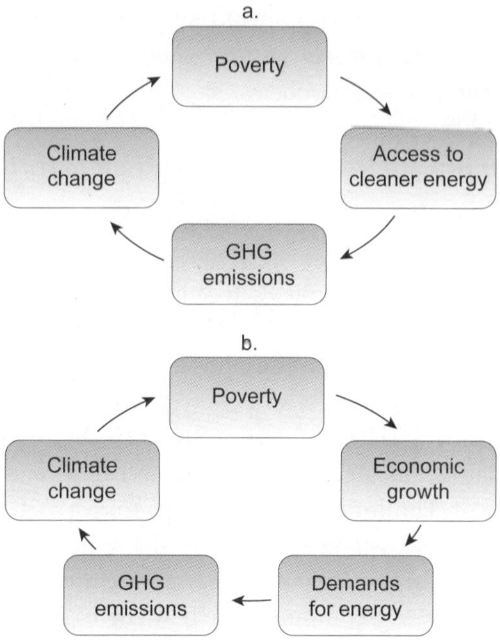

In this positive loop, the system functions like this. The more poverty the region falls into, the less access to cleaner energy (-); the less clean energy and the higher the use of fossil fuel or traditional biomass, the higher the emissions (-) and more negative impacts on health, which worsens poverty; the higher the CO_2 emissions, the greater the climate change impacts (+); the greater the impacts of climate change, the higher the poverty rates (+) (Fig. 2.7a). In attempts to address poverty, economic development via growth (i.e. the Business as Usual model), if unchecked has a similar positive loop, driving the system to collapse: poverty drives economic growth, which increases demands for energy, which leads to more CO_2 emissions, which intensifies climate change, which has impacts on the ecosystem to which the poor are deeply vulnerable (Fig. 2.7b). This poverty loop reinforces itself, unless the structure is changed by removing or changing links.

Regional response to the challenges

Facing these challenges, the national governments of the Mekong countries and regional decision-makers have adopted a series of policies, focusing on energy supplies to meet the demands of economic development to reduce poverty and/or to ensure energy security. This section reviews some of these policies.

Sustainable supply of cleaner energy: National strategies

Energy and climate change have become central policy concerns both at the national and regional level across the region. All the national governments have adopted policies to cope with challenges of energy and climate change to meet the targets of economic development and poverty reduction. These policies emphasize a) enhancing energy access, b) developing renewable energy, and c) improving energy efficiency.

In Thailand, the fifteen-year National Renewable Energies Development Plan 2008–2022 (REDP) seeks to bring renewable energies to 20.3 percent of Thailand's total energy mix by 2022. Enacted in 2009, the Plan aims to increase energy security, use alternative energy sources, encourage high-efficiency energy technologies, and spread green alternatives among communities (IEA 2013). In addition, another significant policy that has been adopted is Thailand's consecutive Energy Strategies; the current Energy Strategy has been effective since 2009. The focus is on building energy security, to reduce energy imports, and increase Thailand's energy independence. Another key policy is the National Energy Efficiency Development Plan with its major target of saving 3–5 percent of energy from 2006–2010, and 5–8 percent for 2011–2015 (Ministry of Energy 2011).

Energy security is also a high priority for Vietnam. The National Power Development Master Plan 2006–2015 aims to enable 100 percent of communes access to electricity by 2015. Renewable energy development and improving energy efficiency are prioritized in this plan. Specific targets have been set accordingly. In 2007 the Government of Vietnam approved the National Energy Development Strategies for up to 2020, targeting to increase the share of renewable energy in the total commercial primary energy to 3 percent by 2010, 5 percent by 2020, and 11 percent by 2050 (IEA 2010). In the same year, the government passed Decision No. 177/2007/QD-TTg on Biofuel Development, setting the targets of increasing the share of ethanol and biodiesel in gasoline and diesel demand to 0.4 percent by 2010, 1 percent by 2015, and 5 percent by 2025 (Phong 2010).

For Cambodia and Laos, the context is somewhat different from that of Thailand and Vietnam. Renewable energy and energy efficiency are less developed, and traditional biomass dominates the energy mix of the two countries. The main reasons are a lack of experience, funds, and data as well as policy barriers, as identified by the energy directors of the two countries, Toch Sovanna of Cambodia and Chantho Milattanapheng of Laos at the subregional Energy and Environment Partnership Forum in 2010. Cambodia has set as a country target achieving 100 percent electrification in rural communities by 2020, using renewable sources (Sovanna 2010). For Laos, the government aims to electrify 90 percent of households by 2020, increase the share of renewable energy to 30 percent by 2025, and make biofuels substitute 10 percent of oil imports by 2025 (Milattanapheng 2010). In terms of energy efficiency, however, both countries are just at the initial stage, building capacity in energy management and energy auditing, mostly under regional cooperation and bilateral projects (Kouphokham 2009; Sovanna 2010).

Myanmar, while having extensive natural gas reserves, has limited domestic consumption. Thus natural gas has become the largest export earner of the country (Thein and Myint 2008). However, renewable energy and energy efficiency are concerns of the government and there are ongoing efforts to increase the usage of renewable energy, especially through regional cooperation. Nonetheless, specific goals or measures have yet to be set up (Thein and Myint 2008).

In 2007, China adopted a "Medium and Long-Term Development Plan for Renewable Energy in China." The plan sets the following targets:

- China will raise its share of renewable energy in total primary energy consumption to 10 percent by 2010, and 15 percent by 2020. This will be achieved by fully utilizing, to the extent possible, technologically mature and economically feasible renewable energy sources, such as hydropower, biogas, solar thermal, and geothermal, as well as by promoting the development of the wind power, biomass power, and solar photovoltaic industries.

- China will also aim to provide electricity to people in remote, off-grid areas and resolve fuel scarcity problems in rural areas through the use of renewable energy, doing so according to local conditions and at the same time effectively protecting the environment.
- China will actively promote the development of renewable energy technologies and industries, building up a renewable energy technology innovation system. The aim was to have achieved the ability to produce domestically the main renewable energy equipment it uses by 2010.

By 2020, local manufacturing capability based mainly on home-grown Intellectual Property Rights will be achieved. The specific renewable energy development targets are summarized in the following table.

Table 2.1 China: Targets for renewable energy power generation (GW)

	2005	2010	2020
Hydropower	117	190	300
Biomass power	2	5.5	30
Wind power	1.26	5	30
Solar PV	0.07	0.3	1.8

To enhance national energy efficiency and mitigate its CO_2 emissions, China set a target of reducing energy intensity by 20 percent in 5 years during the 11th five-year-plan (2006–2010). In Copenhagen at COP16, China announced its Greenhouse Gas (GHG) emissions reduction target—reducing carbon intensity by 40–45 percent by 2020.

Sustainable supply of cleaner energy: Regional strategies

The regional targets of energy development and climate change mitigation have been agreed among the Association of Southeast Asian Nations (ASEAN) member-states, which include the Mekong region countries of Cambodia, Laos, Myanmar, Thailand, and Vietnam.

The ASEAN Plan of Action for Energy Cooperation (APAEC) 2010–2015 aims to "enhance energy security, accessibility and sustainability for the ASEAN region with due consideration to health, safety and environment" (ACE 2009: 12). Seven action plans have been formulated.

The goal of the ASEAN Power Grid (APG) is to integrate the national power grids of the ASEAN member-states in order to meet rising electricity demand and enhance access to energy services while promoting the efficient utilization and sharing of resources (ACE 2009: 13). Other plans under APAEC which will directly contribute to energy and climate sustainability in the Mekong region include those for Coal and Clean Coal Technology, Energy Efficiency and Conservation, Renewable Energy, and Civilian Nuclear Energy. Each of the seven action plans has established objectives and outlined activities; implementation is to be carried out by different agencies. Two objectives agreed collectively by the ASEAN ministers were to reduce regional energy intensity by 8 percent by 2015 from the 2005 level, and to increase the total regional power installed capacity of renewable energy to 15 percent by 2015 (ACE 2009: 22).

Several energy-related policies and actions have been established collectively by the GMS parties. The GMS Intergovernmental Agreement for Power Trade was signed in 2002 and came into force in 2003.

The ADB has facilitated the process of formulating the GMS Energy Strategy for endorsement by the region's governments. Seven priority actions have been identified, including:

1. Natural gas and refined petroleum products need to be considered for cross-border trade and future energy integration.
2. Energy productivity should be enhanced to increase energy security in the subregion.
3. Policy regimes and sector reform improvements and restructuring are needed in a time-bound manner to improve the efficiency of the energy sector.
4. Oil consumption needs to be reduced and technological options, such as coal liquefaction and biofuels, as well as relevant supportive approaches, need to be reviewed in order to decrease dependency on oil imports.
5. More energy efficient and sustainable patterns of transportation need to be considered.
6. There is a need to facilitate private investment in the energy sector.
7. Institutional and human resources development are needed for a sustainable energy future in the GMS.

8. Although some cross-border energy projects have been designed for the GMS, implementation and progress have not been satisfactory. The reasons identified include: the lack of infrastructure to facilitate intraregional energy trading; concerns over externalities which are inherent in energy distribution networks; and an absence of regional frameworks to share costs and benefits from energy trading. Hence, there are still significant barriers to implementing energy trading in the GMS (ADB 2009a).

The way forward: A low carbon economy

The GMS is challenged by poverty and energy inaccessibility. Climate change is intensifying the poverty cycle. Climate change represents a challenge for the GMS due to the following factors. First, the region has quite a large population living below the poverty line, and therefore, conversely, has big potential and a reason to grow out of that poverty. Second, the region has one of the world's fastest-growing economies, with demands for natural resources that could alter ecosystems if sustainable resource management is not adopted immediately (Tyler and Fajber 2009). Third, the heavy dependence of the regional economy on water, forest, and land resources makes the region more vulnerable to climate change. In reality, the region is already suffering from the effects of intensified land erosion, forest degradation, and deterioration in the quality and accessibility of water. Therefore, sustainable development has to be at the very centre of climate change strategies and policies at all levels, in order to improve livelihoods and eradicate poverty.

Mainstream economic development models have created the world's largest economies—while depleting huge shares of the earth's resources—over the last two centuries, leaving developing countries and regions with an unfair reality; there is scarce leeway to practice the same pattern of growth. What is urgently needed in the region is to explore new paths to development that will build sustainable supplies of cleaner energy while enhancing economic resilience to climate change. In the Mekong region, a low carbon economy is proposed to achieve the following goals:

- Reduce poverty by growing economies sustainably.
- Adapt to climate change that is already happening and will increase.

- Develop renewable, cleaner, energy substitutes for fossil fuels.
- Control the increase of carbon emissions to preserve the planet.

A low carbon economy is an integrated practical solution to the complicated challenges of climate change at the regional, national, and community levels. For example, if a rural community's economy is diversified, it would not be wiped out by natural disasters or unexpected economic downturns as is happening now; new crops that are more robust against draughts or floods could be adopted; renewable energy and decentralized power supplies will reduce the exposure of communities to fluctuations in the global energy market while reducing carbon dioxide emissions; recycling and reuse of organic wastes will create community social enterprises and green jobs.

A recent report has identified four categories of opportunities for mitigating climate change as pathways to a low carbon economy: energy efficiency, low-carbon energy supply, terrestrial carbon, and behavior carbon. According to this report, by 2030, a shift of energy supply from fossil fuels to low-carbon alternatives, e.g. renewable energy development to replace conventional fossil fuels, can provide an opportunity of 12 $GtCO_2$-eq (gigatonnes of CO_2equivalent) per year. Energy efficiency can bring an annual reduction of 14 $GtCO_2$-eq. Terrestrial carbon provides an opportunity of 12 $GtCO_2$-eq per year, as forests and soils act as natural sinks for carbon. Stopping deforestation, reforesting marginal areas of land and sequestering more CO_2 in soils through changing agricultural practice will increase carbon sequestration (McKinsey & Co. 2009); this opportunity largely falls in developing countries. Behavior carbon is associated with people's awareness, consumption patterns, and capacities to manage emissions and adapt to climate change.

To achieve a low carbon economy in the GMS, the first task is to examine these opportunities at community, national, and regional levels, with due consideration of its vulnerable economic base and the dilemma of coping with both climate change and expanding energy use for poverty reduction purposes. A low carbon economy is an economy that fits local conditions, balances ecosystem services consumption and regeneration, provides sustainable livelihoods, and also contributes to the well-being of the planet. Communities should be given priority because that is where poverty is located and where the impacts of climate change are most and first felt. In line with the four opportunities identified in the McKinsey report, a low carbon economy should have the following characteristics.

1. Sustainable supplies of low-carbon energy in order to provide:
    - reliable cleaner energy supplies and less fossil fuel dependence;
    - lower cost energy and reduced energy bills; and
    - information and advice about energy savings, including an energy monitoring and auditing system.
2. An enlarged carbon sink:
    - deforestation should be stopped or reduced, and forests conserved;
    - low-carbon land use adopted; and
    - land degradation reduced or stopped.
3. Waste disposal minimized:
    - wastes (organic) recycled and reused;
    - sanitation and health—communities and households—improved; and
    - water pollution stopped or reduced.
4. A diversified economy with green businesses and green jobs created:
    - improved quality of agricultural products;
    - creation of community social enterprises so that money made from the production of renewable energy is used to fund more community projects;
    - creation of low carbon jobs; and
    - diversified and improved sources of income for community members.
5. Improved awareness amongst citizens and community members, particularly youth, of climate change and its impacts on their livelihoods, and improved community capacity and skills to mitigate and adapt to climate change.

These five categories of indicators are connected to support each other as indicated by Fig. 2.8.

Fig. 2.8 Characteristics of community-based low carbon economy

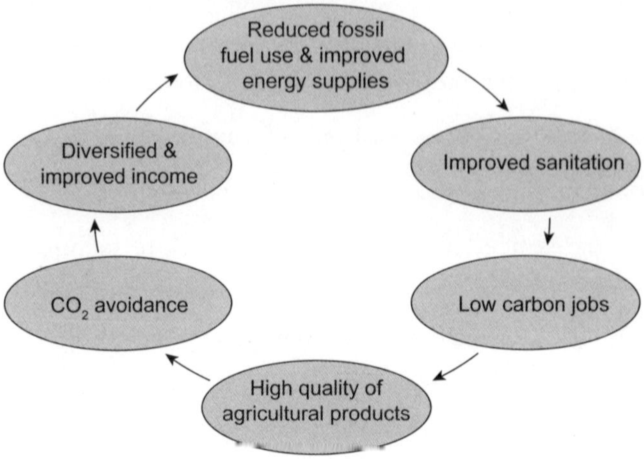

Issues for thought

The suggestions in this chapter are based on the assumption that mainstream economic development will not eliminate poverty in the GMS countries. Although there has been some reduction of the percentage of the population below the poverty line in Thailand, and recently in Vietnam and Laos, the poor are still vulnerable to changes arising from economic, environmental, and political crises. A community-based low carbon economy is suggested as a way to increase the climate resilience of the Mekong region.

To build a low carbon economy, four categories of issues need to be studied. For renewable energy development, it is necessary to find out:

1. How much renewable energy resources are available particularly at national and community levels?
2. What finances are required to mobilize the resources and turn them into energy?
3. What technologies are appropriate and needed?

4. What are the long-term impacts of renewable energy development, e.g. impacts of hydropower on ecosystems?
5. What regional cooperation strategies would benefit renewable energy development—south-south collaboration?

The low energy efficiency in the GMS countries shows great potential for improvement. In order to do so, it is necessary to determine:

1. Cost-effective areas of opportunities for improvement.
2. Economic incentives and instruments to be adopted.
3. Technologies needed, including appropriate technologies and financing mechanisms.
4. Opportunities for technological cooperation, including South–South cooperation.

Climate change adaptation featuring low carbon economies is needed. What regional and national strategies should be developed to build resilient, adaptive, and low carbon economies at all levels? The sustainable management of ecosystems and ecosystem services is an effective approach to climate change adaptation because ecosystems supply energy needed for development and sequester CO_2. An ecosystem-based adaptation strategy should be developed at all levels, based on comprehensive studies of ecosystem services and valuation of tradeoffs.

Note

[1] The Euro area as defined for the purposes of the *Little Green Data Book* (World Bank 2001–2010) comprises of: Albania, Armenia, Azerbaijan, Belarus, Bosnia and Herzegovina, Bulgaria, Georgia, Kazakhstan, Kosovo, Kyrgyz Republic, Latvia, Lithuania, Former Yugoslav Republic of Macedonia, Moldova, Montenegro, Poland, Romania, Russian Federation, Serbia, Tajikistan, Turkey, Turkmenistan, Ukraine, Uzbekistan.

3

Valuing Ecosystem Services in the Mekong Region

Penporn Janekarnkij and Orn-uma Polpanich

Most of the countries in the Mekong region are drawing heavily upon their valuable natural resources to fuel rapid economic growth. The region's major export earners are agricultural and forest-based commodities. The environment is also suffering from increasing development-driven demands for water and other resources. Large areas of terrestrial and mangrove forests have already been destroyed through economic activities, the flow patterns and volume of the Mekong and its tributaries are changing, and floods and droughts are occurring more frequently due to disruptions of the hydrological cycle. Coastal and delta areas are facing saltwater intrusion and inundation from rising sea levels (WWF 2009: 4). These ecosystem changes, along with projected climate change, have been affecting and will continue to severely affect the livelihoods of a large percentage of the region's population.

One reason for the continued degradation of these natural assets is the failure of markets, and society in general, to adequately value their ecosystem services in economic or financial terms. Many crucial ecosystem services such as the flood control or erosion prevention provided by natural forests are not recognized as their values are generally not expressed through market mechanisms. Policymakers need to be able to assess the economic importance of ecosystem services as well as the costs of a policy or action that alters an ecosystem. This chapter addresses the valuing of ecosystem services in the Mekong region through a discussion of the current state and projected trends of some of the major regional ecosystems; how their ecosystem services contribute to the region;

policymaking gaps regarding ecosystem services; and how the Mekong region can maintain its rich ecosystems to enhance economic growth while reducing poverty.

State of the environment

The Mekong River and its tributaries form the Mekong Basin, with a total catchment area of 795,000 km^2. Surface water resources are abundant with run-off amounting to approximately 475 billion m^3 in the rainy season and 78.8 billion m^3 in the dry season. More than 90 percent of the population live in the Lower Mekong Basin, which consists of about 70 percent of the entire area (UNEP 2006). Rural livelihoods are based on the integrated use of a wide range of natural resources adapted to the seasonal changes of flooding and recession (Mekong Wetlands Biodiversity 2004). The remaining terrain is largely mountainous. The coastline of the Mekong Delta alone extends to around 650 km, with about 350 km facing the South China Sea, and 300 km along the Gulf of Thailand. The associated coastal area is characterized by large estuaries, sand dunes, tidal marshes, and mangrove forests.

There has been a continuous decline and degradation of ecosystems due in part to rapid economic development, infrastructure development, and other activities that cause deforestation, loss of spawning grounds for fish, and changes to natural river flow. Most of the provisioning services such as those from aquatic and terrestrial ecosystems from which fish, timber, or wood products are marketed are given some recognition. But it is more difficult to demonstrate the economic necessity of ecosystem services such as flood control, soil erosion protection, or local climate regulation, which are significant for human security and health. The value of these services is poorly quantified and understood, and thus not recognized by society and governments.

Endowed with diverse terrestrial and aquatic ecosystems, the Mekong region has 41 biosphere reserve areas and 19 designated Ramsar Convention wetland sites (UNESCO 2013; Ramsar 2010). A United Nations Environmental Program (UNEP) impact assessment outlines the five major environmental concerns as the loss and/or modification of ecosystems, unsustainable exploitation of fisheries, freshwater shortages, pollution, and global climate change. Of these, the modification or loss of ecosystems was selected as the region's main priority because it depletes the living resources upon which rural communities depend for their very livelihoods (UNEP 2006).

Valuing benefits and costs of ecosystem services

Ecosystem goods and services have been commonly viewed as public goods and used for free. Many of these goods and services are not traded so we do not see their obvious value through the lens of market transactions. Often, ecosystem goods and services are both underappreciated and undervalued. In policymaking, the problem with public goods is that decisions about ecosystem management are complicated by the fact that various types of market failures are associated with natural resources and the environment. But markets do not reflect the full social and environmental benefits of ecosystem goods and services, nor the monetary and non-financial costs of their degradation. A major cause of the excessive depletion of environmental resources is the lack of understanding of the extent of human and economic interdependence on ecosystem services.

Hence, valuing the benefits and costs of ecosystem services is not just an analytical exercise, but an important input for decision-making. Information on the economic value of ecosystem services is necessary for justifying and setting priorities for policies or actions that protect or restore ecosystem services in the complex relationships involving the environment, economic growth, and human well-being. When difficult choices about the allocation of scarce resources have to be made, such valuation enables more informed environmental planning and decision-making that considers the full range of opportunities and impacts associated with particular investments and resource use.

The following section provides a review of the economic value of ecosystems and ecosystem services and how this can be used in policymaking involving a variety of ecosystem services using a range of valuation methods.

Economic value and classification

How can we calculate the overall importance of ecosystem services and obtain robust estimates of their value and benefits for incorporation into accounting systems for policymaking? Value and valuation are approached, viewed, and expressed differently by different disciplines, cultural conceptions, philosophies, and schools of thought (see e.g. MA 2005b; Norton 2012; MA 2005b: 128–9; NAS 2005: 48).

Despite the emergence of the valuing of ecosystems in policymaking, only a minimum economic value that satisfies the buyers and sellers are expressed in market prices, not the total economic value of the transaction. It is therefore important to assess the fundamental life-support services provided by ecosystems and the flow of assigned benefit values. The total system value is always greater than the total economic value that is associated with the direct use of ecosystem products, the indirect use of ecosystem services, or the non-use of species and habitats. As an example, according to the National Academy of Science (NAS 2005), the Irrawaddy dolphin and the elephant have strong cultural values in and of themselves without regard to what they do for humans, so using a value or market-based approach is inappropriate in these cases. This also includes heritage or cultural sources—i.e. globally significant natural and cultural heritage areas in the Mekong region, as well as areas of rich biodiversity—that are worth protecting regardless of their current value to humans, even as these values also remain difficult to quantify.

The measurement of the economic value of a good or service is often based on its utility for livelihood or lifestyle preferences, through the choices and tradeoffs that people make about what it is worth for them, given certain constraints on budget and time. This expresses economic value in units of money in a market-based economy. Thus, economic value is measured by people's willingness to pay (WTP) or willingness to accept (WTA) compensation for changes in their level of use of a particular ecosystem good or service. Accordingly, an analysis of the total economic value of the supply of an ecosystem service requires establishing the consumer and producer surplus in a market although ecosystem services that are not traded may be difficult to analyse (Fig. 3.1).

As depicted in Fig. 3.1, the consumer surplus is the excess of the price which consumers are willing to pay that exceeds the market price for a particular ecosystem good or service, while the producer surplus refers to the excess of the price which producers are willing to accept associated with the marginal costs of production for a cost less than the market price. This values the total economic surplus by summing up the consumer and producer surplus, and can be viewed as the total net economic benefit to people achieved by the production and consumption of a particular ecosystem good or service. It helps to make the case for the efficient allocation of resources based on an economic valuation of ecosystem services.

Fig. 3.1 Valuing an ecosystem service in a market

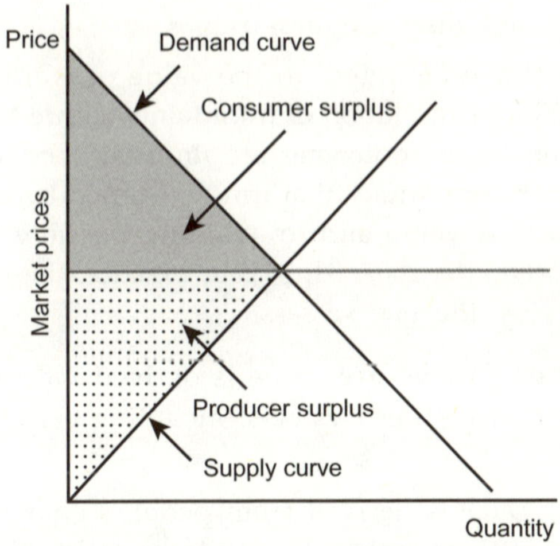

In general, the economic value of ecosystem services can be classified into distinct components according to their use and non use values (Pagiola 2005; DEFRA 2007; TEEB 2010).

Use value is derived from physical interaction with ecosystems that includes direct use, indirect use, option value, and perhaps quasi-option value.

- *Direct use values* are derived from the actual use of ecosystems that are used directly for a given purpose—i.e. production, consumption, and sale—(e.g. timber, fisheries). This can also be in the form of non-consumptive use, which leaves the quantity of ecosystem goods and services undiminished (e.g. recreation and bird watching), but may affect the quality of the good or service, e.g. by pollution or crowding.
- *Indirect use values* refer to benefits derived from ecosystem services which maintain and protect natural and human systems and provide essential global life-support systems, such as the regulation of groundwater recharge, water purification, flood control and storm protection; carbon sequestration; nutrient retention; and microclimate stabilization. Measuring these indirect use values is always difficult.
- *Option values* refer to the value that individuals place to preserve its availability in the future, simply to avoid the risk of not having that option. These future uses may be either

direct or indirect. Option value can also be viewed as a form of insurance, e.g. option of future availability of medicinal plants for drugs and pharmaceutical uses.

- *Quasi-option value* refers to the value of waiting for new information or resolution of uncertainty secured by delaying a decision, where outcomes are uncertain and where there is value or opportunity that can be learned by delaying the decision to develop and to avoiding possibly irreversible changes (DEFRA 2007: 31, 32). In practice, the quasi-option value is very difficult to assess.

Non-use value or passive use value is derived independently from physical interaction with ecosystems, including existence value and bequest value.

- *Existence value* is derived from people's enjoyment of the existence of an ecosystem. People derive satisfaction of merely knowing that certain ecosystems and their services continue to exist, even though they may never actually use or plan to use it, e.g. for spiritual or cultural satisfaction.
- *Bequest value* refers to the willingness to pay for preserving the ecosystem in a given state for the benefit of future generations. The value is derived from the desire to pass on elements of ecosystems intact to future generations because of their symbolic and identification values. Some literatures include bequest value as part of the non-use value (DEFRA 2007: 31; TEEB 2009: 23).

It is important to capture all the values of an ecosystem. Although not all benefits provided by ecosystems are fully translatable into economic terms, valuation can be used to complement other ways of conservation of ecosystem services. The most widely-used valuation method is the Total Economic Value (TEV) framework, which encompasses subsistence and non-market values, ecosystem functions, and non-use benefits as a more complete picture of the economic importance of ecosystems. The valuation can be done in monetary as well as non-monetary terms. The TEV of ecosystem services can be represented by a combination of all use values (direct, indirect, and option) and non-use values that correspond broadly to the internationally accepted Millennium Ecosystem Assessment (MA) framework (MA 2005b), as illustrated in Fig. 3.2. The TEV and the MA frameworks are seen as complementary.

Fig. 3.2 Total economic value of ecosystem derived from its services

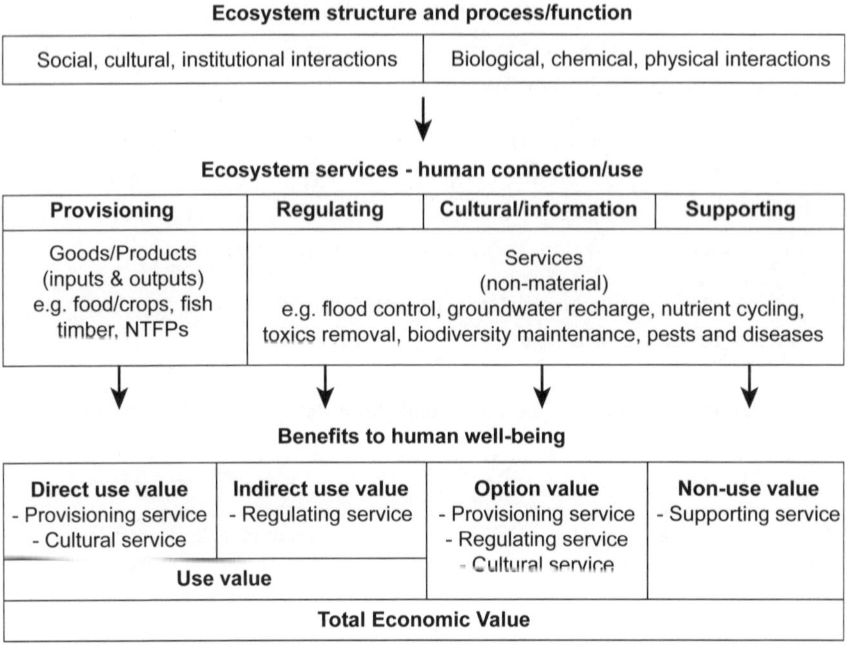

Sources: Adapted from EFTEC (2005) and Turner and Daily (2008).

Ecosystem services valuation methods

In economics, something has value if we are required and willing to give up something to get or enjoy it. Economic valuation is a suite of methods for assigning benefits and costs in monetary units associated with different options, especially services that are not being marketed. It presents a promising approach for highlighting the relevance of ecosystem services and of cost-effective policy instruments for natural resource conservation and impact assessment. Most valuation methods measure the demand for a good or service in monetary units that is consumers' WTP, or their WTA compensation for its loss. Four main sets of valuation methods, monetary and non-monetary, will be discussed here (See Table 3.1).

1. *Market prices method* captures the value of ecosystem goods being traded in conventional markets. Some provisioning services can also be measured if they are inputs in production that contribute to the profits made from final goods. The method includes market price, cost method, factor income, and household production model.

2. *Revealed preference method* measures individuals' preferences for an ecosystem good with their behavior in the surrogate market. This often relies on modeling techniques such as the household production, travel cost, or hedonic pricing.

3. *Stated preference method* is based on a hypothetical market in which people are asked to make choices or to reveal their willingness to pay for a certain ecosystem good/service. The most important approaches are the contingent valuation and contingent choice models.

4. Other methods, including benefit transfer and participatory environmental valuation.

Table 3.1 Economic valuation methods for different ecosystem services

Valuation method		Economic value type	Ecosystem service type
Conventional market/ Market prices	Market price (direct)	Direct use	• Mainly applicable to 'goods' from provisioning services (e.g. timber, NTFP, fish, food) but also some regulating services (e.g. pollination). • Value of ecosystem goods calculated from the gross value of good/ service subtracted by cost of using it, to obtain the net value.
	Market price (related)[a], i.e., avoided cost, replacement/ substitution cost, mitigation/ remediation cost	Indirect use	• Sometimes known as cost method; mainly used to quantify the cost of damage resulting from loss of an ecosystem service, or costs of replacing, or providing substitute from regulating services, e.g.: flood control services derived from the estimated damage if flooding occurs, – groundwater recharge services derived from the cost of obtaining from other sources – storm protection, or erosion protection services derived from the cost of human-made defenses – water purification services derived from expenditure on filtration • fertilizers replaced by nutrients that worms create for soil • damage caused by an oil spill.

Valuation method		Economic value type	Ecosystem service type
	Factor income	Indirect use	• Normally applied to regulating service that serves as input to market products, e.g. water, soil, air quality on crop yields, fish catch, or forestry output and thereby change in income to farmers/fishermen.
Surrogate market/ Revealed preference methods	Household production model, i.e., model of farmer's crop production function, household's damage function, defensive expenditure function, and cost of illness function	Indirect use	• Normally applied to regulating services • Similar concept to factor income, but estimating statistical model/ function, then measuring welfare gain (benefit) or loss (cost) experienced by households affected from ecosystem service change
	Travel cost model	Direct use	• Normally applied to cultural/ information services. • The services contributing to recreational activities are reflected by the amount of time and money that people spend while traveling and visiting the recreational site
	Hedonic pricing model	Direct and indirect use	• Normally applied to regulating and cultural services • Estimating the hedonic price model of market goods or real estate where environmental attributes (ecosystem services) such as air quality, visual amenities, & landscape influence the price.

Valuation method		Economic value type	Ecosystem service type
Hypothetical market/Stated preference methods	Contingent valuation	Use and non-use	• Applicable to all types of ecosystems and their services • Using a survey questionnaire with trade-off scenarios (status quo vs proposed) and hypothetical questions, then ask respondents to elicit their willingness to improve the quality of a particular ecosystem service, data analyzed by statistical model to estimate the willingness to pay for ecosystem service change.
	Contingent choice model (choice modeling)	Use and non-use	• Applicable to all types of ecosystems and services, similar to contingent valuation • Using a survey questionnaire with scenarios and set of choices that include environmental attributes (ecosystem services), then ask respondents to choose the preferred option. Choice set may include bundle of or a single ecosystem service to be valued.
Others	Benefit transfer	Use and non-use	• Apply to all types of ecosystem services • Using data from other sources with similar context (policy, socio-cultural, biophysical) to transfer and adjust to the policy site, can be the value transfer, the function transfer, or the meta-regression analysis
	Participatory economic valuation	Direct use	• More applicable to 'goods' from provisioning services in subsistence economies • Respondents express the value of natural products within the context of their own perception, needs, and priorities by ranking and rating (with scores), the selected numeraire is used to transform the scores to the computed value of each item.

Source: Adapted from TEEB (2010, chap. 3); DEFRA (2007: 37).

Note: [a] The approach must be used with care. Normally, this cost-based approach refers to public projects that could reverse environmental damage, reproduce a lost ecosystem service, or return the damaged ecosystem to its previous quality. The 'engineering costs' are used as proxies for values of ecosystem service change. The costs are related in no particular way to the lost value from natural services. The approach has nothing to do with measuring the welfare effect of households experiencing ecosystem change (Bockstael and McConnell 2007: 326–27).

Integrating ecosystem services value into policymaking

Defining the gaps

Governments across the Mekong region face decisions that involve trade-offs between economic benefits from development, especially hydropower development and their considerable irreversible effects on ecosystems. Development plans are multidimensional, involving social, ecological, economic, and political issues that make decision-making more complex. In many cases, one of the most critical trade-offs involves balancing the benefits of large-scale development projects with the costs that are incurred from resulting ecosystem degradation.

While those who depend on them know that ecosystem services are central to their livelihoods, policymakers often fail to appreciate the multiple functions of ecosystems; and hence undervalue or overlook their importance. The full range of benefits and costs of ecosystem services, particularly those relating to non-marketable goods and services, is underrepresented and lost in the policy debates and decisions. In addition, the benefits of ecosystems are felt differently by people in different places with competing objectives and over different timescales. One reason for such gaps is because the value of an ecosystem is generally not reflected in markets because that valuation is mainly based on the costs of extraction (e.g. labor, equipment, and transport) and raw materials, not taking into account the value of other ecosystem services (e.g. soil fertility, air regulation, and biodiversity conservation).

In this chapter, a few examples will illustrate the difficulties involved in valuing ecosystem services in the Mekong region. One example, in Thailand and Vietnam, is the clearing of coastal mangroves for shrimp farming (aquaculture) to supply the export market, which has caused fishing communities to experience declining fish catches, increased storm damage, and water pollution. Aquaculture has caused a decline in other services provided by mangroves such as being a spawning ground for wild fish, a protective barrier during storms, and filtering pollution. Raganathan et al. (2009: 2) raised the case of the conversion of a mangrove into a shrimp farm at Tha Po village in Surat Thani province, southern Thailand, to demonstrate that the ecosystem services from intact mangroves is much higher than the benefits received from shrimp production. The study by Sathirathai and Barbier (2001) cited by Raganathan et al. (2009) shows that the net income from shrimp at

US$1,164/ha was not compatible with the values foregone from forest products (US$823/ha), fish spawning ground (US$420/ha), and coastline protection (US$34,453/ha) if the mangroves were left untouched. Another example is the dynamic interdependence between linked ecosystems such as that of the Great Lake (Tonle Sap) in Cambodia and the Mekong Delta, where primary production is affected by changes in the Mekong's streamflow and sediment loads. Scientific knowledge of these intimately related ecosystems and their interdependence is necessary to make better informed, sustainable development planning and policy decisions. Another example is that of the unequal benefits received by various people involved in the exploitation of a particular resource. A change in an ecosystem may typically yield benefits for certain groups of people while imposing costs on others (including those living across national borders), who may either lose access to their most critical traditional economic resource or livelihoods or be affected by externalities associated with the change. One illustration of this is the Pak Mun dam hydropower project on the Mun River in Northeastern Thailand.

Much of the data on ecosystem service conditions does not allow for us to discern differences among them, i.e. whether they are in a relatively intact condition, degraded or even impacted by human activities whose effects generally impact an area much greater than a single ecosystem. Although valuing a single ecosystem service shows promise in delivering results that can inform important decisions, the valuation should comprise the entire ecosystem. Measuring the value of a single ecosystem service presents only a partial assessment of natural processes and may provide a false estimate of the total value of an ecosystem and narrow the scope of analysis. A major difficulty is that ecosystem services have no formal market, price or expressed cash value, except for those with the extractive use of component resources (e.g. fish, timber, crops from provisioning services) that are formally bought or sold in the market. Another source of complexity is that various ecosystem services are also spatially linked. Therefore, a workable understanding of the functioning of possibly the entire ecosystem may be required, although the valuation of multiple ecosystem services is more difficult than a single service. In most cases, the economic valuation fails to undertake integrated studies at suitably large spatial scales to fully cover important effects due to difficulties in the quantification and measurement of the functioning of large parts of or the entire ecosystem.

Today, although there is a growing body of evidence and research on the importance of ecosystems using an economic valuation approach to society, there has been a limited use of this evidence by decision-makers. In the Mekong region, the economic value of ecosystem services is not widely accepted or applied, nor is it implemented in economic development policymaking. This slow progress stems mainly from data issues, particularly poor information, and a lack of scientific understanding of linkages between development and ecosystem management, as well as institutional failures.

- *Data issues.* The value of indirect use ecosystem services is not well reflected and factored in policymaking, particularly when decisions on the allocation of public funds are competing with shorter-term market-based commercial and development returns. Within all the Mekong countries, economic value information is often only available for a single ecosystem, with some value domains and in incompatible units, rather than as a dependent and integral component of the functioning ecosystem. Given the lack and inaccessibility of data, policy planners and other stakeholders often make rough, working assumptions about the value of ecosystem services on the basis of poor information. This is partly because assessing ecosystem values is made difficult by patterns of seasonal variation and by complex ecosystem linkages that are only partially understood—with more long-term cross-border studies and data collection required for demonstrating the complexity of ecosystem processes and their contribution to the economy and human well-being. There remains a shortfall of scientific information about the independence of various components and functions of ecosystems; changes in the provisioning services caused by different human actions; the dynamic character, complexity, and interconnectedness of linked ecosystems and human systems over time; comprehensive and reliable data in different temporal and spatial scales; poor articulation or ignorance of uncertainty associated with valuation of trade-offs among sectors. Likewise, there is need for a comprehensive overview of the economic value of ecosystem benefits and costs for the entire Mekong (WWF 2013; UNEP 2012; Johnston and Kummu 2011).

- *Institutional failures*. Economic valuation often remains a largely academic exercise, carried out in isolation from planning processes or only used to show environmental costs retrospectively when implementing policy. Moreover, information about ecosystem costs and benefits are rarely communicated more widely by economists or environmentalists. In general, the percentage of a population whose livelihoods depend on the proper functioning of a particular ecosystem, or ecosystems is not well documented or taken into account in decisions. A lack of clarity in the legal framework leads to uncertainties about resource tenure that encourage destructive or inequitable use because ecosystems cut across administrative boundaries. Thus, the best governance methods and institutions to ensure natural resource conservation and long-term sustainability remain under question. In summary, this systematic under-valuation of ecosystems and the failure to establish effective institutions have resulted in broader economic benefits being under-emphasized in development and conservation policy, planning, and management practice (see UNEP 2013).

Taking the ecosystem service concept into policy setting

The Millennium Ecosystem Assessment (2005a) has warned us that we must value natural systems and their irreplaceable conditions for human well-being, and that the failure to do so has already been a significant factor in their continuing loss and degradation. The first requirement is a recognition of the concept of ecosystem services itself. In recent years, the ecosystem service concept has been widely recognized by various international organizations and in academia (MA 2005a; NAP 2005; Turner and Daily 2008; Ranganathan 2008; Tallis et al. 2008; Dalal-Clayton and Bass 2009), as it takes the MA and TEV frameworks as the foundation of thinking. The concept requires a holistic and people-centered view of ecosystems, recognizing their multiple services and co-benefits, and accepting ecosystems as economic assets required for sustainable development. Fig. 3.3 shows how the ecosystem concept links both development and nature by keeping people at the center.

Ranganathan et al. (2008: 62–64) list a range of policy options for sustaining ecosystem services, and a variety of ways in which ecosystem service considerations can be incorporated into development decision-making, irrespective of limited actual applications. An initial synthesis of the IIED's work with 13 partner countries on "the challenges of the environmental mainstream," provides a map to operationalize environmental mainstreaming (Dalal-Clayton and Bass 2009). While the MA framework stresses the inseparable connection between people and ecosystems, recognizing both its potential and limitations, the TEV framework makes clear *the full range of benefits provided by ecosystems*. This provides a more complete picture of the costs and benefits of altering ecosystem services and trade-offs in order to make a stronger argument for incorporating ecosystem services valuations in policy, program, and project appraisals. In particular, such integration into various development policy domains conveys significant opportunities for poverty reduction while improving local ecosystem goods and services delivery.

A common threat is that most management decisions affecting ecosystem services are made at a local level, but are conditioned by national and international policies. The use of policy tools such as Payments for Environmental Services (PES) and Reducing Emissions from Deforestation and Forest Degradation in Developing Countries (REDD+) aids in the designing of appropriate compensation and institutional mechanisms for tapping opportunities and ensuring benefits to local communities. In addition, integrating the economic value of ecosystem services into regional development cooperation provides a base for informed policy decisions, identifying more efficient and cost-effective alternatives that can achieve multiple benefits linked to human well-being in the Mekong region.

Fig. 3.3 Ecosystem services, human well-being, and development linkages

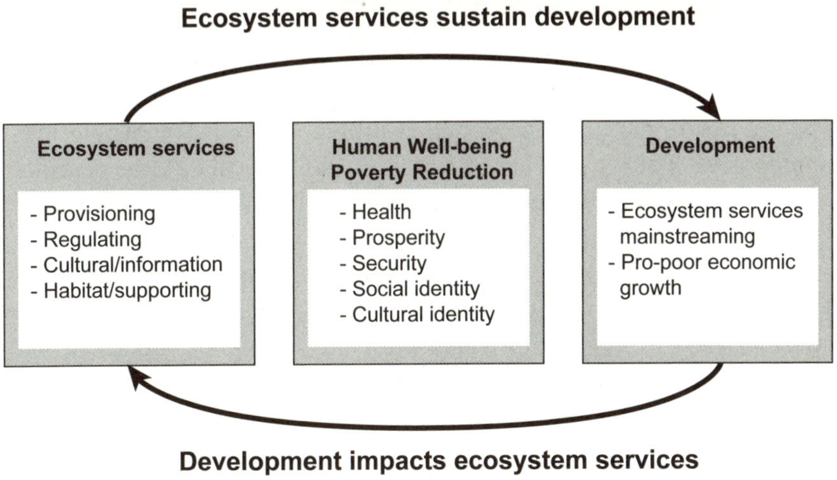

Source: Adapted from Ranganathan (2008).

Pagiola (2004) summarizes different approaches in incorporating ecosystem services values into the policy context (see Table 3.2). The following four approaches are closely related but distinct depending on the policy question being addressed.

First, the value of the total flow of benefits from ecosystems must be determined. Questions such as "How much is a particular ecosystem contributing to economic activity?" can be asked at different levels. At the local level, villagers would like to know how much an ecosystem contributes to them under their protection efforts as a community forest or local fishery. The same question can be asked at the national level, as part of a total ecosystem services accounting. A study by Costanza et al. (1997) provides a good example at a global level. The analysis is typically the sum total value of flow from multiple services.

Second, the net benefits of interventions that change ecosystem conditions should be determined. For instance, a study by Ziegler et al. (2013) reports that more than 85 dams are now proposed to be built on the mainstream and tributaries of the Mekong River. Existing dams have already created undesirable environmental and interconnected outcomes for governments and communities in the Mekong region. They are threatening the ecosystem and food supply by reducing the diversity and abundance of freshwater fish, the major source of protein for 67 million people in the region. If the proposed dams are built, the impact of the entire network could be much more devastating, even threatening human welfare. The study by Ziegler et al. also reveals that the dams would result in changes to the seasonal riparian flow regime—particularly in the Lower Mekong Basin—and have impacts on fisheries, biodiversity, and other ecosystem services. Clearly, this is a good example of the need to demonstrate how ecosystem conditions are being altered as a result of large-scale projects or policies for which the change in the flow of ecosystem services and its impact to different groups of people must be identified. Thus, the value of ecosystem services should be routinely incorporated as part of development appraisals, plans or reviews in order to properly make natural resources and investment choices.

Third, the distribution of costs and benefits of ecosystems should be determined. Equity issues are crucial in the Millennium Development Goals (MDG) to achieve poverty alleviation targets and to protect particular groups such as the poor or indigenous. Policy intervention may impact different stakeholders with very different costs and benefits resulting in ecosystem changes. Examples of the social disparities that can result from development projects are discussed in ICEM's (2010) study of the proposals for more mainstream dams on the Mekong. The question of *who benefits* and *who loses* should be a central concern in development policy, project assessments or programs. From a practical perspective, legal and other institutional regulation over ecosystems play a crucial role in determining benefits and costs against cultural and social aspects. A top-down regime of conservation programs may constitute perverse incentives (disincentives) for local people if they do not perceive that there are any benefits to them from their conservation efforts.

Table 3.2 Different approaches to valuing ecosystem services

Approach	Why do we do it?	How do we do it?
Determining the total value of the current flow of benefits from an ecosystem	To understand the contribution that ecosystems make to society	Identify all mutually compatible services provided; measure the quantity of each service provided; multiply by the value of each service
Determining the net benefits of an intervention that alters ecosystem conditions	To assess whether the intervention is economically worthwhile	Measure how the quantity of each service would change as a result of the intervention, as compared to the quantity without the intervention; multiply by the marginal value of each service
Examining how the costs and benefits of an ecosystem are distributed	To identify winners and losers, for ethical and practical reasons	Identify relevant stakeholder groups; determine which specific services they use and the value of those services to that group (or changes in values resulting from an intervention, such as a change in land use or management practice)
Identifying potential financing sources for conservation	To help make ecosystem conservation financially self-sustaining	Identify groups that receive large benefit flows from which funds could be extracted using various mechanisms

Source: Pagiola et al. (2004: 28).

Finally, there is a need to identify potential sources of financing for conservation. An appreciation of ecosystem services values is not adequate until it leads to investment in ecosystem restoration and protection in order to secure or enhance their services. Valuation can help in identifying the main beneficiaries of conservation efforts and the amount of benefits they receive. Valuing ecosystem services improvement can help design mechanisms such as PES by making the direct beneficiaries engage in the mechanisms and pay some of the benefits following the "beneficiary pays" principle. The pilot site of Lam Dong province in Vietnam provides an example of the implementation of PES that enables service buyers such as the water supply and hydropower companies to pay local communities for water regulation and soil conservation services and secure their watershed protection efforts.

ICEM (2003) synthesized various studies to illustrate how ecosystem services valuation is being used to provide better decision-making in protected area (PA) management in the Mekong region by:

- demonstrating the importance of PAs in national development and economic growth, and showing how the information can be used to influence mainstream development decisions and economic indicators;
- being able to provide strong arguments for PAs as profitable and economically beneficial use of land, resources, and investment funds; and
- increasing the priority of PA protection in economic decision-making by showing the economic cost of failing to conserve PA species and ecosystems.

To this end, Interwies (2010) concludes that applying economic valuation in environmental decision-making is essential when policy choices need to consider the trade-offs between benefits and costs. It also assists in addressing, mitigating, and calibrating the complex interdependence of human well-being and ecosystem services; quantifying trade-offs between ecosystem services, conservation and other priorities; addressing non-linear and abrupt changes; expanding the scope of probabilistic analyses by gaining the most complete picture for possible future development; evaluating interactions of ecosystem services with other determinants of human well-being; and filling gaps in improved understanding regarding human well-being. It is worth noting that generating estimates of the economic value of ecosystem services should be applied with care, particularly on a wider scale. Nevertheless, the value of ecosystem services depends upon underlying ecosystem and market conditions, as well as changes in estimates relating to those conditions. It thus provides a systematic way in which those conditions can be factored into different policy choices. In this regard, quantification of economic values may not be necessary when relating to intrinsic value unless the policy depends on specific criteria used to choose among policy alternatives, e.g., ecological, social, and cultural indicators (NAS 2004).

Discussion

Integrating economic valuation and policy

Undoubtedly, the negative and cumulative impacts of development projects pose threats to the value of biodiversity and ecosystem services, as well as increase the costs of ecosystem degradation and biodiversity loss. There is increasing recognition of the indirect economic value of ecosystems with new efforts to improve impact assessment for ecosystems and to achieve multiple alternative ways of effective use and conservation of natural resources; but often, these measures are taken after ecosystem values have been seriously compromised.

Although findings such as those by MA (2005a) are not available for this region, there is evidence to show that ecosystems in the Mekong region have been degraded due in large part to economic development (see e.g. ICEM 2010; UNEP 2013). The impacts of ecosystem change are long term and not covered in national accounting or reflected in the standard indicators such as GDP. With climate being one of many factors contributing to changes, the Mekong region should prepare itself for the next 20 years to protect and sustain its natural assets while coping with a complex mixture of social, economic, and environmental pressures.

Some policymaking that incorporates ecosystem services has emerged in the Mekong region. At the regional level, the concept has been used for policy evaluation through the strategic environmental assessment (SEA) of mainstream hydropower dam projects. The approach takes into account multiple services provided by ecosystems and links with multiple groups of people. At the national level, many good examples exist. China and Vietnam have invested in measures to reduce the rate of deforestation, increase afforestation, and ensure forest protection for sustainable development. Forestry is also seen as playing a key role in both countries for addressing the negative impacts of projected climate change and loss of biodiversity while achieving international climate adaptation targets. The lessons from their proactive policies and strategies for investment in natural capital need to be learnt by other countries in the region. The institutional reforms for enabling conditions and the policy environment in China and Vietnam should be further investigated.

The initiation of market-based mechanisms such as compensation or payment for ecosystem services in China and Vietnam provide examples of how ecosystem services valuation can be used within the policymaking process. An assessment of the long-term economic returns of ecosystem services as distributed among various affected groups at the local and national level should also be further explored.

Policy challenges

The Mekong region faces certain critical issues that need to be addressed to ensure sustainable development that: preserves the integrity of its ecosystems and their services, creates equitable and non-destructive growth, and benefits the rural majority. Mainstreaming the ecosystem services concept is no longer an option but an obligation for the region's development. In order to balance income while maintaining healthy ecosystems and sustainable rural livelihoods, some key policy challenges emerge.

- *Ecosystem management in economic context.* The ecosystem should be treated as an economic asset that can generate long-term benefits. Investment in the protection and restoration of target ecosystems services can be an effective strategy to secure and sustain the region's economic development. Economic returns drawn from ecosystem services investment should be emphasized to provide sustainable financing for protection and conservation.
- *Ecosystem management in socio-cultural context.* The distribution of wealth generated from ecosystem investment and management should be decentralized to the lowest appropriate level, particularly the poor and indigenous people. In ecosystem investment, pro-poor strategies must be taken to avoid negative social impacts and to improve the livelihoods and well-being of poor rural communities.

> Linking people's well-being with the changes in ecosystem services for any ecosystem alteration is crucial. The profits from ecosystem investment should focus on sustainable livelihoods, especially for marginal groups. A pro-poor policy of ecosystem management requires participatory approaches with community organizations involved in decision-making and livelihood impact assessments.

To link economic and social goals with ecosystem services, policymakers at all levels should focus on growth that benefits the poor. At the local level, mainstreaming ecosystem service concepts can be taken through community-based initiatives, to ensure that the ecosystem services benefits go beyond direct commercial values. At the national level, community-based systems can collaborate with the government for ecosystem management.

Although there is no standardized process or detailed methodological guidance to mainstream the ecosystem service concept in decision-making, many international organizations have launched the concept through country programs such as the poverty environment initiatives and the sub-global assessment under the UNDP-UNEP joint programs. The World Resources Institute (WRI) has produced guidelines to suggest policy options and entry points for mainstreaming ecosystem services into different types of private and public decision-making. The series of publications and website (www.teebweb.org) produced by the Economics of Ecosystems and Biodiversity (TEEB) provide knowledge on the conservation of ecosystems and biodiversity. The main focus of these studies is to evaluate the costs of the loss of biodiversity and the associated decline in ecosystem services worldwide, and to compare them with the costs of effective conservation and sustainable use. TEEB intends to sharpen awareness of the value of biodiversity and ecosystem services and facilitate the development of effective policy, as well as engage with business groups and citizens.

In the Mekong region, a change in policy perspectives is required from treating nature as an economic externality to that of an economic asset, along with greater political will to implement this in practice.

Our conclusions from various studies show that the key policy responses to mainstreaming ecosystem services in the Mekong region are the following (MA 2005; Irwin et al. 2007; Ranganathan et al. 2008; Emerton 2008; Dalal-Clayton and Bass 2009):

- *Restoring the health of ecosystems to development.* In order to reverse environmental degradation ecosystem services must be enhanced as part of the development strategy, and trade-offs across services should be reduced. This strategy will lead to pro-poor economic growth, so that the poorer sections of the population will equally benefit from development. Long-term investment in forest ecosystems, both private and public, can also help to mitigate climate change.
- *Prioritizing ecosystem services trade-offs.* The situation of trade-offs occurs when the ecosystem is limited for competing uses and by increasing one service at the expense of other services. Trade-offs arise from management choices or actions that involve ecosystem alteration in order to achieve narrow economic goals. Examples are coastal development, hydropower infrastructure, demand for biofuel, and converting forests and wetlands for agriculture or aquaculture. Changes in the quality or quantity of ecosystem services normally result in differentiated economic gains and losses among various groups of people. Scenario planning should be used to identify the outcome of ecosystem service trade-offs in the future.
- *Incorporating ecosystem services into the existing strategic environment assessments (SEA).* The SEA approach at sector and national levels guided by the OECD's *Applying Strategic Environmental Assessment: Good Practice Guidance for Development Co-operation* (2006) takes into account the full integration of environmental, social, and economic factors as a holistic sustainability assessment. Its main focus can be broadened to include the dependence of human well-being on ecosystem services. Special consideration should be given to the poor who are especially dependent on environmental assets and are vulnerable to change. ICEM's assessment of hydropower projects on the Mekong mainstream is a good example of how SEA enables multiple ecosystem services to be considered along with social and economic impacts (ICEM 2010).
- *Using economic incentives to protect and restore ecosystems.* Economic mechanisms such as PES and REDD+ in Vietnam's forest sector can be a good example. The mechanisms should

ensure that rural people and communities have secure access to the ecosystem they depend upon for their livelihoods. This can create the financial incentive for them to invest in the long-term health of these natural assets. Other market-based tools such as eco-labeling, business supply chains, and green procurement can bring in private business to engage in mainstreaming ecosystem services. These tools can help to develop and encourage the use of products and methods that reduce dependence and impact on ecosystem services.

- *Relocation of budget subsidies for biodiversity, etc.* More importantly, though it is difficult, governments should gradually relocate budgets by removing the production subsidies for food and fuel and shift towards more protection and restoration subsidies for biodiversity enhancement, and soil and water regulation.

- *Promoting communication and education.* Effective communication of the key messages, evidence, or economic cases for mainstreaming the ecosystem service concept into national development policymaking is essential. There is an urgent need to raise public awareness of the value of ecosystem services.

- *Supporting research concerning ecosystem services.* Urgent research is needed, for example, on the dynamic conditions of key ecosystems in the Mekong region, the consequences of ecosystem changes to individual populations, effective policies to enhance the well-being of vulnerable groups while conserving the ecosystem, and strategies to mainstream ecosystem service concepts in policymaking. However, the initial challenge for mainstreaming the concept of ecosystem services rests on understanding current institutions and governance rather than understanding the conditions and trends of regional ecosystems.

Research requirements

It is common knowledge that the effective management of ecosystems is constrained both by the lack of information about ecosystems and by the failure to adequately deploy the information that does exist. The suggested research agenda for policymakers in the Mekong region is summarized in Table 3.3.

Table 3.3 Suggested research agenda

Research area	Research focus
Ecosystem services mainstreaming	Regional and national ecosystem assessment – Condition and trends of ecosystems and their services – Consequences of ecosystem changes for human well-being Development of ecosystem service indicators – Assessing of ecosystem services indicators in support of policymakers' use of ecosystem service concept – Identifying indicators and their measurement of major linked services that have direct impact on quality of life Ecosystem service valuation reference inventory – Institutional capacity assessment – Survey of mainstream drivers of changes and constraints to influencing them
Ecosystem service investment and enhancement	Gains and losses of ecosystem changes (freshwater, terrestrial, and coastal ecosystems) – Environmental assessment on ecosystem service changes – Economic assessment of benefits and costs at local and national levels – Social and cultural assessment of well-being and livelihoods – Assessing economic feasibility of ecosystem investment – Assessing potential for product development that reduce dependence and impact on ecosystem services
Ecosystem service trade-offs	Assessing ecosystem services trade-offs and impact on vulnerable groups – Determining the balance between extractability and renewability of services – Assessing policy impact of subsidies that create perverse incentives towards ecosystem service deterioration
Ecosystem service market	Assessment of alternative economic instruments and market-based approaches in ecosystem services management – Assessing the potential of PES and other market-based mechanisms at local and national levels under pro-poor growth policy – Determining the economic incentives for consumers as ecosystem service buyers to engage in ecosystem service markets, e.g., carbon finance, user fee, green tax.

There are some key points to be considered in conducting research including:

- using relevant forms of knowledge and information in the assessments, including traditional knowledge;
- incorporating non-market values of ecosystem services in the assessment;
- under the TEV framework, recognizing the bundle of benefits provided by the ecosystem;
- enhancing human and institutional capacity in conducting research, particularly for local communities.

Conclusion

All Mekong countries are at the stage of choosing their long-term development strategies and policies. It is time for them to accept the fact that sustainable economic growth depends on an healthy ecosystem. Under the ecosystem service concept, by integrating equitable growth, the poor, and ecosystem services, development will help to balance pro-growth and pro-poor policies. Although the ecosystem service concept is not new, guidelines for implementing these concepts in decision-making are still at an early stage. As this chapter has pointed out, although the policy tools and examples of their application in actual decision-making are still limited, some tools and examples are already available in the Mekong region.

4

Transboundary Flows of Resources, People, Goods, and Services in the Mekong Region

Louis Lebel, Sopon Naruchaikusol, and Muangpong Juntopas

Places are transformed by flows. In the Mekong region improved transport infrastructure is increasing trade in agricultural and manufactured products. Land ownership and use is shifting in response to external investments, urbanization drivers and demand from newly accessible markets. Enhanced construction capabilities and access to financial services are leading to larger and more complex manipulations of river flows—for irrigation, hydropower, navigation, and flood regulation. Flows have created diverse opportunities as well as risks and burdens to societies.

Places are linked by flows. In the Mekong region there has been an overwhelming, but uneven, increase in interconnectedness. Advances in communication and information technology are changing the way people perceive themselves and others, their aspirations, and how they organize. Modern communities are interacting with states on different sides of borders creating new livelihood opportunities and identities. New dependencies have emerged and old relations transformed. Flows have multiple dimensions with myriad implications for local and regional sustainable development.

Many flows are internal to a country; some go beyond. *Transboundary flows* are defined in this chapter as flows that cross *shared* international borders. Some international flows within the Mekong region are not transboundary as defined in this chapter, for example, between China and

Thailand or Cambodia. Such flows may, however, generate transboundary flows indirectly.

Transboundary flows may be important for several reasons. First, they provide resources that a destination lacks or which are cheaper than those available from local flows. In the latter case they may displace local producers or traders. Second, their dynamics typically lies partly beyond the full control of authorities within one country. International cooperation is typically needed, for example, to deal with issues arising from diversion or storage of river waters by upstream countries, illegal migration, or cross-border trade and investment. Third, when transboundary flows cross borders they are often filtered, tapped, or otherwise transformed. Border institutions—from passport checks and customs duties through patrols to deter and detect 'incursions'—are one of the key ways by which 'states' define themselves and key locations for rent-seeking and corruption.

This chapter explores the main patterns, drivers, and consequences of *transboundary flows* on social development and the environment in the Mekong region. The Mekong region is taken here to include the territory of Cambodia, Lao PDR, Myanmar/Burma, Thailand, Vietnam, and two areas in China—Yunnan province and Guangxi Zhuang Autonomous Region. This geographical area is promoted as the 'Greater Mekong Subregion' (GMS) by the Asian Development Bank (ADB 2008b, 2010b).

Four main types of transboundary flows are distinguished (Table 4.1). *Resource flows* refer to the movement of animals and natural resources. *People flows* refer to the movement of people. *Goods flows* are the movement of agricultural commodities and manufactured products. *Non-material flows* are defined in this chapter as symbolic transactions, in particular, of money, ideas, and information.

Table 4.1 A simple classification of transboundary flows

Types	Examples
Resources	Water, fossil-fuels, timber and other natural resources Fish or wildlife that migrate or move across borders Pests, weeds, and diseases (affecting animals or plants) Regulatory and supporting 'ecosystem services' Air pollution
People	Migrants and refugees Tourists Traders and border communities
Goods	Food and agricultural products Manufactured items – e.g. textiles, automobiles, electronic equipment Narcotics Solid wastes and scraps
Non-material	Financial services such as investments, insurance, loans, and bank guarantees Information—news, science, education, and cultural products Cultural services provided by ecosystems

Resource flows: Energy and water

Resource flows include the movement of animals and natural resources across shared national borders. These can be facilitated as in the case of cargo in a boat, gas in a pipeline or electricity in transmission lines, or be part of natural biophysical processes such as flows of water downstream or migration of fish upstream. Introductions of pests, weeds, and diseases are examples of adverse resource flows.

Fuel and electrical power flows can generate significant foreign exchange earnings for source countries and support electrification and industrialization in recipient countries. Infrastructure and institutions to support these flows are crucial. Thailand already imports natural gas through transborder pipelines from Malaysia and Myanmar. The Trans-ASEAN Gas Pipeline program aims to create a grid of interconnected infrastructure across the ASEAN region by 2020. Construction is underway to expand existing infrastructure with five further lines costing US$7 billion (Clancy 2010). The projects are promoted by oil and gas companies in ASEAN countries.

The level of energy resource flows and their impacts on the environment and development depend, in part, on how well energy markets function and the capacity and willingness of host governments to enforce environmental sustainability standards. More open energy markets are expected to reduce costs. Modeling studies generally show financial benefits of regional energy trade (Watcharejyothin and Shrestha 2009). Energy trade could also reduce emissions of greenhouse gases and other pollutants. But markets are not sufficient: effective regional energy development also needs improved cooperation (Yu 2003), for example, in dealing with the transboundary social and environmental impacts of projects.

One argument against high dependence on single or a few energy sources is that it makes a country vulnerable to politics, natural disasters, and other factors that might affect supply. The security of a natural gas supply may be enhanced by diversifying sources and energy mixes and adopting more flexible transport systems. Transporting liquefied natural gas greatly reduces transport costs and can make more distant sourcing competitive with traditional pipelines. The energy security discourse is frequently used by actors in the Mekong region to justify large-scale energy projects (Simpson 2007). The transboundary consequences of the investments and resource flows from large projects in the energy sector require careful scrutiny as there are both transboundary opportunities and risks.

Hydropower has been widely promoted as a key alternative energy source in the Mekong region (Bakker 1999). The Mekong region is home to five large rivers: three are transboundary—the Nu–Salween, Mekong and Red rivers, and two—the Irrawaddy, the Chao Phraya—have basins largely within one country. Proposals to divert, link, store, and 'improve' rivers abound. Several large projects have been completed; many others are being built or proposed (Grumbine et al. 2012).

The extensive storage, diversion, and regulation of transboundary water flows have important implications for ecosystems and livelihoods. Most evidence indicates that dams on the lower mainstream or major tributaries will have major impacts on fisheries. Floodplain ecosystems are dependent on the nutrients and sediments accompanying the seasonal flood pulse (Lamberts 2006; Kummu and Sarkkula 2008). The largest impacts on migratory fish species will likely come from 11 proposed mainstream dams (Sarkkula et al. 2009) for which mitigation will be extremely difficult given high species diversity. A likely outcome is the loss of freshwater species and ecosystem services following construction (Baran and Myschowoda 2009; Dugan et al. 2010). A large number of the endemic fish species in the Mekong basin are already endangered (Valbo-Jorgensen et al. 2009). The continuing significance of capture and culture fisheries to people living in the Lower Mekong Basin is very high (Baran et al. 2007). Many species are traded in local markets and a few are an important part of regional and extra-regional value-chains (Loc et al. 2010). Fisheries also generate employment in processing, trade, and retail (Bush 2004; Glemet et al. this volume).

Although the likely adverse impacts of dams on fisheries are recognized by most promoters of hydropower energy development, this trade-off is largely viewed as acceptable. This peculiar position can be explained by a prevalent view that the regional fisheries have little future and little role to play in poverty alleviation or modern development (Friend 2009; Sneddon and Fox 2012).

People flows: Migrants and tourists

People flows include voluntary and temporary movements such as tourism and temporary legal and illegal labor migrants as well as involuntary movements of people such as refugees, trafficked women or children, and international criminals. People are different from other things that flow because they can actively respond through resisting or facilitating further flows. People are also conduits for information, values, and beliefs.

Migrants

In the Mekong region the largest migrant flows are to Thailand. Transboundary flows of migrant workers into other countries are smaller. The typical patterns of migration vary among pairs of countries, but economic factors figure prominently (ADB 2009c).

In 2005 it was estimated that there were about 1.8 million foreign workers in Thailand with about 38 percent unregistered (Martin 2007). A survey by the Ministry of the Interior carried out in early 2007 suggested that the number of irregular or unregistered migrants from Myanmar, Laos, and Cambodia was around 2.8 million (Pholphirul and Rukumnuaykit 2010). Thailand's policies and border practices swing back-and-forth depending on labor needs in agriculture, fisheries, and industry (Martin 2007; Kaur 2010). Irregular registrations of illegal laborers is done to meet quotas or justify repatriations (Brees 2008). Illegality in migration is widely believed to make migrants vulnerable to trafficking (Segrave 2009).

Fig. 4.1 Migrant workers registered in Thailand from the three main source GMS countries

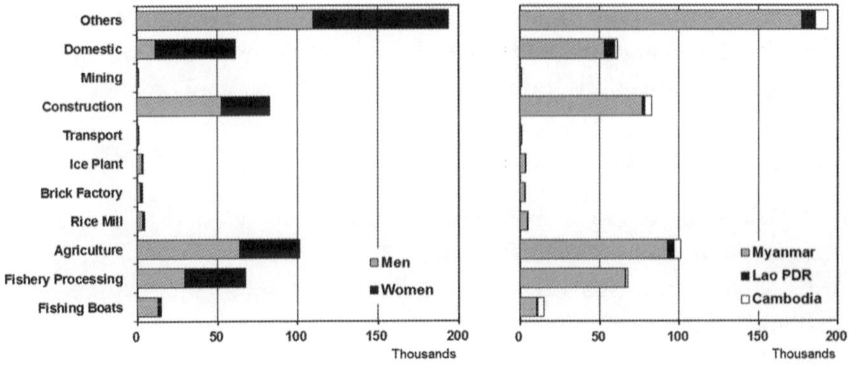

Typical concerns in destination countries are competition for jobs, transmission of disease, and criminal activity. In general, evidence for these impacts is modest compared to the amount of attention they receive—a pattern easiest to explain as prejudice and discrimination. Most jobs taken by migrant workers are those which local workers do not want or are complementary (Caouette et al. 2006; Chalamwong and Prugsamatz 2009).

Not much has been written about the benefits migration brings to destinations and host communities although these may be substantial (ADB 2009b). In 2007 migrant workers made up between 5–10 percent of the Thai labor force (Caouette et al. 2006, Martin 2007). More than half of the country's domestic workers are estimated to be migrants (Martin 2007). Thailand is now highly dependent on migrant labor, and in the future may become more so as demographic shifts unfold. Estimating the net economic contribution of migrants is complex as there are several factors that should be taken into account including: relative skills and their impacts on productivity; and the effects of migrants on national wage levels (Pholphirul and Rukumnuaykit 2010).

Migration also has impacts on source locations. Remittances play an important role for maintaining households (Deelan and Vasuprasat 2010). Most funds are transferred back through informal channels like friends or family members. Women send back more money than men (Deelan and Vasuprasat 2010). Overseas migrants not only bring back money when they return—they also bring back experience and different views of what constitutes a good life (Plungsricharoensuk 2009).

Formal bilateral cooperation on migration between Thailand and source countries—Cambodia, Laos, and Myanmar—has been established but implementation has been slow. Regional arrangements are underdeveloped and international frameworks and principles are not fully accepted or followed by countries within the Mekong region.

Tourists

In 2011 the number of tourist arrivals (in millions) was: Thailand (19.1), Vietnam (6.0), Cambodia (2.9), Laos (2.7), and Myanmar (0.8). The intra-ASEAN share of arrivals was: Laos (80 percent), Cambodia (38 percent), Thailand (29 percent), Vietnam (14 percent), and Myanmar (12 percent) (ASEAN 2012). More detailed information for 2010 illustrates that the significance of shared borders is not two-way. In the first three-quarters of 2010 Thais dominated (60 percent) tourist arrivals to Laos; second were Vietnamese (17 percent), and third, Chinese (7 percent) (Mantanarat 2010). In contrast Laotians made up only 4.3 percent, Vietnamese 2.5 percent, Cambodians 0.9 percent, and Burmese 0.6 percent, of the 15.8 million tourist arrivals to Thailand in 2010 (ATF 2011).

Chinese made up just under half of Asian tourist arrivals to Vietnam in 2007 (Truong and King 2009). Prospects are for greatly increased transboundary flows of tourists from China to other countries in the Mekong region, perhaps via Yunnan (Kaosa-ard 2007). Yunnan is itself an important domestic tourist destination with ethnic minorities a key element of the tourist product (Doorne et al. 2003). In key locations like Xishuangbanna, however, domestic tourists continue to greatly outnumber international tourists (Kiyoshi 2002). One of the dilemmas in managing tourism in which local culture is an important part of the tourist product is maintaining the appearance of authenticity in the face of growing demands for reproducible and commodifiable experiences (Shamshub 2010).

The contribution of tourism revenues as a percentage of GDP in 2011 in the Mekong region countries was high for Cambodia (15 percent), intermediate for Thailand (8.2 percent), Laos (6.8 percent), and Vietnam (5.4 percent), and lowest for Myanmar (0.2 percent) (WEF 2012). Tourism in the region has high potential given its diversity, central location in Asia, and relatively low prices, but inadequate infrastructure and environmental protection are key limitations.

In general the distribution of tourism income is not as equitable as that from agriculture as it tends to favor those with capital to invest;

but it can be better than that from other non-farm sources (Kaosa-ard 2007). Environmental and social costs are not easy to measure, but can be substantial; it is also possible to undertake measures to improve environmental quality and create jobs for low-income households through tourism (see: Kumar et al. this volume). A study of the tourism industry in Chiang Mai suggests a high level of domestically retained value, implying significant contributions to local development (Kaosa-ard 2007). But this varies a lot: a significant fraction of the income generated by tourism can flow back out or *leak* from a destination as when hotels and other services are owned by foreign companies or tourists purchase primarily imported goods and services. Such leakage can be as much 20–40 percent in the poorer Mekong countries (ADB 2005).

Regional cooperation on tourism development can be relevant to handling crises, attracting greater numbers of visitors overall and facilitating tourist movements among sites in different member states (Anastasiadou and de Sausmarez 2006; Wong et al. 2010). The GMS Tourism Sector Strategy outlines areas of collaboration among Mekong countries in developing tourism destinations and logistics as well as targeting specific zones for tourism development in each country (ADB 2005). The strategy followed on from the 2000–2002 Mekong Tourism Development Project through which ADB made available low interest loans for tourism development in the Lower Mekong Basin (Laws and Semone 2009). One of the seven core strategies was to streamline cross-border tourism. Key ideas are to improve and expand border checkpoints with visa-on-arrival and visa extension services. The idea of a GMS-wide visa was also proposed. Common visa arrangements—as found in the EU for example—have not yet been implemented in ASEAN or the GMS. In 2009 and again in 2010 GMS countries cooperated in the "Visit Mekong Year" campaign (ADB 2009c; Chheang 2010). The Mekong Tourism Coordination Office promotes tourism through a coordinated website.

Goods flows

Goods flows include trade in agricultural produce and manufactured goods as well as adverse goods flows like exported industrial wastes and narcotics. Trade in the Mekong region has a very long history from Yunnan caravans to commerce along the Mekong River (Giersch 2010; Tapp 2010).

Trade

Many free trade agreements (FTAs) were signed by Mekong region countries between 2000 and 2010: China (11), Thailand (9), Vietnam (6), Cambodia (5), Myanmar (5), and Laos (4). Others are being negotiated. Most are bilateral although lower-income countries have often worked through ASEAN (Kawai and Wignaraja 2010) and may increasingly do so after the formation of the Asian Economic Community. As much as 80 percent of Lao PDR's and Myanmar's country trade is under FTAs and around 50 percent for other Mekong region countries. Most trade in the Mekong region has been in manufactured goods. Agricultural products have been largely or partly excluded from these agreements as a result of lobbying or social concerns—an issue which has also impacted negotiations and schedules for the launching of an ASEAN Economic Community.

The sum of the total value of imports from and exports to GMS member countries has grown substantially since 2000 for all countries (Fig. 4.2). In terms of total value China, Thailand, and Vietnam dominate transboundary flows (Fig. 4.2). But these countries also have larger economies and levels of trade with the rest of the world, and in the case of statistics from China, other provinces which are not strictly part of the GMS. Assessing the significance of intra-regional trade for member countries and provinces necessitates a closer look.

Fig. 4.2 Total value of intra-GMS trade

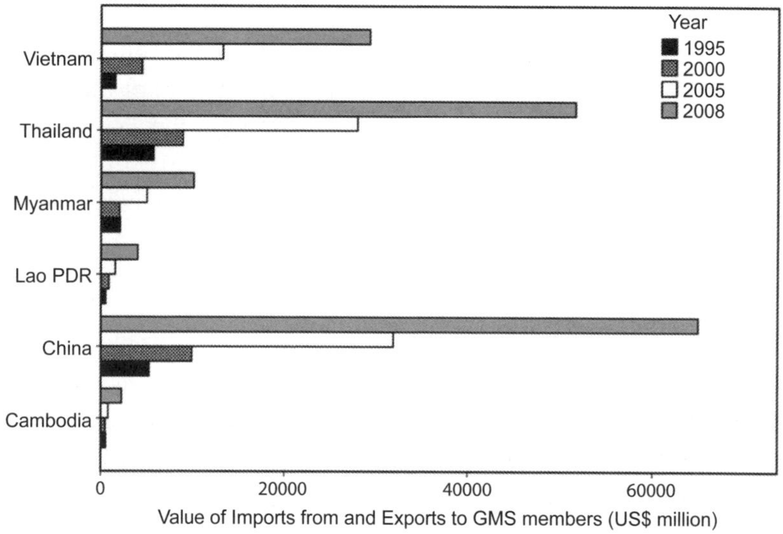

Source: Redrawn from UN Comtrade data cited in AusAID (2010).

Overall exports from the Mekong region grew on average at 12.7 percent a year during 1992 to 2007 (Table 4.2). Within the Mekong region the volume and value of trade across shared borders remains particularly important to Laos, Myanmar, and Yunnan (Than 2006). In 2004, for example, 60 percent of Laos's imports were from Thailand (Sussangkarn 2006). Over the period 1994 to 2006 exports from within the Mekong region to China grew annually by 22 percent (Jha et al. 2010). Overall, between 2000 and 2006, intra-GMS trade grew by almost 27 percent per year. At the same time intra-region trade remains modest in many areas and relative to overall trade for most countries; the sum of intra-GMS exports to total exports was just 6.5 percent in 2006 (Table 4.2).

In the case of Yunnan, studies suggest a trend towards greater trade with more developed countries beyond GMS (Poncet 2006). Trading more broadly appears to have been beneficial at times, for example, during the Asian financial crisis, where sectors like the Cambodian garment industry which focused on trade with the United States and the European Union being much less impacted than those caught up in the local contagion.

Table 4.2 Transboundary trade in the GMS (% per year over the observed period)

Countries	GDP Growth	GMS–World (1992–2007)		GMS–Asia (2000–06)		GMS–GMS (2000–06)		Intra-GMS Exports to Total Exports (2006)
		Export Growth	Import Growth	Export Growth	Import Growth	Export Growth	Import Growth	
Cambodia	8.9	20.3	20.2	4.6	22.9	2.9	30.5	2.5
Lao PDR	6.5	17.6	16	21.8	17.7	18.8	18.2	51.7
Myanmar		13.6	10.5	17.4	11.3	38.1	22.9	50
Thailand	4.1	11.7	9.5	15.4	12.8	22.5	24.3	4.7
Vietnam	7.6	19.9	22.8	14.9	24.6	16.4	26.5	3.1
PRC–Yunnan	10.7	16.3	23.2	20.4	20.5	28.7	32.6	30.8
PRC–GZAR		9.7	19.2					
Total GMS	5.5	12.7	11.7	15.8	16.2	26.7	26.9	6.5

Note: Statistics from GMS–Asia and GMS–GMS do not include Guangxi Zhuang Autonomous Region (GZAR).
Source: Based on ADB (2008).

Reorganization and relocation of production within multinational organizations can have complex impacts on transboundary flows within the Mekong region. In the communications technology sector, for example, China has grown rapidly as a major exporter to the United States and Japan, overtaking other Southeast Asian countries—a pattern that in part reflects relocation of production facilities to China (Xing 2010). At the same time China imports a substantial amount of components from countries such as Thailand and Malaysia. Although China is often viewed as a labor-intensive exporter, as this case illustrates, it is also an importer of manufactured components for final assembly (Coxhead and Jayasuriya 2010).

Agricultural products

Agricultural exports to China from Mekong region countries have grown substantially in the last decade and are likely to grow much further in coming decades (Jha et al. 2010). There are likely to be both opportunities and challenges (Johnston et al. 2009). Thailand's FTA with China in 2003, for example, impacted the competitiveness of shallots grown in northeast Thailand (Boossabong and Taylor 2009). Investment and trade are linked, with China and Thailand in particular already playing important roles in neighboring countries. Differences in monetary and exchange rate policies among countries contribute to price differences, creating opportunities across borders especially for agricultural commodities affected most by local prices (Russell 2008). But, volatile currency exchange rates result in lower exports of agricultural commodities from Thailand as exporters tend to avoid risk (May 2010).

Sugar in Thailand is a good example of a highly regulated agricultural commodity (Sorapipatana and Yoosin 2011). Under the 1984 Sugar and Cane Act a fund is used to help stabilize prices. A state body also oversees sugar exports. Traders are frequently accused of hoarding; illegal trade with neighboring countries is rife since prices are often double those in Thailand (*Bangkok Post* 2010a). Sugarcane and its byproduct molasses are increasingly used for bioethanol production and Thai firms are investing more in feedstock production and mills in neighboring countries in anticipation of growing regional and global biofuel markets. Khon Kaen Sugar, for example, has a mill in Cambodia, along with a 90-year farming concession for 20,000 ha in Koh Kong, and a 30-year concession for 10,000 ha that will supply a mill in Savannakhet, Laos (*Bangkok Post* 2010b). Mitr Phol Sugar, Thailand's largest company, registered in Laos under the name Mitr Lao, has a further 10,000 ha in Savannakhet (Nolintha 2011). Most sugarcane for biofuel is grown under land concessions or contract farming arrangements to ensure feedstock supplies. Contract farming appears to have more benefits for small farmers than concession arrangements (Voladet et al. this volume). Most of the sugar from Laos is exported to the European Union after passing through roads and ports in Thailand, underlining the importance of transboundary flows for this commodity.

The case of rubber is particularly instructive as it combines elements of goods, resource, and people flows. Chinese demand for natural rubber is the key driver of trade within the Mekong region. Investments in rubber have also been justified by the Chinese and Lao governments as an opium-replacement policy (Cohen 2009). To secure supplies, Chinese firms have invested directly in rubber production, especially in Luang Namtha and neighboring provinces in northern Laos (Ziegler et al. 2009). Chinese migrant labor plants and taps trees in concessions granted by the Lao government or through contract farming arrangements (Shi 2008). Latex is exported back across the border into China; indeed the entire production–consumption system is transnational, from seedlings through to exports. Thai and Vietnamese firms are investing in southern Laos. The environmental and livelihood consequences of the very rapid expansion of rubber in Laos have been uneven and profound (Cohen 2009). Large-scale projects with associated land concessions (Schoenweger and Üllenberg 2009), in particular, have increased land insecurity of small farmers and not provided the poverty alleviation benefits expected (Baird 2010). On the other hand, there is also evidence that smallholder rubber farmers in Xishuangbanna have created informal cross-border networks that successfully established rubber in northern Laos to the benefit of Akha and Tai farmers on both sides of the border (Sturgeon 2013).

Non-material flows: Money and information

Non-material flows are symbolic rather than material transactions. Such flows are often more flexible than flows of physical or material goods although they are often closely linked.

Money

Inbound transboundary investment flows and trade in related financial services varies greatly in importance among countries in the Mekong region. The absolute size of foreign direct investment inflows is much larger for Thailand and Vietnam than the other countries (Fig. 4.3). Relative to size of their economies, inflows are important for all Mekong countries, but especially Cambodia, Laos, and Vietnam, where FDI has grown a lot in recent years.

Fig. 4.3 FDI flows in the Mekong region

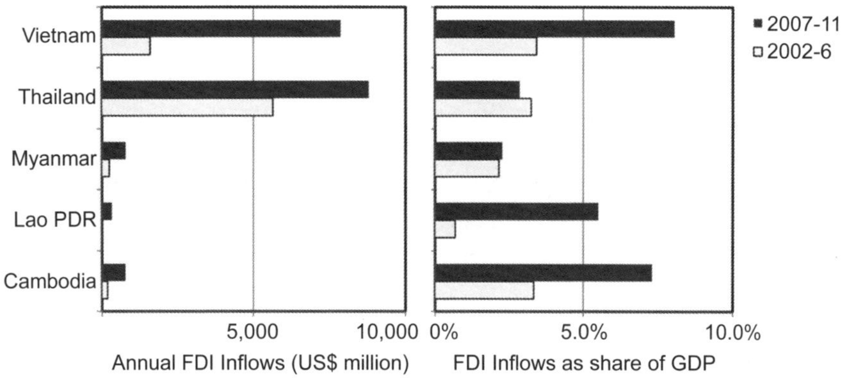

Source: Calculated from data in UNCTAD (2013).

The top three investor countries in Laos between 2001 and 2008 were, in order, Thailand, China, and Vietnam. Most FDI was directed at mining and hydropower (Rutherford et al. 2008). China's trade with, and investments in, Laos substantially influences manufacturing, agriculture, and extractive industries in that country (Goto 2011). Investments at the local level sometimes trigger discontent, even for what might appear at first as uncontroversial projects like a bridge on the Thai–Lao border (Lin and Grundy-Warr 2012).

A study completed in 2009 estimated as much as 13 percent of the land area of Laos has been granted as concessions to foreign companies (Schoenweger and Üllenberg 2009) for mining and large-scale agro-industries. The lack of transparency in granting concessions and leasing contracts facilitates corruption and sometimes results in loss of access rights for local people. A recent report gives examples where the same area of land was granted to different firms, concessions were granted under agreement to a village head without consulting villagers, and misleading descriptions allowed mature forests to be cleared for agriculture (Schoenweger and Üllenberg 2009). Foreign investment in land concessions in Cambodia is also widespread with most dedicated to wood or agro-fuel and food production (Üllenberg 2009). Officially, economic land concessions cover about 5.5 percent of Cambodian territory, with foreign investors involved in more than half of the 59 concessions larger than 1,000 ha and covering an area of almost 300,000 ha in total. Legal frameworks exist on paper, but implementation is weak. While concessions have led to significant increases in employment, wages, and transport infrastructure, they have also had negative impacts on indigenous farmers displaced without adequate compensation (Üllenberg 2009).

Myanmar is highly dependent on its Mekong region neighbors as a source of foreign investment. Thai businesses, including chambers of commerce and business associations, have actively shaped Thai foreign policy relations with Myanmar since 1988 (Pongyelar 2007). As a consequence, Thai investment in Myanmar grew strongly in many areas including banking, tourism, manufacturing, retail, and natural resource sectors. Transboundary flows of goods and investment to and from Thailand grew despite trade sanctions by the United States and allied countries and disputes over refugee and migrant flows. In 2002, Thailand was the third highest source of FDI in Myanmar, after Singapore and the United Kingdom (Pongyelar 2007). An outstanding but largely ignored issue is how to improve the corporate social and environmental responsibility practices of Thai, Chinese, and other foreign businesses investing in Myanmar (Middleton et al. 2009). China's hydropower industry, for example, is investing in overseas markets with the support of the Export–Import Bank of China, and other financial institutions, but has yet to have adopted internationally accepted standards for large infrastructure development (McDonald et al. 2009).

Information

In the last decade there have been dramatic shifts in the organization and scale of information and communication technology (ICT) services across the Mekong region. Whereas most services were at one time supplied by state enterprises with little competition or independent bodies (UNESCAP 2004), ownership patterns and regulations are now much more complex.

Transboundary flows of investment and information have expanded with the scaling up of activities of firms and cooperation among governments. The Vientiane Plan of Action for GMS Development 2008–2012 included strategies for improving cooperation in telecommunications (Chheang 2010), which has led to the notion of a GMS Information Superhighway Network (ADB 2010a). Greater private sector involvement is promoted. Access to telephones and the Internet has grown rapidly, but starting from very low baselines in the least developed countries of the region (UNESCAP 2004; Lim and Wi 2005). A special feature of the region is the now relatively high coverage of mobile phone subscriptions in all countries apart from Myanmar (Fig. 4.4). Internet use remains comparatively low in Myanmar, Laos, and Cambodia (Fig. 4.4). According to comparisons against the ICT-opportunity index in the Mekong region, countries are ranked as follows, from lowest to highest: Myanmar, Laos, Cambodia, Vietnam, Thailand, and China (Emrouznejad et al. 2010).

Fig. 4.4 Telecommunication indicators (per 100 inhabitants) in GMS countries, 2009

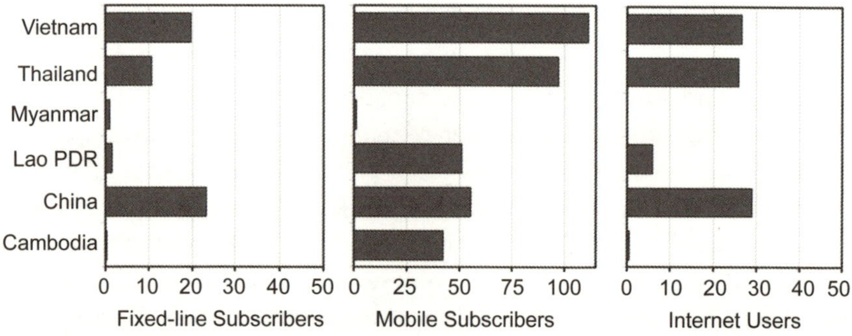

Source: ITU 2010.

The flow of ideas across borders can change how recipients view critical issues, such as environmental change or democracy and public participation. Governments in the region have at times been very concerned and continue to exert significant influence over information flows through mainstream communication channels.

The mass media has become increasingly privately owned and commercialized, but content is still highly controlled (Garden and Nance 2007). Coverage of regional issues in respective mainstream national media is modest and usually with a strong 'national interests' slant. A remarkable exception has been the cohort of journalists created through the "Imaging Our Mekong" program run by the Inter Press Service (IPS) and Probe Media Foundation. The program has trained more than 200 journalists since 2001 through a series of annual fellowships directed at developing comparative and cross-border news stories (e.g. IPS 2006).

One of the most contentious domains of transboundary flows are those related to popular culture. A common thread in studies of globalization in the Mekong region are concerns about the impacts of popular or mass culture on traditional values and systems of belief. Theeravit (2003), for example, sees the key impacts of globalization as shifts towards capitalist values of individualism, materialism, and consumerism. Laotian youth appear to be as keen to experience the culture they hear and see on Thai mass media as their government is to control and criticize it (Pholsena and Banomyong 2006). While some commentators fear the generation gap, others see an opportunity to escape past forms of patron–client relationships and military–elite political organization. Transboundary information flows have multiple and complex consequences and as a result are simultaneously promoted and resisted by different elements in society.

Another type of non-material flow that is 'information-like' is the notion of virtual flows or the resources that go into making a product (Ma et al. 2006). Transboundary virtual flows of water, for example, could be calculated based on the volume of water it took to produce a certain volume of agricultural commodity traded across a border. Although the water does not physically flow from one country to another—just the product—in some sense the water was allocated (or assigned) to consumers in the destination rather than source country. As a source of influence on policy and practice they might constructively be thought of as a type of information flow. Not much analysis of virtual flows has been carried out so far in the Mekong region. Thailand as a major food exporter, for example, exports large amounts of virtual water—and much of this green rather than blue water (Chapagain and Hoekstra 2010).

Discussion

The transboundary flows of goods and investment within the Mekong region are often not as large as those promised in political discourses promoting regional economic integration or counter-discourses expressing concerns over globalization. Domestic and external flows are often large relative to inter-regional flows and depending on scales and units of analysis may overwhelm the latter. This is especially true for the most open and industrialized economy of Thailand, whereas for Laos or Myanmar intra-region transboundary flows are much more significant. Differences in competitiveness among countries add another layer of explanation for differences in levels of transboundary flows found within the region.

Transboundary flows of people vary widely over time, by boundary, source, and destination conditions. People move across borders with varying degrees of freedom. Movements out of Myanmar into Thailand from border areas to flee war and suppression include an involuntary element arising out of lack of options. Tourist movements are at the other extreme. The largest migrant and tourism flows are to Thailand. Fluctuations in economic conditions have major impacts on both types of flows. Changes in government policies influence the fraction of flows which are legal and documented. Lives in the borderlands are impacted by different forms of people flows, including, for instance, conflict within Mynamar, broader economic conditions that effect tourism, and state efforts at regulating trade and movements through and around borders. Migration has repercussions, negative and positive, for migrants and their families that stay at home. Tourism has often been promoted as a way to alleviate poverty. In practice, the impacts of tourism development are mixed with examples of both increases in local income sources as well as high leakage and adverse social and environmental impacts.

For transboundary flows of resources like water or clean air, it is alterations in flows and changes in quality rather than just quantity which are of concern to destinations. For mobile resources such as fish, separating local versus transboundary flows is challenging, but there is sufficient evidence to be concerned about transboundary impacts on fisheries from dam cascades in the mainstream of the Mekong. Regional gas pipelines and power distribution grids also raise environmental and social concerns around construction projects and their real effects on energy security. Alongside infrastructure are issues of institutional development that can facilitate markets and trade for energy and other natural resources.

Investment and financial service flows have received a lot of attention in analyses of the drivers and consequences of regional economic integration because of the impacts they have on resource extraction and the movements of people and goods. Other types of non-material transboundary flows such as those related to ideas, technical skills or culture have received less attention. These latter flows are more indirect but profound, influencing the evolution of other flows as they shape perceptions, beliefs, and expectations. National and regional perspectives are in a dynamic struggle over the merits and adverse impacts of transboundary flows.

Transboundary flows are not independent of each other. This is a key rationale for considering diverse phenomenon like trade, migration, rivers and foreign investment together under a common framework around the notion of flows when trying to understand regional development processes. International trade and investment agreements, cooperation on conservation and exploitation of natural resources, introduction of standards, and national policies to protect particular sectors or promote exports, influence flows in diverse ways. Patterns in transboundary flows, for instance, are strongly influenced by roads and what goes on at borders. International regions are formed, defined, and reinforced by these dense and complex networks of transboundary flows.

Transboundary flows create opportunities, risks, and burdens for social development, economic growth, and environmental sustainability. Separating out impacts of transboundary flow from other internal and external flows is not straightforward and has rarely been done carefully. Nevertheless, there is evidence for some adverse direct and indirect effects. Benefits may not go to low-income groups because they are losing access to agricultural or forest land as a result of transboundary investments in plantations, commercial crops, and other activities that result in land concessions being made to foreign firms. Likewise jobs that are created may not go to those in low income groups because they do not have the required skills. At the country level the challenge for smaller, less developed states like Laos and Cambodia is to maintain some control over their natural resources as they become integrated into the other larger economies in the region.

Transboundary flows are not universally important, especially when their magnitude and impacts are compared with internal and global flows. Nevertheless, the interactions stimulated by flows across shared borders are significant to well-being and sustainability in particular places—sources, border areas, and destinations. Moreover, transboundary flows often take on a life of their own in public discourse with competing narratives of regional economic integration on the one hand and loss of identity and security on the other. For the most part these perceptions are not backed by careful study or clear evidence. More critical research on transboundary flows—their drivers, consequences, and interactions is needed. The case studies in the rest of this book address some of these key questions and raise many others.

5

Urbanization and Sustainable Development in the Mekong Region

Carl Middleton and Dusita Krawanchid

In the Mekong region, a growing proportion of the population lives and works in urban areas. At best, urban areas, as centers of economic growth, employment, education, and innovation can offer opportunities for economic and social development and cultural enrichment. At worst, urban areas can be centers where there is a lack of basic services, employment opportunities, and decent housing—places of poverty and environmental degradation.

Ensuring that urbanization is sustainable and fair is one of the greatest challenges facing policymakers and the public in the region. Urbanization to date has been shaped by rapid economic growth, population growth, expanding industrialization, and deepening integration into the global economy. The process of urbanization has also transformed the region's rural areas and deepened their linkages with urban centers, including across borders.

Our Common Future offered the most widely cited definition of sustainable development, namely to "meets the needs of the present without compromising the ability of future generations to meet their own needs" (WCED 1987: 43). It broadly captures the economic, social and environmental dimensions of sustainability, and has been both lauded for its flexibility and criticized for its amorphousness. In Peter Evan's book, *Livable Cities: Urban Struggles for Livelihood and Sustainability*, the livable city is characterized as one where work is located near decent housing and has wages that are commensurate with rent, basic services are accessible to all, and the urban environment is both healthy and has sufficient green space (Evans 2002: 1–2). Meanwhile, for the city to be sustainable, it must also not create an ecological footprint that degrades its hinterland through resource consumption and waste generation. Also important is intergenerational justice, both in the form of ecological sustainability, and ensuring livelihoods and quality of life for the present and future generations.

This chapter maps out the opportunities, challenges, and prospects for urbanization in the Mekong region. We highlight how the experience of urbanization differs by socioeconomic groups, and that creating livable cities for all inevitably is, and will continue to be, an intensely political process. Managerial and institutional deficits in urban planning are often identified as important reasons for environmental and social shortcomings in the region's urbanization processes (Yuen 2009: 17, 34–40), where influential international and domestic investors and large urban landholders have often bypassed planning processes and regulations (Sheng 2010: 139). Counterbalancing these interests in the pursuit of livable cities requires democratized, deliberative, and decentralized decision-making, accountable government, and the participation of empowered citizens. The chapter furthermore argues that creating livable cities and ensuring sustainable urbanization can only be attained in the context of ensuring sustainability across the wider region. Therefore, it is crucial that the interconnectedness between urban areas and rural hinterlands—for example the flows of people, finances, resources, and waste—are recognized and accounted for in decision-making, including those that are principally considered as urban.

The next section identifies important trends in urbanization, including population growth and migration, urban–rural connectedness and peri-urban expansion, economic growth and the impact of globalization, and urban poverty. The following section considers in greater depth the relationship between urbanization and the environment. It first conceptualizes the urban–environment nexus, before examining current challenges to the urban environment in terms of brown, grey, and green environmental issues, and then the linkages between urbanization and regional environment sustainability. The fourth part of the chapter discusses the prospects for democratizing urban governance, and the final section offers conclusions and explores the future of urbanization in the region.

Urbanization in the Mekong region

Urbanization, population growth, and migration

Urbanization is a process which leads to a higher proportion of a country's total population living in cities and towns as a result of population growth, migration, and incorporation of rural areas into urban areas. In the Mekong region, this process has been a partially planned and partially organic process. The region's estimated total population was 223 million people in 2010, and is expected to reach 229 million people by 2025 (UNDESA 2012). At present, 32 percent of the region's population lives in urban areas; this is expected to rise to 40 percent by 2025 (Table 5.1).

Table 5.1 Urbanization in mainland Southeast Asia

	Urban population ('000) (% of total in brackets)			Rural population ('000)			Urban annual growth rate (%)		
	1990	2010	2025	1990	2010	2025	1990–95	2005–10	2020–25
Cambodia	1,482 (15.5)	2,801 (19.8)	3,975 (23.8)	8,050	11,337	12,711	5.32	1.79	2.49
Lao PDR	647 (15.4)	2,054 (33.1)	3,563 (48)	3,545	4,147	3,866	5.06	5.30	2.95
Myanmar	9,664 (24.6)	15,388 (32.1)	21,777 (40.9)	29,604	32,575	31,417	2.23	2.45	2.12
Thailand	16,793 (29.4%)	23,315 (31.1%)	29,704 (40.8)	40,279	43,486	43,108	1.46	1.62	1.63
Vietnam	13,591 (20.3)	26,700 (30.4)	39,837 (40.1)	53,510	61,149	59,499	3.78	3.26	2.30

Source: UNDESA (2012).

Large cities, where populations exceed 5 million people, are becoming increasingly commonplace; whilst in 1950 there were only 8 cities in the world with a population greater than 5 million, by 2000 there were 42 (Cohen 2006). In the Mekong region, Bangkok and Ho Chi Minh City have already surpassed 5 million people, with Yangon likely to follow soon (UNDESA 2012). It is not city size alone, however, that represents a challenge to sustainable urbanization, but also the rate of growth; for example, whilst it took London 130 years to grow from 1 million people to 8 million, it took Bangkok only 45 years (ADB 2008a).

Urbanization throughout the region is furthermore characterized by the primacy of major cities over secondary cities and towns. In 2011, Phnom Penh, Bangkok, Yangon, Ho Chi Minh City, and Hanoi had estimated populations representing 55 percent, 36 percent, 28 percent, 23 percent, and 11 percent of each country's total urban population respectively (UNDESA 2012).[1] Analyses of the social and environmental consequences of urbanization have tended to focus on the growth of these large cities, yet secondary cities and towns are also growing rapidly (Sheng 2010). The urbanization of secondary cities and towns probably represents greater social and environmental challenges as these areas generally have weaker basic services, higher poverty rates, and local governments that are institutionally weaker (Tacoli 2003; Satterthwaite 2006).

Migration from rural to urban areas may be either domestic or cross-border, and can be permanent, seasonal or temporary. Trends in migration to urban areas also reflect a rapid transformation in agrarian rural livelihoods. There has been a shift from subsistence to larger-scale commercialized agriculture, and livelihoods and poverty are becoming delinked from the land and farming as rural occupations diversify outside of agriculture (Rigg 2003, 2006; De Koninck 2012). As rural–urban economic and social connectedness deepens (see below), people living in rural areas are becoming more mobile and their livelihoods delocalized; remittances are playing a growing role in rural household incomes. These wage remittances can pay for children's education, better housing, and enable investment in agriculture, and rural migrants working in towns and cities are central to the region's urban workforce. On the other hand, labor required at key times of the agricultural calendar may be lost, and a loss of skilled rural entrepreneurs can result in long-term rural economic and environmental decline.

In the Mekong region, Thailand is the main destination country for international migrants, where approximately 3.1 million people from Myanmar, Cambodia, and Lao PDR are working, the majority of whom are unregistered, irregular migrants (Huguet et al. 2011). They principally work in dirty, difficult or dangerous jobs—the so-called 3Ds, including those in construction, manufacturing, plantation, fishing, domestic help, and the entertainment and sex industries. Information on domestic migration is comparatively sparser. In Vietnam, during 1994–99, 36 percent of inter-provincial moves were rural–urban. Deshingkar (2006), and Dang (2003) conclude that rural-to-urban migration has been important to the surge of urbanization in Vietnam. In Thailand, rural-to-urban migration has increased since 1985, especially to Bangkok from Northeast Thailand, where in the mid-1990s remittances accounted for almost a quarter of household income (Dang 2003). Long-term migration has seen a net loss in population in the Northeast, whilst Bangkok and its periphery in the central region have been the main area gaining migrants (Huguet et al. 2011). In Cambodia, around 35 percent of the total population is considered as internal migrants by the government, but most of these internal movements are intra-provincial and very short range (IOM 2006). Phnom Penh, however, has seen its population double between 1998 and 2008 from 567,860 to 1,237,600 residents, and according to the Ministry of Planning (2012) 80 percent of this growth was due to net migration into the city.

Urban population growth is largely due to natural population increase within urban areas themselves, and the incorporation of surrounding residential and farming areas (i.e. peri-urbanization). Net migration into urban areas can also make a contribution (Yuen 2009: 26). The temporary status of a significant number of urban residents, who migrate for seasonal employment into urban areas and who may not have legal resident status, can make predicting urban population growth and stability uncertain, with implications for urban policies and planning. Urban population change, is also shaped by more unpredictable considerations, such as economic growth or decline. Yet, in the face of this uncertainty, the 'readiness of the receiver' has implications for the sustainability of any particular urban area.

Urban–rural connectedness and peri-urban growth

It is now widely recognized that the concept of an urban–rural divide has become a "misleading metaphor… that oversimplifies and even distorts realities" (Tacoli 2003: 3). The interaction of rural and urban is most apparent in the landscapes of peri-urban areas, but readily extends into urban hinterlands via production systems, markets, service provision, and livelihood strategies. For example, demand for food and other natural resources in urban centers for local consumption or subsequent sale to international markets shapes rural land use and agricultural production via market mechanisms. Conversely, manufactured products and imported goods flow from urban to rural areas. In addition to the migration of people, other interactions include: information flows, including on market prices and employment opportunities; services, for example, of secondary schools, higher education, government offices, and hospitals; and financial flows, including remittances, and also investments and credit from urban-based institutions (Tacoli 2003).

Whilst the relationship between urban and rural areas is complex and place- and context-specific, it is deepening, resulting in both positive and negative interactions, opportunities and risks (Hardoy et al. 2001; Rigg 2003; Tacoli 2003). The experience of urbanization in both urban and rural areas differs by class, gender, ethnicity, and other socioeconomic groupings. Household, local, national, regional, and global economies overlap, shaped by multi-scaled governance mechanisms and a panoply of actors ranging from local and national governments, to domestic and transnational businesses, and civil society.

Peri-urbanization, which blurs the spatial distinction between rural and urban areas, presents a particular set of challenges and opportunities to sustainable urbanization. Villages and prime agricultural land have been transformed by housing, roads, and industry, changing landscapes, and modified hydrological and biogeochemical cycles. The complex process of peri-urban expansion may be partly a result of government policies including tax holidays, loans and market support, and zoning regulations, as well as flows of foreign direct investment (FDI) into export-oriented industries and manufacturing. There has also been an expansion of residential, commercial, and leisure investments, facilitated by domestic capital partnerships (Parnwell and Wongsuphasawat 1997; Goldblum and Wong 2000; Marcotullio 2003; Sajor and Ongsakul 2007). All this taken together results in an increasingly intense competition over land use, even as peri-urbanization remains mostly poorly planned or regulated across the region (Hall et al. 2011).

Mega regions of urbanization have emerged, also known as extended metropolitan regions or *desakotas* (Indonesian, lit., 'village-towns'), that stretch along transportation corridors between large city cores surrounding national capitals and major cities (McGee 1991). Bangkok is a prime example of expansive *desakota* development (Hung and Yasuoka 2000). Overall, Bangkok's area has expanded from 67 km^2 in the 1950s to 426 km^2 by the mid-1990s; this growth was only partially planned, and has been fundamentally shaped by the needs of FDI. The surrounding 'city region' of Bangkok, called the Bangkok Metropolitan Extended Region (BMER), incorporates the five surrounding provinces, covers an area of 7,761 km^2, and has a registered population of 10 million, which is projected to grow to 30 million by 2020 (Yuen 2009). In Vietnam, Ho Chi Minh City and Hanoi are likewise rapidly spreading over surrounding agricultural land: between 1991 and 2004, the urbanized area almost doubled, and the population increased by almost 1.5 times (Duong 2008).

In these low-density and extensive areas of growth, competition for use of land and other natural resources is intensifying, urban infrastructure and basic service provision lags seriously behind the rate of growth, and pollution and environmental degradation is an increasing problem. Sajor and Ongsakul (2007), for example, examine evolving mixed land use in the Rangsit Field, Pathumthani province in the BMER, and implications for peri-urban water governance. Until the 1970s, up to 90 percent of the area was used for monoculture rice cropping, supported by canal irrigation systems developed in 1900. Since the mid-1980s, agricultural practices changed towards orchards and vegetable farms and an increased proportion of land was turned over to built-up areas, including housing projects and factories. These developments placed the low income rice farming residents in the area in competition with the new users of irrigation canals, who also extracted the canal water and contributed towards water quality degradation through waste disposal. Sajor and Ongsakul (2007) also identify considerable conflict between the old and new users of land and water resources that has been predominantly to the detriment of low income farmers, who are politically disadvantaged in terms of access to powerful politicians in comparison to developers. Governance deficits are furthermore mediated by administrative separatism, ambiguity and multiplicity in the functional jurisdiction of official water-related bodies, and the general lack of opportunity for public participation.

Economic growth, globalization, and urbanization

The forces of globalization concentrate in cities and determine to a significant degree the direction and nature of urbanization, for example, through shaping policies on economy, trade, investment, industry, and agriculture. Global cities in Southeast and East Asia, such as Singapore and Hong Kong, and their aspiring Mekong region peers such as Bangkok, Ho Chi Minh City, and Hanoi, are often perceived to compete for investment with each other within a global marketplace (Rimmer and Dick 2009). These cities have become important nodes in global financial and production networks. Despite the global economic crisis since 2008, the region's economies remain strong and likely to continue growing in the long term (Yuen 2009).

Industrialization in the Mekong region is predominantly an urban phenomenon, closely linked to the region's integration into the global economy both for investment and export markets (Giok-Ling 2009). Industrialization first and most intensively occurred in Thailand since the 1980s, and then in Vietnam since the early 1990s. Governments often perceive a trade-off between protecting environmental quality and attracting FDI, and have therefore permitted poorly planned and weakly regulated industrialization. Bangkok, for example, did not have an official city plan in operation until 1992, and even then could only maintain limited effective use of land control (Setchell 1995, Krongkaew 1996, Plumb 1999, cited in Sajor and Ongsakul 2007). Whilst industrialization is associated with rapid national economic growth, it has been accompanied by new and serious environmental problems in urban areas and their hinterlands, including high levels of air and water pollution, contaminated soils, and damaged ecosystems. This is a major threat to public health, and has led to acute and chronic illness amongst workers and local communities, including those who cannot afford to move away or who consider that they have no other viable choice but to work there (HEI 2010).

Large cities are centers of political power and are considered engines of economic growth (Sheng 2010). Primarily as a result of the industrial and service sectors, urban areas generate the greatest share of GDP (Table 5.2), although such statistics should be considered in the context of the interconnectedness of urban and rural economies discussed above.

Table 5.2 Estimated urban share of national GDP (%)

	1990	1995	2000	2005	2006
Cambodia	49.9	48.6	62.1	65.8	70.4
Lao PDR	33.8	45.0	47.4	55.2	53.2
Myanmar	42.7	40.0	42.8	46.9	47.4
Thailand	85.6	89.2	91.0	89.8	89.3
Vietnam	61.3	72.8	75.5	79.1	78.3

Source: Yuen (2009: 10).

National development policies have been biased towards, and hence privileged, urban economic growth, enabling and supporting industrialization, through, for example, prioritizing investment in urban infrastructure and social services. Whilst fostering economic growth, this investment and policy bias has, over time, been an important factor in (re)producing uneven economic and social development between urban and rural areas, even as they remain interconnected through flows of resources, people, and finance. Phongpaichit and Benyaapikul (2013:15, citing World Bank 2012), in the context of Thailand being caught within a middle-income trap, highlight widening economic and social inequalities including between regions, noting that "Bangkok accounts for 17 percent of population and 26 percent of GDP, but receives a 72.2 percent share of total public expenditure. On the other hand, the Northeast region accounts for 34 percent of population and 11 percent of GDP and receives only 6 percent share of total public expenditure." Urban bias has privileged the needs of powerful economic actors, such as transnational corporations, financiers, and domestic business interests, that coordinate economic production, and labor and resource extraction, including in the hinterlands. Walsh and Amponsrira (2013: 888) reveal how big business has also pushed out small traders across the region, for example "as traditional markets are replaced by air-conditioned shopping malls."

Whether or not the interaction between market and other institutional structures versus the power of social actors' agency shapes positive or undesirable outcomes continues to be debated. Evans (2002), for example, drawing in particular on the work of Manuel Castells (1989, 1996), conceptualizes an interaction between the "spaces of places" and "spaces of flows": "Spaces of places" are largely physically self-contained spaces where ordinary urban citizens interact with each other and their environment, and where local politics predominate; in contrast, "spaces of flows" are the terrain of transnational networks through which information and resources flow, where economic and political power resides, and is shaped by business and political elites while largely excluding ordinary citizens. Evans (2002: 11) subsequently explores the potential of transnational nongovernmental organizations and political parties—termed "translocal intermediaries"—as agents that can link and scale up place-based struggles for social and environmental justice to the "space of flows."

Economic globalization in the Mekong region has also been shaped since the early 1990s by the ADB's Greater Mekong Subregion (GMS) program (Kaosa-ard and Dore 2003). The program has promoted regional economic integration between Cambodia, Lao PDR, Myanmar, Thailand,

Vietnam, and Yunnan and Guangxi provinces of China (ADB 2012).[2] A key strategy of the GMS program has been to reduce regional transport costs, and to facilitate liberalized trade and investment between major cities and urban areas by promoting 'economic corridors' (ADB 2012). The GMS program has facilitated the construction of large-scale cross-border infrastructure such as highways, railways, high-voltage power transmission lines, and hydropower dams to physically interconnect urban centers and their hinterlands. The ADB promotes a model of regional development that in many ways emphasizes urban-led economic growth with an intended goal of poverty reduction. Others have highlighted the costs of economic regionalization, namely the environmental and social impacts of resource extraction in the hinterlands, especially for the marginalized (Guttal 2006; Cornford and Matthews 2007), including increasingly prevalent land-grabbing across the region (Hall et al. 2011). Oehlers points towards the risks of an "entrepôt" mode of development emerging, "defined by increasingly complex corridors linking poles of [urban] activity, but with very little else within or between," where a lack of sufficient emphasis on improving productivity risks "a low productivity, low wage trap" for the majority (Oehlers 2006: 467, 472).

Urban poverty

In the cities of the Mekong region, a relative minority are wealthy, well-educated, and mobile, and fully enjoy urban amenities, living near the center, or in gated communities within well-serviced and relatively green spaces. There is also a growing middle class in the suburbs, who enjoy relative material abundance where services meet most needs, although they may face other challenges, such as long commutes to work. At the other end of the socioeconomic scale, there are the urban poor who live in slums and squatter settlements, and survive largely within the informal economy (Savage 2006; Seabrook 2007).

Urban poverty is multidimensional and relates not only to low levels of income, employment, and assets, but also a lack of access to basic services and to local political and bureaucratic systems that are unable or unwilling to meet the needs of the urban poor (Elliott 2006; Sheng 2010). Whilst levels of poverty remain higher in rural than urban areas in Southeast Asia (Balisacan et al. 2005), within urban areas there is also great inequality in opportunity, wealth, and access to resources. For example, the gini coefficients in Phnom Penh, Hanoi, Bangkok, Ho Chi Minh City, and Chiang Mai are estimated to be 0.37 (2004), 0.39 (2002), 0.48 (2006), 0.53 (2002) and 0.58 (2006), respectively (Yuen 2009). Many urban poor survive within the informal economy through small-scale, unregulated, and semi-legal activities that typically rely on local internal resources, family labor, and traditional technology (Schneider 2002; Park 2005; Elliott 2006).

The urban poor live in settlements and areas that are often hazardous and detrimental to their well-being. Poverty and a lack of development interact closely with slum dwellers' relationship to and impact upon the local urban environment. Slums may be characterized as urban settlements with inadequate access to safe water, sanitation and infrastructure, poor quality housing, overcrowding, and insecure land tenure (UNHABITAT 2003), and are pervasive throughout mainland Southeast Asia; UNHABITAT reports that the percentage of the urban population living in slums in 2005 in Cambodia, Laos, Myanmar, Thailand, and Vietnam was 79 percent, 79 percent, 46 percent, 26 percent, and 41 percent, respectively (UNHABITAT 2009). Difficulties faced by people living in slum and squatter settlements include the risk of eviction with no legal recourse, a lack of basic public and emergency services, environmental and health risks, and a lack of access to formal credit; these impacts are often gender-differentiated (Elliott 2006). Furthermore, slum residents pay more for basic service provision than other urban residents, such as water supply or electricity, but receive poorer service.

The vast majority of slum dwellers do not choose to live in such conditions by choice. Many are crowded out from the formal land market, because land has become very expensive in the city, reflecting an inequitable political economy of urban land ownership and the real estate market. Key policies required to improve the circumstances of the urban poor should address the issue of land tenure security, support peoples' ability to own land and build their own housing, and put in place controls on land development and real estate speculation to mitigate rising land prices. Legal land ownership, especially for high-value urban real estate, often remains tenuous and tends to marginalize the claims of the urban poor who may have occupied land for generations without ever having officially claimed legal ownership. Whilst land reform legislation is ongoing, contested verdicts typically privilege the interests of a powerful minority (Yuen 2009).

Urbanization and the environment

Conceptualizing the urban–environment nexus

Conceptual approaches seeking to understand how ecological processes interact with people and economic, social, and political systems are diverse, and increasingly multi-disciplinary across the physical and social sciences (Kennedy et al. 2012). At the same time, fundamental questions of the ontological and epistemological status of nature and "how best to understand the production of 'socionature' in an age of globalized capitalism" remains heatedly debated (Braun 2005: 635). The urgent challenge of conceptualizing the relationship between urbanization and the environment—and translating these into policy recommendations—remains at a comparatively early stage, but is increasingly the focus of urban scholars and practitioners (Haughton and McGranahan 2006).

Urban areas are simultaneously physically built areas, natural ecosystems, and environments that are socially and culturally constructed. Urban areas and their environment are dynamic, produced and reproduced by socio-natural processes. Heynen et al. (2006: 1) describe cities as "dense networks of interwoven socio-spatial processes that are simultaneously local and global, human and physical, cultural and organic." The ecological changes resulting from the process of urbanization occur within urban and peri-urban areas themselves, and beyond into their hinterlands, which may stretch from nearby areas to a global scale.

The habitat of urban people, their pets and gardens, adapted animals, and other flora and fauna, such as birds and pests (rats, weeds), all constitute the intra-urban ecosystem (UNU/IAS 2003). Given the high concentrations of humans in urban areas, they dominate the intra-urban ecosystem. Intra-urban ecosystems are linked to and buffered by the ecosystems of hinterlands, on which they depend for support in the form of energy, water, and other material (UNU/IAS 2003). As urban areas and the intra-urban ecosystems grow, so does consumption and waste production, expanding hinterlands in the process. Sheng and Lebel (2009) note that whilst urban areas can be places of intense resource consumption associated with higher levels of wealth, equivalent levels of wealth not spatially organized within relatively compact urban areas would result in larger resource footprints.

Adopting the metaphor of an *urban metabolism*, the intra-urban ecosystem, which extends to the extra-urban ecosystem, can be imagined as a consumer of resources and a creator of waste products that produces the urban environment and that flow through networks of sewerage and water pipes, electrical cables, waterways and transportation routes, and within buildings, factories, parks and gardens (Robbins 2004; Heynen et al. 2006). The concept of an urban metabolism has been taken up by multiple disciplines, including industrial ecology, urban ecology, ecological economics, political economy, and political ecology (Broto et al. 2012). The notion of a metabolic relationship between society and nature in which urbanization processes produce local and global environments reveals the centrality of nature to urbanization (Braun 2005). Such insights render visible how urban areas are produced from and embedded within nature, and challenge the dualisms of urban/rural and city/nature (Braun 2005); as David Harvey famously wrote, "…there is nothing *unnatural* about New York City…" (Harvey 1996: 186).

Intra- and extra-urban ecosystems are interconnected through multiple feedback mechanisms between places and across scales from local to regional to global. These scalar relationships are increasingly recognized as relational across scales rather than linearly hierarchical in favor of urban areas, and include relations between networks of cities and towns, and how urban and rural ecosystems reflexively shape one another, dissolving the artificial notion of their separation (Haughton and McGranahan 2006). Indeed, the concept of scale often appears as central when the relationship between urbanization and the environment is described by various actors, thus also implying the potential existence of a politics of scale (Haughton and McGranahan 2006).

Urban areas themselves are differentiated spaces of uneven development, as are the hinterlands within which they are embedded. As such, politics often embroil the urban environment and its impacts upon hinterlands. Braun (2005) highlights that although the production and reproduction of urban nature is itself an intensely political process, this cannot be separated from wider political processes in society.

Intra-urban environments: Brown, grey, and green urban environmental issues

This section outlines key intra-urban environmental challenges across the Mekong region in terms of 'brown,' 'grey,' and 'green' environmental issues. It seeks to highlight how people's experience of the urban environment is differentiated, reflecting power and political voice (Evans 2002), and how environmental and social inequality are often closely correlated, in many instances, reproducing and reinforcing each other.

For poorer citizens, who constitute the majority living in the Mekong region's urban areas, 'brown' environmental issues predominate. These include poor quality and overcrowded housing, lack of basic services (waste water removal, sanitation, water supply), hazardous pollutants in urban air and waterways, and accumulation of solid wastes. These have immediate impacts, including on human health, and act predominantly at a local scale. Brown environmental burdens disproportionately affect the urban poor, due to their low income and means to purchase services and closer dependence on urban ecosystems, such as canals, waterways and areas of wasteland turned to productive use. Brown environmental burdens also serve to amplify economic inequalities; this reflects not only the fact that wealthier citizens are able to afford basic services that mitigate brown issues, but also because the weaker political voice of the urban poor is less heard by authorities, and public resources are not directed towards meeting their basic needs. Savage (2006) contends that 'brown' environmental issues are the most urgent in Southeast Asian cities, and to avoid the specter of 'sick cities' governments should prioritize investment in brown infrastructure, such as garbage disposal, sewerage issues, clean water, and efficient and clean energy systems.

For example, the provision of treatment facilities for human sewage, grey water, and industrial pollution within urban areas, whilst gradually improving, remains inadequate. Water-based sewerage systems remain out of reach for most urban areas due to its cost, making septic tanks the most common technology employed (Sheng and Lebel 2009). Yet, poor construction and maintenance means that wastewater is often inadequately treated, and groundwater and waterways remain seriously polluted representing a major public health risk. According to reporting for the Millennium Development Goals, in 2008 the total coverage of improved sanitation[3] in urban areas in Cambodia, Laos, Myanmar, Thailand, and Vietnam was 67 percent, 86 percent, 86 percent, 95 percent, and 94 percent, respectively (WHO and UNICEF 2010). However, the ADB noted in 2004 that in Phnom Penh, only 41 percent of the city was covered by the city's sewerage program, and approximately 12 percent of households have no toilet facilities; while in Ho Chi Minh only approximately 12 percent of the city had sewerage network coverage (ADB 2004, cited in Marcotullio 2007). In Bangkok, only 2 percent of households were connected to the city's sewerage system, 25 percent rely on septic tanks, and the rest use pit latrines or other means (ADB 2004 cited in Marcotullio 2007). In 2005, however, a number of major large-scale waste water treatment plants were commissioned to address this issue (WQMO n.d.).

Solid waste production is also growing in both volume and diversity from residential, commercial, industrial, and construction sources (UNEP 2004). Estimated averaged rates of waste generated per day in the late 1990s and early 2000s ranged from 0.45 kg/person/day in Myanmar to 0.64 kg/person/day in Thailand, and are anticipated to at least double by 2025 (Uyen and Schnitzera 2009). Collection and safe disposal of solid waste is a major logistical challenge: it is estimated that between 50 percent and 80 percent of urban solid waste is collected each day, although in small towns it may be significantly less (Sheng and Lebel 2009; Uyen and Schnitzera 2009). Uncollected waste is usually burned or dumped, creating local public health risks. Most collected waste is disposed of in landfills or dump sites which are generally poorly operated and maintained, such that contaminated leachate is a common problem. Furthermore, suitable landfill sites are becoming more difficult to find as urban areas expand. Rates of recycling are low across the region; formal and informal waste separation and recycling of materials is estimated to recycle between only 8 and 12 percent of the solid waste generated (Uyen and Schnitzera 2009).

All people living in urban areas are increasingly affected by 'grey' environmental issues, which include the impacts of air pollution from industrialization and motorization (Lebel et al. 2009). Grey environmental issues are a threat to both health and ecological sustainability, although the onset of their impact tends to be delayed. Whilst various sources of grey pollution exist in many if not most urban areas, some people are more exposed than others; e.g., industrial estates are typically located close to poorer neighborhoods, either because workers prefer to live close to their workplace or because they cannot afford to move away.

Most major urban areas across the region have experienced serious air pollution, including particulate materials of 10 micrometers or less (PM–10) and sulphur dioxide (SO_2), causing respiratory and other serious health problems such as cancers in urban and peri-urban areas (UNEP and TEI 2007; Lebel et al. 2009). The main emitter of PM–10 is vehicle traffic, especially diesel engines. Growing vehicle use, together with low emission standards, poor road networks, and outdated technology, contribute to the high level of vehicle-sourced air pollution. Thailand has the largest fleet of vehicles in the region by far; and this has more than doubled in size since the early 1990s. Yet all countries are witnessing rapid increases in the number of vehicles on the road (UNEP and TEI 2007). Whilst cars may become more fuel efficient and cleaner over time, environmental problems associated with motorization, including increasing use of fossil fuels, air pollution, traffic congestion, and contribution towards climate change remain unavoidable. Improving the coverage and efficiency of public transport, such as bus services, together with the provision of accessible and affordable mass transit systems, such as subways and urban railways, are critical for the creation of sustainable and livable cities.

For a relative minority but growing group of wealthier citizens, the brown environmental problems are largely overcome in their urban places. Their concerns are more related to 'green' environmental policies to create a pleasant city environment, including the creation of city parks, the planting of trees along the roadsides, and spaces for urban agriculture. It should be noted, however, that often the brown environmental issues may be displaced, rather than resolved, and threaten hinterland ecosystems; for example, sewers may carry waste waters away from wealthier areas but then discharge it untreated into waterways, and collected solid wastes dumped in poorly maintained landfills outside the city. Furthermore, the high levels of consumption and waste generation associated with wealthier lifestyles can result in resource depletion, producing inter-generational threats to sustainability.

The relationship between economic growth and environmental impact is hotly contested; it has variously been argued that environmental protection is reinforced with economic growth, that growing affluence corresponds to a growing environmental burden, and that the environmental burden will rise and then fall following an "environmental Kuznets curve" (Bai and Imura 2000; McGranahan 2007). McGranahan (2007) concludes that the track record of urbanization suggests that economic growth does result in a growing environmental burden, although this relationship is in fact politically mediated. In many early industrializing cities of the North, it was not until local groups and government organized and mobilized that living conditions were improved (Szreter 2005, cited in McGranahan 2007).

Urbanization and linkages to regional environmental sustainability

As areas of concentrated production and consumption, activities in urban areas draw upon their hinterlands for domestic consumption, industrial processes, and for export globally, risking what McGranahan (2007) terms the spatial displacement of environmental burdens. Demand for natural resources include: energy, such as electricity and gas; agricultural products, such as food and biofuels; and primary materials, such as mined minerals and timber originating from forests and plantations. The distance between the point of consumption and point of production in urban and rural areas, respectively, can hide from consumers the environmental and social costs of their consumption. Weak governance associated with the exploitation of valuable natural resources risks—and often results in—the unequal sharing of environmental burdens and benefits, and the creation of environmental injustice (Walker 2012).

Export-orientated industrialization and economic growth has transformed the region, including through: construction of large infrastructure, such as roads, hydropower dams, and fossil fuel-fired power stations; intensification of land use and growth of agro-business models of agriculture; and the spread of industrial facilities and mining industries (Nevins and Peluso 2008; Lebel et al. 2009; Hall et al. 2011). Whilst indicators of human development are improving for many people, development remains uneven and increasing demands are being placed on the region's natural resource base that is vulnerable to over-exploitation. Region-wide, habitat and biodiversity loss, fragmented and polluted ecosystems, land degradation, forest cover and wetland loss, river degradation, and fish stock depletion threatens to undermine the ecological foundations upon which long-term sustainable development depends (UNEP and TEI 2007). At its most extreme, land grabbing (Hall 2011) and water grabbing (Matthews 2012) has occurred.[4] Even as rural and urban areas are tied together, given that the region's majority rural population still depend significantly upon this natural resource base for their livelihoods and well-being, this mode of development threatens local livelihoods and economies even as national GDP continues to grow.

Meeting the region's growing demand for electricity, for example, epitomizes the linkages between consumer and industrial demand mostly in urban areas, and the "spatial displacement of environmental burdens" to the hinterlands. Thailand's government estimates that electricity demand in Thailand will approximately double to 65,000 MW by 2030 (EGAT 2010), and Vietnam's government predicts demand to triple by 2020 to 75,000 MW (Socialist Republic of Vietnam 2011). Across the region, the construction of new large-scale energy-generating projects, including coal-fired power stations and hydropower dams, both domestically and through the regional power trade, are high on all governments' agendas (Middleton et al. 2009). A reliable and sufficient supply of electricity is important to many dimensions of development, of course. Yet, the track record across the region to date reveals significant social costs and environmental damage experienced by affected rural communities who do not participate adequately in the decision-making, nor receive enough compensation or restitution (Hirsch 2010; Middleton 2012). Civil society groups, especially in Thailand and Vietnam, have sought to reform electricity planning processes, including by promoting the potential that energy efficiency measures, renewable energy, and decentralized energy options could play, especially in the more industrialized cities of the region in Thailand and Vietnam (Greacen and Footner 2006; VUSTA 2007). They have argued that existing plans mostly serve the interests of the state-owned electricity utilities, energy companies, and the construction industry, rather than the needs of the regions' electricity consumers and ensuring social justice and environmental sustainability (Greacen and Greacen 2012). Such reforms would also redefine the relationship between urban electricity consumption and its impact on hinterlands.

The complexity of urban–rural linkages—via flows of natural resources, wastes, finance, people, ideas—makes it challenging to distinguish the environmental impact of 'urban' centers alone. That said, a number of factors, ranging from consumerist urban lifestyles to inefficient urban planning and design and high levels of energy intensity, all place urban areas as loci of consumption. Cities, for example, are major contributors to climate change; globally, it is estimated that cities produce close to 80 percent of all carbon dioxide and other greenhouse gas emissions (ADB 2008a). The Mekong region is at present not a major emitter of greenhouse gases in comparison to more industrialized regions, yet its emissions are rising, and some urban areas, such as Bangkok, have comparable carbon intensity to industrialized countries (UNEP and TEI 2007). Furthermore, national greenhouse gas emissions are growing at a rate that is faster than GDP growth, meaning that economic growth is becoming more rather than less carbon intensive. Climate change is anticipated to affect the region profoundly (ADB 2009a). Already, urban areas in low-lying delta areas, such as Bangkok and Ho Chi Minh, are facing serious challenges from flood risks caused by extreme weather events and rising seas levels (Chan et al. 2012). Addressing climate change will require fundamental reforms to the ways urban areas are planned and built, including reducing energy demand, for example, through improving transportation within cities, and making buildings more energy efficient (Yuen and Kong 2009).

Democratizing urban governance

Urban areas in the Mekong region are becoming increasingly complex, a result of their physical growth and their increasingly diverse and technologically-advanced economic activities. Yet many urban areas lack a well-formulated process and clear urban planning framework, including land use zoning, environmental regulations, and building codes. Overall urban governance is weak, with poorly funded, insufficiently skilled, and overly-centralized structures of urban government (Sheng 2010).

Sheng (2010: 134) defines governance as "the quality of relationship between the government and its citizens," where the measure of quality can be gauged by measurements of "inclusiveness, participation, transparency and accountability, equity, predictability, adherence to rule of law, and subsidiarity." Others have emphasized the importance of sustainability and ensuring environmental justice (Haughton 1999; Agyeman et al. 2003). Within globalization, many actors with asymmetrical power relationships are now involved in shaping urban decisions, spanning the local and global, including community representatives, local and national governments, public officials, transnational and domestic and local businesses, and civil society.

It is now widely acknowledged that 'master planning' urban development has largely failed to meet the needs of many rapidly growing cities and their inhabitants (UNHABITAT 2009; Yuen 2009). In recognition that sustainable neighborhoods within urban areas are the building blocks for building sustainable urban environments, there is a growing trend globally for decentralization and greater participation of urban citizens. Kenworthy (2006: 81) describes this as a move away from "predict and provide", and towards "debate and decide" processes.

Over the past two decades, with the exception of Myanmar, the governments of the Mekong region have had clearly articulated policies towards decentralization and privatization (Sheng 2010). This has been partly in response to a growing urban middle class who are increasingly educated, politically aware, and vocal with their demands of urban governance. In practice, however, these policies have struggled to be implemented, both in terms of delegating authority from national to local governments and enabling public participation (Yuen 2009; Sheng 2010). Furthermore, where decentralization has occurred, the capture of benefits by local elites and patronage networks at the expense of ordinary people and the urban poor has been unfortunately common (Sheng 2010). On the other hand, Goh and Bunnell (2013) emphasize how processes of decentralization have been accompanied by new forms of urban grassroots activism and social movements in both primary and secondary cities.

A particular tension in the Mekong region lies in the significant influence of the market, which is dominated by a powerful private sector and the lack of capacity or willingness on the part of government to regulate or plan urban areas and land use towards ensuring the wider public interest. The urban political economy is too often shaped by national politicians influenced by the interests of powerful transnational and local investors in real estate (the "space of flows"), insufficiently counter-balanced by local government, civil society, and local people (the "space of place").

The tendency towards privatization across the region also raises important public interest questions towards the provision of public services, including security and public spaces, and investments in public infrastructure, ranging from roads to water supply. The extent that state versus market versus community action is required continues to be debated. Overall, however, the private interest remains privileged—despite the end of overtly pro-privatization state strategies—and has a disproportionately large say in the form and reproduction of the region's urbanization. Important decisions affecting the public are taken beyond the scrutiny of the public itself. Whilst the urban middle class may be able to purchase the services provided by the private sector, this can reduce the political pressure upon and resources of local governments to provide such services to the remainder of the population (Sheng 2010). Evans (2002: 6) concludes that:

> Rejecting the market out of hand works no better than blind faith in their efficacy. Markets must be taken seriously without being taken as 'natural' or exogenous. Normally, the coalitions of private and

> public actors that construct markets have socially minimalist goals .... Replacing these 'minimalist markets' with ones whose rules take livability into account is at the core of any quest for more livable cities.

The degree to which such urban and regional sustainability challenges can be met depends, to an important extent, on the quality of urban governance, regardless of the level of overall economic development (Ooi 2007). Democratized urban decision-making cannot be separated from wider factors, including literacy, political rights, and civil liberties. Public pressure, lobbying, and political competition are all important factors, beyond economic growth alone, that determine how environmental burdens are addressed and apportioned, and whether sustainability and livable cities will be created (McGranahan 2007).

Conclusion: Towards sustainable urbanization

Throughout the Mekong region, the process of urbanization will continue for the foreseeable future. If the region's expanding urban areas are the engines of its economic growth, they unfortunately remain at present dirty and inefficient. Living conditions for the majority of the urban poor are difficult; they often suffer poor quality housing, land tenure insecurity, a lack of basic services, and unavoidably encounter a locally degraded environment that places health at risk and reinforces poverty. There is also, however, a growing urban consumer class who have benefited from the economic growth that urban areas bring, and who enjoy more of the amenities urbanism may offer; as they are able to afford access to basic services, they are largely buffered from direct local environmental impacts. And yet through higher levels of consumption, they are creating growing ecological footprints that threaten long-term sustainable development. These environmental and social challenges are produced and reproduced by both the major decisions and small everyday decisions of business, politicians and people—for example, how people travel to work, where and how they shop, and how much waste is recycled (Haughton and McGranahan 2006).

Ensuring that urbanization and urban lifestyles are sustainable is a central challenge to ensuring sustainability in the wider Mekong region. Urban areas and their hinterlands are connected through flows of goods, services, investment, finance, people, and knowledge. Haughton (1999: 234) writes "[A] sustainable city cannot be achieved purely in internal terms: a sustainable city is essentially one which contributes effectively to the global aims of sustainable development,... it is futile and indeed virtually meaningless to attempt to create a 'sustainable city' in isolation." Governance mechanisms are required that link across scales from the local to the national to the global, between rural and urban areas, and that holistically integrate economic, social, environmental, cultural and political dimensions of sustainability. Complexity and temporal and spatial scale mismatches between current urban governance, policy and management practices and the need to account for regional and global sustainability considerations, however, remain a challenge (Bai et al. 2010).

Evans (2002: 2) defines a livable city as "providing livelihoods for its citizens, ordinary as well as affluent, in ways that preserve the quality of the environment." The scope of this challenge ranges from reducing consumption, energy, waste, and pollution intensity in urban areas whilst maintaining economic growth, through to providing the necessary basic services and physical infrastructure, to committing to democratic governance and participatory urban design practices that build sustainable communities from the bottom up. Whilst addressing the brown environmental issues should be a priority to making urban life bearable for all, having addressed these issues, urban areas can and should be pleasant places to live and work. Understanding of the processes of urbanization is deepening, for example, as viewed through the lenses of urban metabolism (Ramaswami et al. 2012) or ecological footprints (Newman 2006). There is also no shortage of potential solutions, ranging from creating sustainable production–consumption systems (Lebel and Lorek 2008) to principles for sustainable urban design (Lebel et al. 2007; Newman and Jennings 2008) that seek to decouple gains in well-being from rising intensities of resource consumption.

Creating sustainable urban areas, however, is primarily a political task. Decisions about economic policy, industrial policy, land use, urban planning, environmental quality, and other related policies that determine forms of urbanization and associated environmental burdens, reflect existing power structures and political struggles (Evans 2002). Whilst power inequalities and the degree of transparent, accountable and democratized decision-making remain a significant challenge, given that urban areas are places of intense interaction, they hold the potential for transformation through the types of struggle that have always catalyzed change.

Notes

The authors would like to thank Edsel Sajor, Steven Johnson, Louis Lebel, Orn Uma Polpanich, Muanpong Juntopas, John Soussan, Nantiya Tangwisutijit, Albert Salamanca, and Rajesh Daniel for their support and advice in the process of preparing this chapter. All remaining errors remain the responsibility of the authors.

1. Data for Yangon and Ho Chi Minh City is for 2010.
2. More recently, the ASEAN Economic Community, has pursued a comparable mode of regional economic liberalization, in which it considers the GMS as a subregion of the wider ASEAN region (ASEAN 2008).
3. Improved sanitation facilities: ensure hygienic separation of human excreta from human contact. They may make use of the following facilities: Flush/pour flush to piped sewer system, septic tank, or pit latrine; Ventilated improved pit (VIP) latrine; Pit latrine with slab; or composting toilet.
4. Land grabbing is understood as agriculture-driven resource grabbing, whilst water grabbing refers to the appropriation of water resources.

II

Cross-border Collaborative Research

6

Vulnerability of Livelihoods in Flood-prone Areas along the Cambodia–Vietnam Border

Nguyen Duy Can, Seak Sophat, Vo Hong Tu, Sok Khom, and Chu Thai Hoanh

The Lower Mekong Basin (LMB) is considered prone to flood hazards and affected by upstream Mekong flow patterns as well as global climate variability. Although the average annual changes in water level rises and flow velocity in the LMB are not as great as in upland areas, floods in the Delta region usually last many days or even many months every year. Seasonal monsoonal floods bring benefits such as abundant aquatic and fisheries resources and improved soil fertility, but severe floods also pose a major threat to lives and property (Nguyen and Nguyen 2008). Such floods have damaged vast areas of agricultural land and affected the livelihoods of a great number of people. In particular, over 4 million ha in Cambodia and about 3 million ha in Vietnam were inundated by floods in recent years (MRC 2006, 2009b). Many thousands of households in both Cambodia and the Mekong Delta of Vietnam are reported to be vulnerable to floods (CFSC 2011). Given projected climate change with possibly higher rainfall intensity and variability, the heavily populated megadeltas are expected to be at the greatest risk of increased river and coastal flooding (Bates et al. 2008; Wassmann et al. 2004; Smith et al. 2013).

To estimate climate change vulnerability, a number of studies have applied the Livelihood Vulnerability Index (LVI), which uses multiple indicators to assess human exposure to natural disasters (including floods) and climate variability, social and economic characteristics of households that affect their adaptive capacity, and current health, food, and water resource characteristics that determine their sensitivity to climate change impacts (Hahn et al. 2009). The LVI analysis was first used in Mozambique (Hahn et al. 2009), then adapted in other countries such as Nepal, Ghana, and Trinidad and Tobago (Lamichhane 2010; Urothody and Larsen 2010; Khajuria and Ravindranath 2012; Etwire et al. 2013).

This chapter aims to assess the sources of livelihood vulnerability and improve our understanding of conditions in the flood-prone provinces along the Cambodian–Vietnamese border. This study is based on an empirical study of livelihood assets and coping strategies and includes a review and analysis of secondary data on flood heights and impacts. The chapter concludes with recommendations on how to strengthen adaptive capacities at the household, community, and provincial levels.

Methodology

Sustainable livelihood framework

This study adopts the Sustainable Livelihood Framework (SLF) to guide the assessment of livelihood vulnerability to floods (Birkmann 2006). SLF analysis involves a critical examination of multiple elements, encompassing five livelihood assets (natural, human, physical, social, and financial capital), their vulnerability context (shocks, trends, and seasonality), and the influence of transforming structures and processes for livelihood strategies and outcomes (Chambers and Conway 1992; DFID 2000).

The vulnerability context was assessed from census data and a literature review of general indicators to study flood situations and their impacts on agricultural production. Livelihood assets were assessed through focus group interviews, participatory rural appraisal surveys using different tools, and a household questionnaire. For assessing livelihood vulnerability, we adapted the LVI to estimate the different impacts of floods on communities in two provinces of Cambodia and Vietnam. Policy interventions and gaps were evaluated using a literature review, a workshop with boundary partners, and interviewing key informants.

Study area

The study was carried out in two provinces, the first one in An Giang, Vietnam, and second in Kandal, Cambodia (Fig. 6.1). Two villages in the flood-prone area of each province were selected for survey. In An Giang, Phu Huu village (An Phu district) and Ta Danh village (Tri Ton district) were chosen, while in Cambodia two communes, Prek Khmeng (Lvea Em district) and Kaam Samnar (Luek Dek district), were chosen.

Fig. 6.1 Map showing the research sites in Cambodia and Vietnam

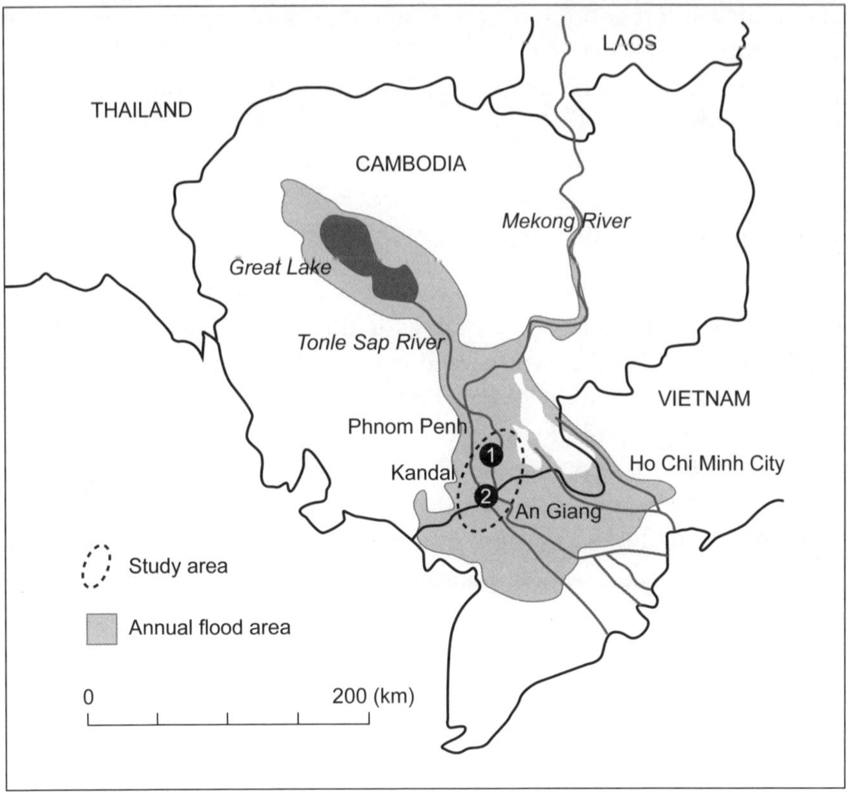

Note: (1) Kandal: 2 villages in flood-prone areas (Prek Khmeng, Kaam Samnar); (2) An Giang: 2 villages in flood-prone areas (Phu Huu, Ta Danh).

Data collection techniques

Approximately 240 households in each province were surveyed. The villages selected for survey were based on suggestions by local authorities (during an inception workshop), and representatives of the areas vulnerable to floods in these provinces. Interviews were conducted by field staff from two universities (Cantho University and the Royal University of Phnom Penh) and by local partners. Household surveys consisted of five sections: demographic information, livelihood assets, flood situations and impacts, resilience and adaptive capacity measures, and policy interventions. Key informant interviews were conducted with 20 staff at district and provincial levels working in the field of flood protection and mitigation. We also conducted 10 focus group discussions in each country with each group including 7–10 household heads from different social groups.

Constructing the LVI

Using the SLF, the vulnerability context is a major determinant of the sustainability of livelihood assets as it directly influences livelihood strategies, institutional processes, and livelihood outcomes of communities (Chambers and Conway 1992; DFID 2000). The level of vulnerability under the impact of extreme floods and climate variability in both countries was calculated by applying the LVI. Indicators, or sub-components, of community vulnerability to flood impacts are grouped into 10 major components (Table 6.1). These components are classified under five different livelihood assets of households (HHs) in the SLF. Each major component includes several indicators or sub-components developed based on available data collected through household surveys on flood impacts in An Giang and Kandal provinces (Table 6.1).

Table 6.1 Livelihood vulnerability index (LVI): Types of capital and components

Capital	Major component	Sub-component/ Indicator
Human	Health	HHs with family member with illness (%)
		HHs with family member ill due to floods (%)
	Livelihood strategy	Ave. agri. livelihood diversification [1/ (no. of agri. livelihood activities + 1)]
		HHs dependent on agri. as major source of income (%)

Capital	Major component	Sub-component/ Indicator
		HHs with family member engaging in non-farm activities (%)
		HHs with no jobs (during flood season) (%)
	Knowledge & skills	HHs: head unlettered (%)
		HHs: head with only primary school education (%)
		HHs: head not trained to cope with flood (%)
Natural	Land	Landless HHs (%)
		HHs with small land (0.1–0.5 ha) (%)
	Natural resources	HHs that did not cultivate 3rd crop (%)
		HHs that depend on (exploiting) natural resources (%)
		HHs that depend on fishing during flood (%)
	Natural disasters & climate variability	Ave. no. of severe floods in the past 10 years (%)
		Ave. no. of deaths or injuries caused by severe floods in the past 10 years
		HHs did not receive flood warning (%)
		Mean standard deviation of monthly ave. water level in Tan Chau (earlier flood zone; years: 2007–11)
		Mean standard deviation of ave. precipitation by month (ave. 5 years)
Social	Socio-demographic	Dependency ratio
		Female HHs head (%)
		Ave. family members in HHs
		Poor HHs (%)
	Social networks	HHs received help because of flood (%)
		HHs without members of any organizations (%)
Physical	Housing & production means	HHs with temporary house (%)
		HHs with housing affected by flood (partially to totally submerged) (%)
		HHs reporting no access to production means (%)
Financial	Finance & income	HHs borrowing money (%)
		HHs with net HH income lower than US$1,000 per annum (%)
		HHs with non-income within flood season (%)

Calculating the LVI

The formulas used for calculating LVI and LVI-IPCC (Intergovernmental Panel on Climate Change) were adapted from those used by Hahn et al. (2009).

Results and discussion

Recent flood patterns

Floods in the Lower Mekong Basin including Kandal province in Cambodia and An Giang province in Vietnam can be characterized as seasonal floods; they are also affected by what happens upstream on the Mekong River (MRC 2006). The Mekong River flows into Cambodia with over 95 percent of its water. During the flood season, water flows back up the Tonle Sap from the Mekong mainstream into the Great Lake. When the water decreases in the mainstream, water flows out of the Tonle Sap down into the Mekong mainstream and this causes seasonal floods downstream in the Mekong Delta (MRC 2006; Nguyen and Nguyen 2008). Records from our survey confirmed that seasonal floods in Kandal province begin in early July and end in October, while in An Giang, the flood occurs later, during mid-July (upstream) and peaks in September or October.

The literature review showed that the floodwaters arrive fairly slowly, but last for a long time. Annually about 60 to 70 percent of An Giang province is inundated for a period of 3 to 6 months at depths of 0.5–4.0 m (Nguyen and Nguyen 2008; CSFC 2010). Similarly, about 40 percent of Kandal is also inundated for 2 to 4 months each year (CDM 2010). Result from surveys in two representative villages in An Giang showed that the duration of floods is 3.5 and 3 months (August to November) for Phu Huu and Ta Danh villages, respectively. The duration of floods in the two surveyed villages in Kandal was the same and both villages remained inundated for 3 months (from July to October).

In the Mekong Delta, floods are classified into three types, based on flood data recorded at Tan Chau gauging station in An Giang: 'big' or damaging floods, where the flood level is more than 4.5 m, 'normal' floods from 3.5 to under 4.5 m, and a 'small' flood when the water is less than 3.5 metres deep. Recorded data showed that the flood level varied from year to year unpredictably and was affected by La Niña, but indigenous knowledge believes that the biggest floods usually happen in the 'Dragon years' of the lunar calendar, such as in 1964 and 2000 (Ngoc 2011; Hong 2012). In the Mekong Delta, out of a total of 19 years with 'big' floods, 11 years coincided with the La Niña phenomenon, while 1904, 1928, 1952, 1964 and 2000 were Dragon years that coincided with La Niña (Hong 2012). Besides, the 'big' flood in 2011 that caused widespread devastation in Thailand, Cambodia, and Vietnam coincidentally also continued into 2012 (the most recent Dragon year). There was a trend on an average of a 'big' flood occurring 3–4 years and causing extreme damage. Fig. 6.2 shows the record of 'big' floods in the Mekong Delta of Vietnam over a period of 50 years (1961–2011) with 14 'big' floods, of which five occurred in a cycle of 3–4 years (Fig. 6.2). Kandal province in Cambodia, bordering the Mekong Delta of Vietnam, was also affected by floods, as was An Giang province.

Fig. 6.2 Big floods recorded at Tan Chau gauging station, An Giang, 1961–2011

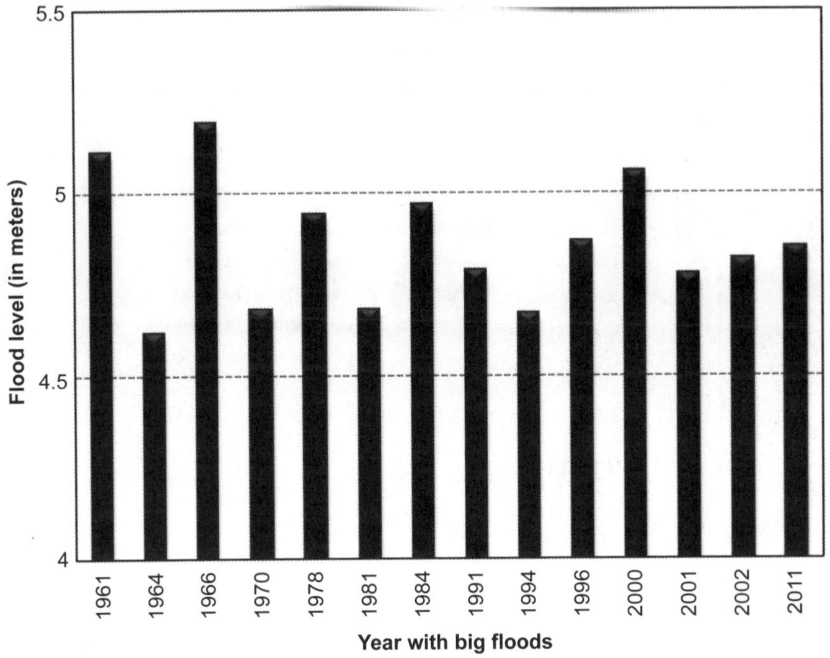

Source: Statistical data from 1961 to 2011.

Impact of floods on economy and agriculture

The floods bring necessary water and nutrients for supporting local livelihoods (e.g. fish, and sediment for agriculture), but they also cause severe damage and disruption in terms of human life, housing, and infrastructure, transportation, health, education, services, and agricultural production (MRC 2005, 2006). Figures 6.3 and 6.5 show the impacts of floods on human life, economy, and agriculture in An Giang and Kandal. In An Giang, the floods in 2000 and 2001 killed many people: in 2000, the floods killed 134 people (of whom 94 were children) and in 2001, 135 people, (of whom 104 were children) (Fig. 6.3). In the study sites in Kandal, a lower proportion of deaths (1 percent) was recorded in the household surveys, and this proportion has reduced from year to year, except for some 'big flood' years. This is attributed to the considerable efforts of the government, such as in constructing safe houses, as well as the local communities' flood adaptations.

Fig. 6.3 Fatalities and damage to houses by flooding in An Giang and Kandal, 2000–2011

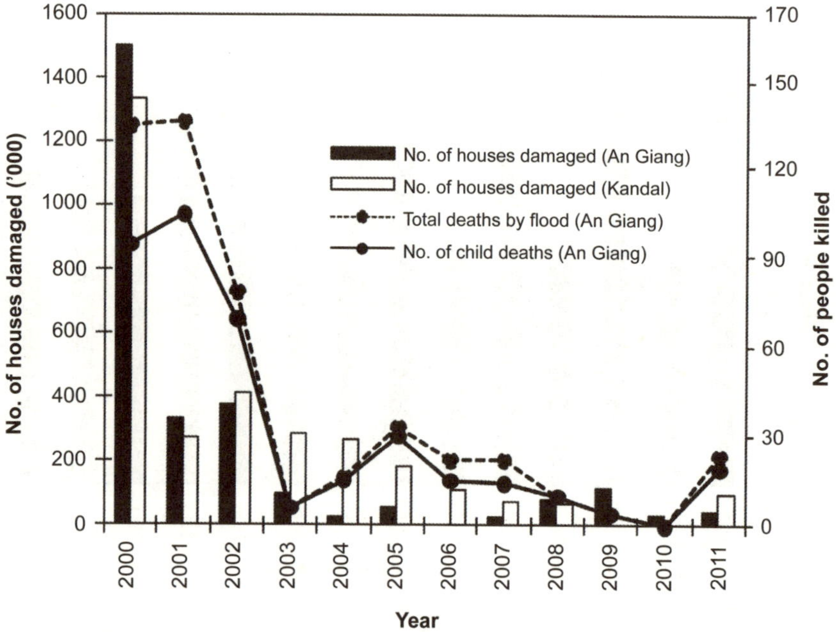

Source: Data from An Giang CFSC; Kandal CDM.

Housing damage due to floods is a big problem for households in flood-prone areas. The big flood of 2000 in An Giang, Vietnam, destroyed 151,867 houses, although damage to housing has been reduced over the years. The floods in 2011 damaged 55,554 houses (see Fig. 6.3). The floods in 2000 were the worst in Kandal with 131,136 houses being affected, of which 3,074 were completely destroyed. Much of the provincial infrastructure such as roads, bridges, schools, and canals was also damaged. Rice and other crops were totally submerged.

Fig. 6.4 presents some information on flood impacts in terms of economic value in the flood-prone areas of An Giang and Kandal provinces. Literature reviews and secondary data showed that the total economic damage in An Giang province from floods was higher than that in Kandal province, especially during the big flood of 2000 (MRC 2005, 2006). In particular, in An Giang, the flood in 2000 caused damage valued at about US$42 million, and in 2011, flood-related losses were estimated at US$49 million (CFSC 2001, 2011). Whereas in Kandal, the flood losses in 2000 were estimated at US$15.4 million (CDM 2001), while estimates for 2011 were not available. Yet, despite many efforts by local governments, including investments in dikes, for instance, in the Delta, flood losses are still high. The damage to agriculture in both An Giang and Kandal was also high. For instance, in An Giang, it was estimated that the floods damaged 10 percent of the total economic output in 2000, whereas in Kandal, it was 44.2 percent. Inundation of rice cropping areas in 2000 in An Giang was over 4,900 ha, and over 5,400 ha in 2011. In Kandal data for the damage to rice-growing areas was not available.

Fig. 6.4 Economic damage caused by floods in An Giang and Kandal, 2000–2011

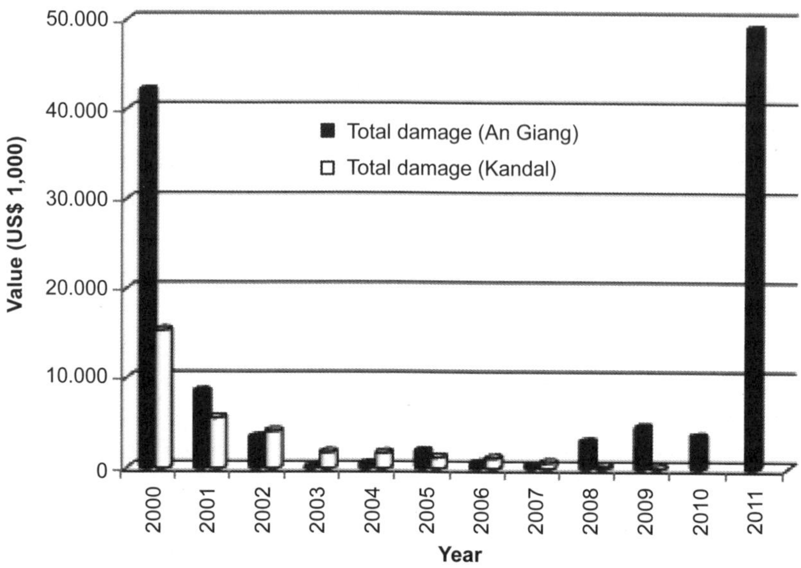

Source: Data from An Giang CFSC and Kandal CDM.

Vulnerability assessments of livelihoods

Table 6.1 presented the summary of the LVI results for all 31 sub-components, 10 components, and 5 types of capital. Fig. 6.5 also illustrates the comparative vulnerability levels of the 5 types of capital. The overall LVI of Kandal was found to be 0.432, which makes Kandal's livelihoods moderately vulnerable (Overall LVI ranged from 0 to 1; 0 denoting least vulnerable and 1 denoting most vulnerable) to the possible impacts of flood and climate variability, higher than the overall LVI of An Giang, which was 0.362. Results from vulnerability assessments for all 5 types of capital and respective components are discussed below.

Table 6.2 Summary LVI results for all types of capital and components, An Giang and Kandal

Capital	Major component	Sub-component	Unit	Observed value		Vul. Index	
				An Giang	Kandal	An Giang	Kandal
Human	Health	HHs with family member with illness	%	3.1	1.5	0.031	0.015
		HHs with family member ill due to floods		0	15	0.000	0.150
	Health vulnerability (h1)					**0.016**	**0.083**
	Livelihood strategy	Ave. agri. livelihood diversification [1/(number of agri.. livelihood activities + 1)]	1/# crops	0.25	0.25	0.063	0.063
		HHs dependent on agri. as major source of income	%	60.2	65.3	0.602	0.653
		HHs with family member doing non farm activities	%	20.8	17.4	0.208	0.174
		HHs with no jobs (during flood season)	%	34.3	0	0.343	0.000
	Livelihood strategy vulnerability (h2)					**0.304**	**0.222**
	Knowledge & skills	HHs head unlettered	%	13.9	26.8	0.139	0.268
		HHs head just passed primary school	%	64.1	76.4	0.641	0.764
		HHs head no training to cope with flood	%	95.9	86.7	0.959	0.867
	Knowledge & skills vulnerability (h3)					**0.580**	**0.633**
Weighted average of h1, h2, h3: Human vulnerability (H)						**0.327**	**0.309**
Natural	Land	HHs who are landless	%	33.2	26.9	0.332	0.269
		HHs with small land (0.1-0.5 ha)	%	23.9	18.7	0.239	0.187
	Land vulnerability (n1)					**0.286**	**0.228**
	Natural resources	HHs that not cultivate the 3rd crop	%	98.2	98	0.982	0.980
		HHs that depend on (exploit) natural resources	%	40.9	50.7	0.409	0.507
		HHs that depend on fishing during flood	%	33.9	37	0.339	0.370
	Natural resources vulnerability (n2)					**0.577**	**0.619**

Capital	Major component	Sub-component	Unit	Observed value		Vul. Index	
				An Giang	Kandal	An Giang	Kandal
	Natural disasters and climate variability	Ave. no. of most severe floods in the past 10 years	no.	0.4	0.5	0.040	0.050
		Ave. no. of deaths/ injuries as result of most severe flood in past 10 years	no.	40	80	0.296	0.593
		HHs did not receive a warning about flood	%	0	6.7	0.000	0.067
		Mean standard deviation of monthly of ave. water level in Tan Chau (earlier flood zone; years: 2007–11)	cm	114.5	114.5	0.441	0.441
		Mean standard deviation of ave. precipitation by month (ave. 5 years)	mm	100.2	90.5	0.562	0.419
	Natural disasters and climate variability vulnerability (n3)					**0.268**	**0.314**
Weighted average of n1, n2, n3: Natural vulnerability (N)						**0.364**	**0.388**
Social	Socio-demographic	Dependency ratio	Ratio	12.3	12.1	0.123	0.121
		female head HHs	%	5.4	16	0.054	0.160
		Ave. family members in a HHs	persons	4.11	5.35	0.422	0.670
		poor HHs	%	25.5	33.3	0.255	0.333
	Socio-demographic vulnerability (s1)					**0.214**	**0.321**
	Social networks	HHs received help due to flood	%	9.1	80.6	0.091	0.806
		HHs not members of any organizations	%	87.7	80	0.877	0.800
	Social networks vulnerability (s2)					**0.484**	**0.803**
Weighted average of s1, s2: Social vulnerability (S)						**0.304**	**0.482**
Physical	Housing and production means	HHs with temporary house	%	19.8	9.3	0.198	0.093
		HHs with housing affected by flood (partially to totally submerged)	%	16.3	38	0.163	0.380
		HHs that report no access to production means	%	51	45	0.510	0.450
	Physical vulnerability (P)					**0.290**	**0.308**

Capital	Major component	Sub-component	Unit	Observed value		Vul. Index	
				An Giang	Kandal	An Giang	Kandal
Financial	Finance and incomes	HHs borrowed money	%	52.1	40.7	0.521	0.407
		HHs with net HHs income lower US$1,000	%	29.6	86.2	0.296	0.862
		HHs with no income during flood season	%	46.5	23	0.465	0.230
	Financial vulnerability (F)					**0.427**	**0.500**
Livelihood Vulnerability Index (Weighted average of H, N, S, P, F)						**0.362**	**0.406**

Notes: Index values were interpreted as relative values to be compared within the study sample only. The LVI is on a scale from 0 (least vulnerable) to 1 (most vulnerable).

Vulnerability in terms of human capital

Kandal showed greater vulnerability on the health component index than An Giang (0.083 versus 0.016, respectively). Kandal respondents reported a higher proportion of households with family members who became ill due to the floods (15 percent for Kandal and 0 percent for An Giang). In terms of livelihood strategy, An Giang showed a higher vulnerability index on this component (0.304) than Kandal (0.222). The major contributing factors for this are that An Giang had higher vulnerability scores for two of the livelihood strategy indicators than Kandal; in particular, more households in An Giang reported having family members doing non-farm activities compared to Kandal (0.208 vs. 0.174 respectively). Non-farm activities here were considered a vulnerability because farmers had stopped farming due to the floods and opted for non-farm work outside their village. Besides, a higher percentage of households in An Giang reported having no jobs during floods (households with no jobs index: An Giang 0.343, Kandal 0.000). These two factors made the overall livelihood strategy score higher for An Giang than Kandal.

Regarding knowledge and skills, Kandal showed greater vulnerability than An Giang (0.633 vs. 0.580). This was because the household heads illiteracy index was higher for Kandal (0.268) than An Giang (0.139); the household heads with primary school index was also higher for Kandal (0.764) than An Giang (0.641). Illiteracy or low education levels are important indicators of knowledge and skills, and households with illiteracy and/or low education levels may be constrained in finding means of livelihood and could therefore be more vulnerable. For example, many non-literate household heads reported that they could not read well or understand information in the media, including flood-related warnings or announcements, and therefore were not prepared.

Vulnerability in terms of natural capital

Land is an important household asset and indicator of wealth. An Giang showed greater vulnerability on the land ownership index than Kandal (0.286 vs. 0.228). Landless households often stay in the flood zone and earn their living by fishing during the flood season, and therefore become more vulnerable. Data from household surveys showed a higher proportion of landless households for An Giang (33.2 percent) than Kandal (26.9 percent). The households that depend fully on natural resources also become more vulnerable as they remain in the flood zone. Kandal showed higher vulnerability in this component index than An Giang (0.619 versus 0.577 respectively). This was due to a higher proportion of households that depend on natural resources in Kandal (50.7 percent) than An Giang (40.9 percent). To measure vulnerability to natural disaster and climate variability, only four indicators were used and Kandal showed greater vulnerability than An Giang (0.314 vs. 0.268). The major contributing factors for this was that Kandal had higher values for indicators of the number of dead and percentage of households that did not receive flood warnings. Also, after the 2000 flood, An Giang received some investment from the government that helped reduce the impact of natural disasters. The overall natural vulnerability index was estimated at 0.388 for Kandal and at 0.364 for An Giang.

Vulnerability in terms of social capital

As illustrated in Fig. 6.5, the overall social vulnerability index was identified as being higher for Kandal (0.482) than for An Giang (0.304). This was attributed to the higher socio-demographic and social networks indices. The socio-demographic component index was estimated at 0.214 for An Giang and 0.321 for Kandal. For the social networks component, Kandal showed greater vulnerability than An Giang (0.803 vs. 0.484). Better relationships help to minimize risks from floods. Results from the focus group discussions proved that the communities in An Giang had strong grassroots organizations such as the Women's Union, Farmers

Association, Red Cross, and close relationships with village authorities, all of which helped local people to minimize their risks.

Fig. 6.5 Comparative vulnerability of five types of capital, An Giang and Kandal

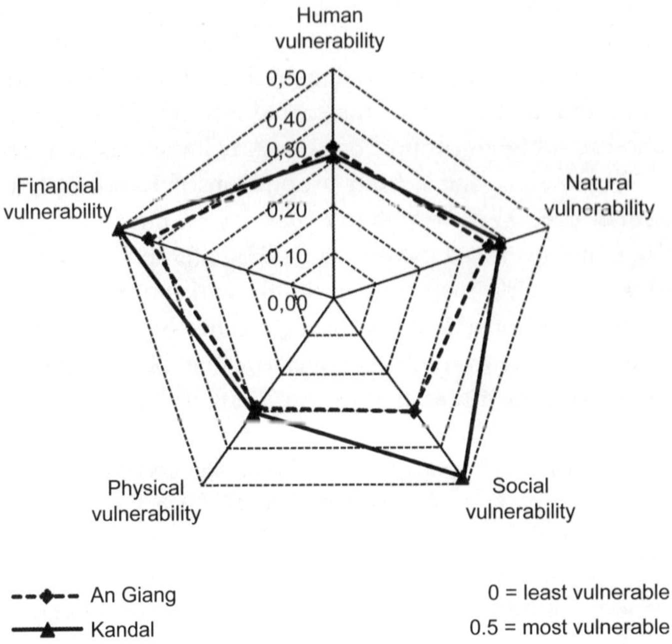

Vulnerability in terms of physical capital

In order to measure physical capital vulnerability, only three indicators were used. The indicators of the percentage of temporary houses and percentage of submerged houses measure the level of vulnerability of households while the percentage of households with no access to production means reflects how people cope with floods. Results of the survey showed An Giang with a higher value for the indicator of temporary houses than Kandal, while Kandal showed higher values for indicators of houses affected by flood and households with no access to production means. The reason for the high proportion of Kandal's households with no access to production means was due to their poor living conditions so they lacked basic production facilities. These three indicators contributed in making the overall physical vulnerability index for Kandal at 0.308, which was higher than that for An Giang (0.290).

Vulnerability in terms of financial capital

The financial capital indices refer to how people cope with floods. Three indicators were used to measure financial vulnerability. An Giang showed a higher value for the indicator of households that borrowed money during floods than Kandal (52 percent vs 45 percent). Similarly, An Giang also showed a higher value for the indicator of households with no income during the floods compared to Kandal (46.5 percent vs 23 percent), but for indicators of net household income of less than US$1,000, Kandal showed a higher proportion than An Giang (86.2 percent vs 29.6 percent). This means that a high proportion of Kandal's households earn less than US$200 per capita per year, which puts them close to the current Vietnamese official poverty line (US$240). These three indicators, households borrowing money, households with no income during the flood season, and households with net income of less than US$1,000, contributed to making the overall financial vulnerability index for Kandal 0.500, which was higher than that for An Giang (0.427).

Fig. 6.6 Vulnerability diagram of the major components of LVI for An Giang and Kandal

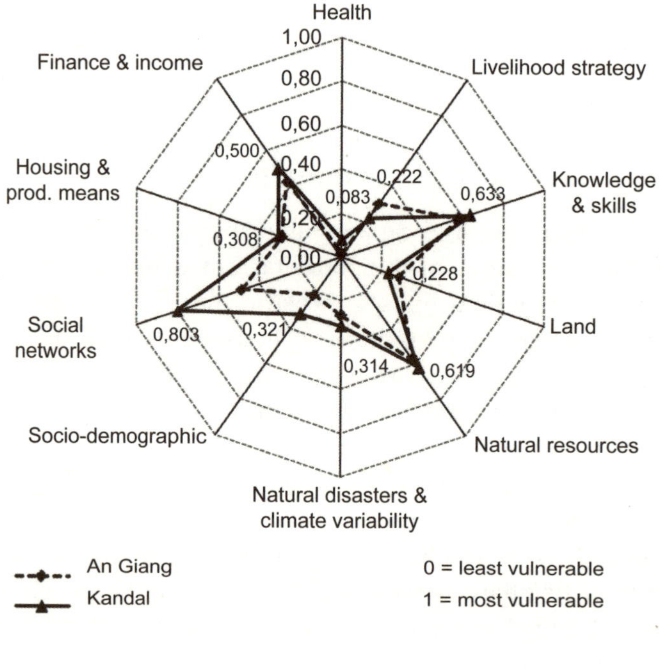

As a whole, Fig. 6.6 serves as a summary figure for all types of vulnerabilities and the differences in vulnerability indices between the two provinces. Kandal appears to show a higher vulnerability than An Giang in terms of social networks, natural resources, knowledge, and skills, and finance and income indices. All these factors indicate people's capacity to cope with floods; they should be taken into account when framing strategies to reduce vulnerability to floods.

When examining LVI values for different social groups, it was apparent that the economically poor groups were most affected by floods and the better-off group was the least for both An Giang and Kandal (Table 6.3). This can be explained by the better-off group experiencing outmigration, having jobs and higher levels of knowledge and skills, while the poorer households face financial deficits, are mostly landless, have less access to production means, and possess lower quality physical capital (housing) that is also more prone to damage by floods.

Table 6.3 Flood effect indicator values by household capital and social groups in An Giang (AG) and Kandal (KD)

Household assets	Effect index for social groups in AG			Effect index for social groups in KD		
	Better-off	Medium	Poor	Better-off	Medium	Poor
Human	0.275	0.327	0.329	0.370	0.381	0.407
Natural	0.348	0.364	0.365	0.376	0.388	0.424
Social	0.247	0.304	0.310	0.466	0.482	0.514
Physical	0.277	0.290	0.299	0.308	0.308	0.345
Financial	0.373	0.427	0.490	0.417	0.500	0.534
Overall effect index	0.320	0.362	0.371	0.413	0.423	0.466

LVI-IPCC: An Giang versus Kandal

Using similar indicators, the LVI-IPCC analysis yielded consistent results (LVI-IPCC: An Giang −0.015, Kandal −0.029) (Table 6.4). Fig. 6.7 shows a vulnerability triangle, which plots the contributing factor scores for exposure, adaptive capacity, and sensitivity [with the function LVI-IPCC = (Exposure − Adaptive capacity) x Sensitivity]. The triangle indicates that Kandal may be more exposed (0.314) to flood and climate variability impacts than An Giang (0.268).

Fig. 6.7 Vulnerability triangle of LVI–IPCC factors, An Giang and Kandal

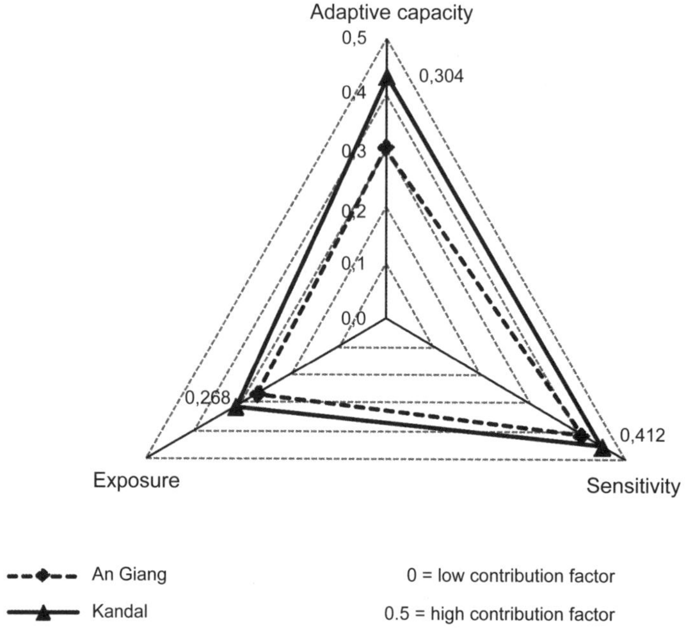

Table 6.4 illustrates the major components for the respective contributing factors of LVI–IPCC (exposure, adaptive capacity, sensitivity) and the resulting indices for An Giang and Kandal. Based on socio-demographics, livelihood strategies, and social networks, Kandal showed a higher adaptive capacity than An Giang (0.378 versus 0.304, respectively). When taking into account health, knowledge, and skills; land and natural resources, and financial assets, Kandal may also be more sensitive to flood and climate variability impacts than An Giang (0.452 versus 0.412). The overall LVI–IPCC scores indicated that Kandal households may be more vulnerable than An Giang households (–0.029 versus –0.015, respectively).

Table 6.4 Factors contributing to LVI–IPCC for An Giang and Kandal

Major components	An Giang	Kandal	Contributing factors	An Giang	Kandal
Natural disasters and climate variability	0.268	0.314	Exposure	0.268	0.314
Socio-demographic	0.214	0.321	Adaptive capacity	0.304	0.378
Livelihood strategies	0.304	0.222			
Social networks	0.484	0.803			
Health, knowledge & skills	0.354	0.413	Sensitivity	0.412	0.452
Land and natural resources	0.461	0.463			
Financial	0.427	0.500			
LVI-IPCC = (Exposure - Adaptive)*Sensivity				**-0.015**	**-0.029**

Practical implications of assessing livelihood assets

In practical terms, the assessment of livelihood vulnerability is too complicated to be fully covered because there are many aspects, dimensions, and factors relating to livelihood vulnerability—economic, political, demographic, etc., that cannot be included (Carney 1998: Adger et al. 2001; Sullivan 2002). This chapter only focused on some major risk factors (major components) that influence the flood-related vulnerability of households. The sub-components we used to construct the LVI in this study were selected based on the available data mainly from household surveys for our particular study sites and may not apply to other communities, and other sub-components, where these indicators may not be present.

Coping strategies and adaptation options

Current coping strategies

The household surveys showed that farmers in An Giang have a number of strategies to reduce their flood vulnerability such as: strengthening their houses, upgrading residential grounds, stocking up food, etc. Similarly in Kandal, the major methods to cope with floods are: preparing necessary drugs for diseases, sanitizing residential areas, strengthening houses, moving children to safe areas, etc. (Table 6.5).

Table 6.5 Household perceptions of common coping strategies to reduce vulnerability, An Giang and Kandal

An Giang, Vietnam		Kandal, Cambodia	
Strategies	% of respondents	Strategies	% of respondents
Strengthen houses	100.0	Strengthen houses	45.3
Prepare necessary food	40.0	Prepare necessary medicines and sanitary measures	51.3
Upgrade residential grounds	90.0	Protect wells, keep clean water	18.0
Send kids to childcare center	10.0	Take children to a safe place	20.0
Fishing during flood	15.0	Seek work during flood	10.0

To assess the level of flood disaster manageability and coping methods, a total of ten communities with about one hundred households in each province were asked whether they cope with floods or not. Three criteria were used to assess households: "cope well", "just cope," and "don't cope." Table 6.6 shows the results from the PRA of the coping capacity, including strategies to reduce vulnerability. Better-off and medium-income households were associated with the "cope well" capacity as they had solid housing or could make repairs or strengthen their houses in advance, had good preparedness plans, and good coping methods such as upgrading dikes and roads to prevent intrusion of floodwater. However, poorer households were only able to "just cope." This explains why poor households are more vulnerable to floods than better-off and medium households.

Adaptation options to reduce vulnerability to floods

On the basis of results from these two case studies in Cambodia and Vietnam, we identified a number of coping strategies (current practices and future strategies) relating to different aspects of vulnerability and flood disaster risk management. These coping strategies were closely linked to the livelihood of households and are aimed at seeking appropriate policies to address or reduce vulnerability. This research showed that local communities and households have used their available local resources in coping with floods (Tables 6.5, 6.6). Meanwhile, a number of coping strategies, policies, and local innovations for coping with floods have been acknowledged (MRC 2009a). Table 6.7 summarizes current practices and future strategies for flood vulnerability reduction. It includes the matrix of key actors, and their responses and strategies with regard to flood disaster management. These were developed based on research findings, existing recommendations or policies from literature reviews, and interaction with authorities.

Table 6.6 Methods of coping with floods in An Giang and Kandal

Criteria	Coping capacity	
	Rich and Medium HHs (Cope well)	**Poor HHs (Just cope)**
Location: An Giang flooded zones		
Housing condition	Having solid housing/ strengthened house in advance	Semi-solid housing or temporary housing
Preparedness plan:	Preparations of food for flood season	No food plan
	Have budget, saved enough money for flood season	No budget, just borrow money from neighbor
	Have health care plan, store enough of clean water during flood season, enough medicine	Have no plan for health care, just community health care service
Coping methods:	Upgrade dikes, roads	Migrate to safe place
	Exploit natural resources	Migrate elsewhere for employment
	Own facilities (nets, boat) for fishing	Daily fishing (fish, snails) for consumption
Location: Kandal flooded zone		
Housing condition	Having solid housing/ strengthen house in advance	Semi-solid housing or temporary housing
Preparedness plan:	Relocate to safe place during flood period	Lack of safe place, just move to land higher than the flood level
	Prepare food and medical kit for flood season	No food or medical plan
	Save money for flood season	No budget plan, just borrow money from neighbor or micro-finance firms
Coping methods:	Upgrade safe place/construct safe places (hill)	Migrate to existing safe areas in the commune or temporarily stay at relative's house
	Change crops pattern and use short duration varieties	Use short duration varieties
	Improve and construct more irrigation systems	Rehabilitate existing irrigation system to secure paddy/crops
	Seek alternative jobs	Migrate elsewhere for employment
	Possess facilities (nets, boat) for fishing during flood season	Daily fishing (fish, snails, rats) for consumption

Source: Field Survey 2011.

Table 6.7 Matrix of scale-dependent actors, responses, and strategies to reduce flood vulnerability

Scale of interest	Key actors	Institutional response	Current practices and future strategies to reduce vulnerability
Vietnam (An Giang province)			
Provincial, district	Provincial/ district authorities, CCFSC, DARD, DoNRE, banks	Direction, policies and programs; Funding, relief, loans; Emergency response	• Propagation of 'Living with floods' model • Construction of 'flood protected residential cluster' (FPRC) • Propagation of the 'four on-site principle' model • Construction and upgrading of dikes, canals for triple rice crop • Setting-up of early warning system • Allocation of funds for flood disaster management
Village/ Community	Village authority, mass org., village CFSC, companies, households	Follow and implement policies; Social safety networking; revolving loans; relief efforts	• Propagation of childcare centre model • Strengthening and repairing infrastructure (dikes, embankments, roads, school, etc.) • Propagation for raising people's awareness on flood impact and management • More training on swimming skills for children • Management of FPRC and provision of jobs
Household	Individual	Family; Relative networks	• Adjust cropping system and pattern • Strengthening and repairing infrastructure (dikes, roads, houses, etc.) • Resettle or temporarily move to FPRC • Migrate for employment
Cambodia (Kandal province)			
Provincial, district	Provincial/ district authorities, CDM, DAFF, DWRM, CRC	Policies & programs; Funding, relief, Emergency response	• Early warning system • Planning and providing aid relief • Flood management and mitigation program • Mobile phone program • Flood damage and needs assessment • Build infrastructure

Scale of interest	Key actors	Institutional response	Current practices and future strategies to reduce vulnerability
Commune/ Community	Commune authority, commune CDM, NGOs, HHs, associations	Response to policies; Social safety networking; Relief	• Improve irrigation • Construct safe place • Provide filtered water and build reservoir • Training about agri. techniques, health care • Community-based disaster preparedness
Household	Individual	Family; Relative networks	• Upgrade and strengthen house • Relocate to safe place • Protect wells for domestic water use • Change crop variety and cropping pattern • Seek more jobs for income generation

Notes: 1. In Vietnam: CCFSC= Central Committee for Flood and Storm Control; DARD= Dept. of Agriculture and Rural Development; DoNRE= Dept. of Natural Resources and Agriculture.
2. In Cambodia: CDM= Committee for Disaster Management; CRC= Cambodia Red Cross; DAFF= Provincial Dept. of Agriculture, Forestry and Fisheries; DWRM= Dept. of Water Resources Management.
Source: Authors, based on field surveys 2011.

Policy interventions

Results from PRA surveys showed that since the floods in 2000, the strategy of 'living with floods' was widely applied in many provinces and became a large campaign at different levels. In An Giang, the program 'living with floods' was used in all districts and villages in flood-prone areas. The 'four on-site principle' concept has also started to be widely used in An Giang since the floods in 2000. The 'four on-site principle' strategy consists of four elements: (i) command on-site, (ii) means on-site, (iii) forces on-site, and (iv) logistics on-site. This concept is actually one of decentralization to local authorities and communities and the enhancement of public awareness of flood response and control. Accordingly, local authorities and communities can make decisions and use their own means and logistics to prevent and control flooding, and to mitigate flood damage, especially in emergency cases (Nguyen and Nguyen 2008). The construction of a 'flood protected residential cluster' (FPRC) was the most successful program in An Giang. From the group discussion in two of the study villages, the FPRC program shows some positive impacts as perceived by local people. For example, this program is helping vulnerable people to resettle to safer places. Households in the village that are poor and most vulnerable to floods can resettle or temporarily stay in these places during the flood season.

In Kandal, PRA surveys showed the major intervention program that both national and local government authorities have put in place or are in the process of carrying out. These programs have several dimensions but give priority to: early warning systems and emergency announcements; safety service teams; relocation of people to safe places; flood management mitigation; mobile phone programs; flood damage and needs assessment; and the rebuilding of existing safe places.

Results from PRA surveys also showed some gaps between policy and practice. In many cases, both in Kandal and An Giang, policy interventions to reduce vulnerability are often in the form of 'general statements', such as "need to reduce damage from flood" or "need to improve their livelihood," but do not provide any concrete measures or suggestions on how such measures could be applied in the practical context. Respondents (officials) in both An Giang and Kandal reported that the relationship between policy development and vulnerability assessment is weak, as most of the assessments are conducted at the national and regional levels, which leads to the policymakers having limited understanding of local situations, especially at the district and village levels. In addition, they also reported that decision-makers have not paid much attention to research on flood management. These information and communication gaps need to be addressed in order to support accurate understanding of methods to reduce livelihood vulnerability to flood and climate variability among different stakeholders and communities.

Conclusions

Based on the two case studies in An Giang (Vietnam) and Kandal (Cambodia), we presented the results of assessing the socioeconomic and livelihood vulnerability of communities to floods and climate variability. In this study, the LVI and LVI–IPCC were used as alternative methods for assessing the relative vulnerability of livelihoods to flood impacts. The results showed huge damage by floods in both countries in recent years. The livelihoods of many people who live in the flood-prone areas have suffered. Kandal had a higher LVI than An Giang (0.0362 vs

0.406, respectively), indicating relatively greater vulnerability to floods and climate change impacts due to Kandal's lower level of household livelihood conditions. Vulnerability to flood of households varied amongst social groups, with poor households suffering more from floods and climate variability as they had insufficient resources to cope. A number of current practices and institutionally planned strategies to reduce vulnerability from floods was captured at different levels, and these strategies were closely linked to livelihood contexts.

Recommendations

- This study provides the LVI and LVI–IPCC as alternative methods for assessing the relative vulnerability of livelihoods to flood impacts, and these approaches could be applied for other purposes such as evaluating the effectiveness of a particular potential program or policy by introducing scenarios into the LVI model for baseline comparison.
- In both provinces, livelihood-based approaches need to be considered for flood management and climate change adaptation strategies. Results from this study showed relative high LVI for all five livelihood assets, showing that livelihood vulnerability for flood-prone areas should be taken into account in strengthening assets. Flood and climate variability measures should pay more attention to improving the livelihoods of poor people as they are the most vulnerable.
- Current practices and future strategies to reduce vulnerability were identified in this study. The strategies that are useful in reducing vulnerability are public awareness campaigns about living with floods, the construction of FPRCs, propagation of the '4 on-site principle' model, construction and upgrading of dikes, canals for triple rice-cropping in flood-prone areas, setting-up of early warning systems, and allocating funds for flood disaster management in Vietnam and Cambodia.

Note

The authors thank SUMERNET for its support for this research. We also thank Dr. Chu Thai Hoanh who was the research mentor for his advice, Ms. Orn-uma Polpanich for her early inputs to proposal development, and Dr. Louis Lebel for his valuable comments.

We wish to acknowledge the generosity and support of all researchers and institutions involved in this research, especially the leaders of the An Giang Provincial Committee for Flood and Storm Control (CFSC) and the Kandal Provincial Committee for Disaster Management (CDM), and many local leaders and farmers, for without their support this research project could not have been successful.

7

Increasing the Benefits from Contract Farming to Rural Households in the Mekong Region: Insights from Studies of Rice and Sugar

Saykham Voladet, Yanyong Inmuong, Aung Win Htut, Winston Set Aung, Men Prachvuthy, Louis Lebel, Graham Eagleton, Pippa J. Featherston, and Angela L. Bush

Governments in the Mekong region view contract farming as one way to assist the entry of subsistence farmers into the commercial mainstream (Cai et al. 2008; Sriboonchitta and Wiboonpoongse 2008; GoL 2009; RGoC 2010; Manoram et al. 2011; U Ye Min Aung 2011). The Ayeyarwady-Chao Phraya-Mekong Economic Cooperation Strategy (ACMECS) has also identified cross-border contract farming as a means of enhancing trade between neighboring countries and paving the way towards integration into the ASEAN Economic Community (ADB/M4P 2005b; ADB 2009c; Thaung 2011).

Contract farming is "an agreement between farmers and processing and/or marketing firms for the production and supply of agricultural products under forward agreements, frequently at predetermined prices" (Eaton and Shepherd 2001: 2). Contract farming models have had an important influence on the development of agribusiness in the GMS, and have been applied to several commodities, although the details of such agreements vary greatly (Manarungsan and Suwanjindar 1992; Prowse 2012) and as yet, relatively few farmers in the region are direct participants (NSOT 2008).

In Thailand, contract farming evolved in the early decades after World War II influenced both by ad hoc practice and indirect government patronage, and was only partly regulated (Sirisambhand 1985; Ekasingh et al. 2007). Over time it has become established in agribusiness, covering a range of commodities: fresh and processed fruit and vegetables; poultry, pork, and aquaculture products; hybrid seed; and organically grown rice and sugarcane (Manarungsan and Suwanjindar 1992; Naritoom 2000; Setboonsarng et al. 2006). In the sugar industry contract farming developed gradually, aided by rural credit and export policies, but in the late 1970s disputes erupted between millers and cane farmers. The latter formed associations to lobby for better benefits. Government intervened with mediation that continued until 1984, with the passing of the Cane and Sugar Act. Since 2004, policy settings that promote biofuel production in Thailand have emerged in the National Alternative Energy Development Plans, increasing the demand for cane and affecting farmers as well as industry (Gonsalves 2006).

In Vietnam, unofficial local reforms in the agricultural sector, driven by market incentives and eventually sanctioned in 1986 under *doi moi* (economic renovation), led to changes in the collective system. Contract farming was one of the mechanisms that led to this change. Luveco, a company set up in the Red River Delta in 1986, began contracting individual farmers and their cooperatives to supply canned fruit and vegetable products mainly to Eastern Europe (ADB/M4P 2005a). In the south, the Charoen Pokphand Group from Thailand commenced operations in 1992 in Dong Nai province and became an important influence on subsequent similar developments. The commercialization of agriculture was an explicit goal of Government Decree 80 (June 2002), which advocated contract farming, and today there is considerable diversity in contract farming practices nationally (ADB/M4P 2005a).

Market-oriented agriculture in Laos and Cambodia had its beginnings in the reorientation of state policies in 1986. Along with Vietnam, both countries implemented reforms promoting the expansion of agribusiness, including contract farming with cooperatives and individuals through state-owned enterprises and foreign investment (Yapa and Jacobs 2010).

Rice production represents 78 percent of the total cultivated land in Cambodia, and by the 1990s the government was considering rice as a potential export earner and not just a basic commodity for domestic food security. This opened the way for niche export companies to engage in rice contracts with farmers. Government policy in 2010 set a target to achieve a paddy surplus of more than 4 million tons and at least one million tons of milled rice exports by the year 2015 (RGoC 2010). Now, contract farming competes with investment in a range of other tradable commodities (Sari 2010; Cai et al. 2008).

In Laos, contract farming developed in the early 1990s, largely out of cross-border trade with buyers in neighboring countries dealing in commodities such as rubber, maize, and vegetables (LEAP/FAO 2007; Manoram et al. 2011). It has now become one of the preferred platforms for FDI in the agricultural sector (LEAP-FAO 2007; Manoram et al. 2011; Leebouapao and Voladet 2011). In the case of sugarcane, contract farming began in the north of the country in the 1990s as part of cross-border trade with China; in the south, it is a more recent development closely connected with the sugar industry in Thailand.

Contract farming is at a fledging stage in Myanmar, though the number of contracting firms is increasing. In the Ayeyarwady Delta it materialized in its current form to help rice farmers affected by Cyclone Nargis (FAO 2008b) and as one way to link the long-recognized potential of Myanmar's agricultural sector to valuable international markets (U Ye Min Aung 2011). The government encouraged companies to re-establish devastated rice-growing areas through loans, technical assistance, and eventual development of exports (AFMA 2011; U Ye Min Aung 2011). In 2009, the Myanmar Rice Industry Association (MRIA, now the Myanmar Rice Federation, MRF) and Rice Specialization Companies (RSCs) were established. The RSCs contract farmers at the township level in rice-producing areas. The government helps to facilitate partnerships between companies, traders, processors, and growers.

Reported benefits of contract farming

Contractual arrangements can help farmers who have land, labor, and local knowhow gain access to: markets (especially high value international markets); credit; technical support; material inputs; and hence new opportunities for raising cash income (Naritoom 2000; M4P 2005; Setboonsarng et al. 2006; LEAP 2007; Cai et al. 2008). For the firms involved, the paramount concern is to ensure a regular supply of marketable commodities of pre-agreed quality and price (Manarungsan and Suwanjindar 1992; Eaton and Shepherd 2001; Setboonsarng 2008).

There are doubts in some quarters, however, about the consequences of contract farming for the economic, social, and environmental well-being of rural communities (Delforge 2007; Panchamlong 2010; PEI-Lao PDR 2011; Boulay et al. 2012). In our recent investigation of such impacts (Prachvuthy et al. 2013), we found that contracted farmers are more likely than other farmers to have experienced recent increases in land holdings or profit per hectare. Contracted farmers appeared twice as likely to increase their profits if they had to produce on schedule, had received seeds, agro-chemicals or cash loans, or benefited from training from the contracting firm; but appeared half as likely to have increased their profits if they had to repay credit on schedule or were forced to purchase certain fertilizers or chemicals. The majority of farmers reported that contract farming leads to better market knowledge and crop management, except in Laos. There was no evidence to suggest that cultivating crops under contract had greater negative environmental impacts than farming carried out without contracts.

Variations observed amongst the countries studied raises the question of why particular contract arrangements benefit some farmers more than others. This chapter draws on new empirical evidence from our comparative study of contract farming under different conditions in Myanmar, Thailand, Laos, and Cambodia and a review of relevant policies and previous research to analyze how contract farming could be made more beneficial to rural households in the GMS.

Methods

The field research was carried out in four GMS countries between September 2011 and September 2012. In Cambodia and Myanmar, rice was the focus; in Laos and Thailand, it was sugarcane (Fig. 7.1). Rice and sugarcane involve major cropping systems; they are both also significant export commodities and hence enable us to present recent contract farming trends within the respective countries.

The farmers surveyed were involved with several firms in each location, namely: in Cambodia a commercial company, Angkor Kasekam Roongroeung Co. Ltd (AKR) and a social enterprise, the Cambodian Center for Study and Development in Agriculture (CEDAC); three rice specialist companies in Myanmar, Gold Delta, Khittayar Hinthar and Zalon Ayeyar; two foreign companies in Laos, Yingmao Sugar (from China) in the north and the Thai-owned Mitr Lao in the south; and in Thailand the mills of two companies, Mitr Phol Sugar Group and Khon Kaen Sugar Co.

Fig. 7.1 Map showing five field study locations: Kampong Speu province (Cambodia); Ayeyarwady and Bago regions (Myanmar); Luang Namtha and Savannakhet provinces (Laos) and Khon Kaen province (Thailand)

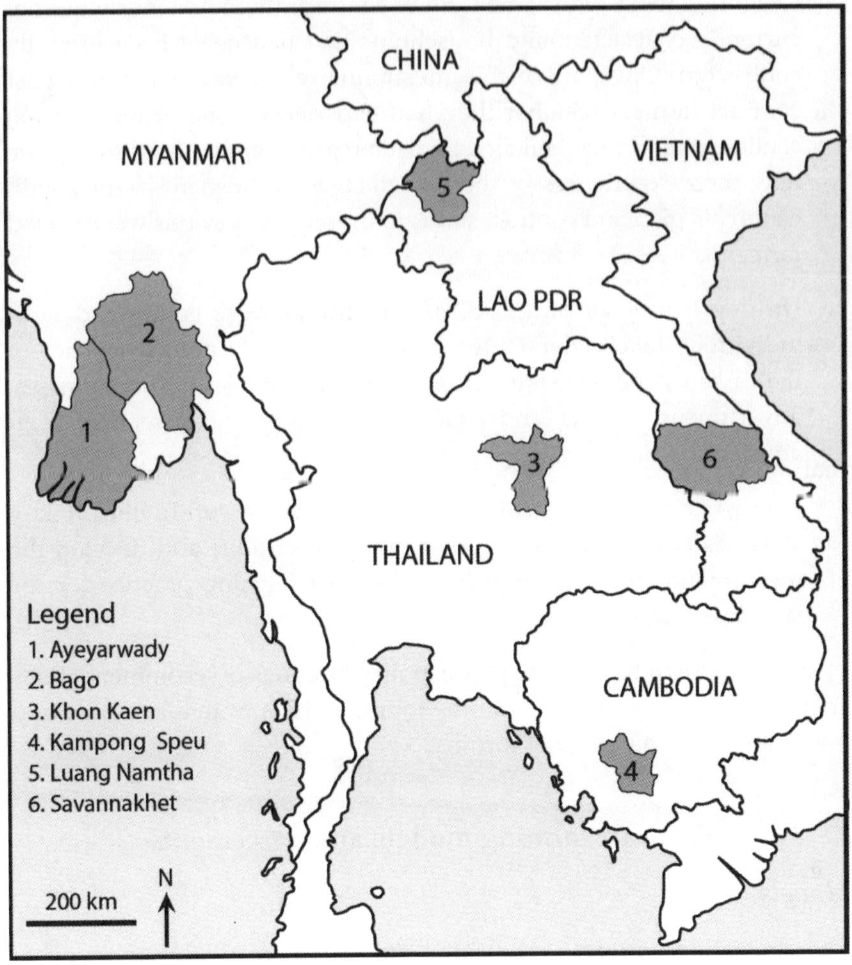

The investigative methods used within each of the four countries are briefly described below (for more details, see Prachvuthy et al. 2013).

1. Published research, government documentation and grey literature relevant to the selected crops, study locations, and models of contract farming, were reviewed.

2. Approximately 200 farmers per country ($198 \leq n \leq 239$) from representative rural communities were surveyed using a structured household questionnaire. Farmers were grouped into one of three categories: those with some form of contract; those previously but not currently contracted; and households that had never been formally contracted. One part of the questionnaire asked current and past contract farmers whether they had experienced one or more of ten challenges and nine benefits. Logistic regression was used to explore how their experiences of these challenges and benefits varied with country/region and contract status (currently vs previously contracted farmers).

3. In-depth interviews (23–32 per country) were conducted with individual stakeholders. Interviewees included farmers from each of the three categories; staff of contracting companies and enterprises; government officials; and NGOs, researchers, and personnel from advisory bodies.

4. One to three focus group discussions were conducted at key stakeholder workshops in each country to critique and develop the recommendations for improving contract farming practice for the benefit of rural households.

Subsequent to the inquiry phase, policy briefings of recommendations arising from the research within the four countries were formulated and reviewed within a variety of forums.

Contract farming models and agreements

Models

The centralized model (see Easton and Shepherd 2001) of the Thai sugar industry features strong vertical coordination; the contracting firm purchases the output of thousands of small-scale farmers with a system of quotas distributed at the beginning of the season, usually with prearranged pricing, credit, and input options. In the nucleus estate model, like sugarcane production in Savannakhet, Laos, the contracting firm manages a plantation of its own connected to its processing factory, but purchases additional supplies by entering into contracts with small farmers, often with credit, inputs, and technical advice as part of the package. The multipartite model—similar to sugarcane production in Luang Namtha, Laos, and rice contract farming in the study areas of Cambodia and Myanmar—involves government, statutory bodies, and private companies all engaging with farmers in different roles, for example, provision of credit or marketing.

Contracts

The details of contract agreements vary substantially among study sites, reflecting different cultural, historical, and political realities (Table 7.1).

Table 7.1 Selected features of contract models with different firms in the four countries

			Available to Contracted farmers[b]							Required of Contracted farmers[b]				
Country	**Contractor**	**Contract type[a]**	**Credit**	**Fertilizer**	**Stock/seed**	**Ag Chemicals**	**Training**	**Cultivation/Planting**	**Harvest/Delivery**	**Scheduled delivery**	**Exclusive supply**	**Quality**	**Quantity**	**Set pricing**
Cambodia	CEDAC	S & I	r	–	–	–	r	–	r	r	–	c	–	c
	AKR	S & I & P	r	–	c	–	r	–	o	–	c	c	–	c
Lao PDR	Yingmao	S & I & P	r	r	c	–	r	–	r	–	c	c	c	c
	Mitr Lao	S & I	r	r	c	r	r	c	r	c&n	r	n	o	c
Myanmar	Zalon Ayeyar	S & G	r	a	–	–	–	–	–	–	–	–	–	r
	Gold Delta	S & G	r	r	o	–	r	o	–	–	–	–	–	r
	Khittayar Hinthar	S & G	r	r	o	–	o	–	–	–	–	–	–	r
Thailand	Mitr Phol	S & I	o/a	o	o	r	o	o	o	c&n	–	n	c	c
	Khon Kaen	S & I	o/a	o	o	o	o	o	o	c&n	–	n	c	c

Notes: [a] Codes for contract types: S=signed contract; I=individual contract; G=group contract; P=government or local authority partners counter-sign or witness individual or group contracts.

[b] Codes for provisions and requirements: r=routinely available; c=compulsory input; o=optional input; a=input from alternative provider supported; n=negotiated requirement.

Contracts produced by the Cambodian company AKR consist of a signed agreement between the firm and the farmer witnessed by a village or commune chief, together with a group contract between the firm and village/commune authority requiring a minimum of 10 contracted households. In 2011/2012 CEDAC changed from using verbal agreements, with farmers listed in a monitored membership book, to signed individual contracts. Our study reveals that many Cambodian farmers were afraid of the word 'contract'. Therefore, it was strategic for CEDAC to introduce informal contractual arrangements before moving to written contracts.

Recent contracts drawn up by each of the three rice specialist companies studied in Myanmar's Ayeyarwady and Bago regions are of the same general type. Agreements are made for groups of 5 to 12 farmers, with each choosing their own leader. The contract records details of loans received and farmers sign a receipt. The original document is retained by the firm; no copy is made available to the farmers. Instead, farmers receive a copy of the loan receipt and repayment details. If one of the group defaults, the others are responsible for repayment. To be eligible for loans under this arrangement, farmers must present a land tax slip verifying the area of their rice land.

The district contract farming committee in Luang Namtha facilitates negotiations between the farmer association and Yingmao Sugar Factory in an annual meeting, the outcome of which is a framework for contract farming for the local region. The framework outlines contractual conditions such as the support available and the crop price.

Mitr Lao Sugar Factory is the only sugarcane processing firm in Savannakhet province that engages in individual formal contracts with farmers. Local government authorities are observers but not active participants in this process. The agreement covers the costs of land development, so contracts entail a twelve-year commitment.

In Khon Kaen, Thailand, extension officers from sugar milling firms convene with farmer groups to discuss requirements for supplying sugarcane, some of which result in smallholders being excluded. Officers monitor properties and check legal documentation before approving a farmer's entry into a signed contract. In certain cases, farmers unable to meet the quota from their land holdings alone need to demonstrate their ability to meet their obligations by purchasing supplementary sugarcane from other small farmers to make up the shortfall (Inmuong et al., in prep.).

In the following sections we outline for each study area key contract features (including details of inputs, training, market access, quotas and prices) and the pros and cons of these for farming households. We then describe and critique laws and policies regulating contract farming and make specific recommendations for improvement.

Inputs

Material and financial

In all study areas, certain agricultural inputs or sometimes credit for purchase, are routinely available from firms and/or through alternative low-interest bank loans with the support of firms (see Table 7.1). Logistic regression reveals how contracted farmers' experience of challenges and benefits varies with country/region and contract status. Farmers who have exited a contract were more than twice as likely as currently contracted farmers to have experienced a constrained choice of agricultural inputs, a challenge listed by 28 percent of current and past contracted farmers (Table 7.2).

Table 7.2 Challenges and benefits of contract farming: A regional comparison

Challenges & Benefits	Total %	Country/Region: Odds Ratios					Status: Exited vs. Current
		Thailand	Cambodia	Myanmar	S. Laos	N. Laos	
Inputs							
A constrained choice of inputs	28	1	0.01	0.10	0.22	0.62	2.24
Transport costs too high	20	1	0.01	0.04	0.07	0.00	0.96
Tight repayment schedule	23	1	0.20	0.36	0.93	0.73	1.26
Better quality inputs	56	1	0.01	0.14	0.21	3.40	0.35
Reliable availability	67	1	0.00	0.20	0.29	1.74	0.24
Access to credit	68	1	0.01	0.28	0.01	0.01	0.19
Lower interest rate	54	1	0.24	7.02	0.05	0.08	0.16
Access to equipment	39	1	0.02	0.37	0.83	1.61	0.56

Challenges & Benefits	Total %	Country/Region: Odds Ratios					Status: Exited vs. Current
		Thailand	Cambodia	Myanmar	S. Laos	N. Laos	
		Knowledge & techniques					
Insufficient farm mgmt. knowledge	25	1	0.13	0.57	**2.31**	1.00	0.61
Tight production schedule	19	1	0.11	0.02	0.38	0.36	0.82
Technical support & training	71	1	0.05	0.02	0.02	0.09	0.32
Increased yield	61	1	0.02	0.02	0.02	0.03	0.61
		Markets					
Low price	19	1	0.04	0.12	1.52	0.08	0.86
Quality standards difficult to meet	25	1	**3.64**	0.81	0.63	0.45	0.70
Quantity quotas difficult to meet	19	1	0.49	0.15	0.45	0.43	0.85
False quality claims	8	1	0.35	0.26	0.35	0.83	0.38
False quantity measurements	17	1	0.06	0.33	**9.83**	1.12	0.56
Better market access	77	94%	86%	46%	96%	99%	0.47
Sale price guarantees	641	100%	93%	26%	60%	89%	0.55

Notes: Values in left-hand column are overall frequencies reported by current and past contract farmers and those in right-hand columns are odds ratios, except in two cases where countries could not be compared because of insufficient variation and the entries are simple percentages. Odds ratios significantly different from 1 are in bold, with Thailand as the comparator. Status is the odds ratio for farmers who have exited contract farming arrangements versus those who are currently under contract.

Both Cambodian contract firms have made available small-value, interest-free loans. CEDAC has given each farmer about 20 percent of the total rice value that they expect to produce in the contract. In contrast, farmers contracted by AKR usually have not taken up the loans (amounting to about US$50 per farmer per season) because they have been required to travel about 50 km to the company's premises to do so. Both contract firms assist with harvest and delivery of produce as part of their contracts for organically grown rice varieties. AKR also provides their farmers with seed for the area under contract. At the end of harvest, farmers must return an equivalent quantity of premium quality planting seed to AKR or be penalized at the high rate of US$7 per kg for the shortfall.

Low interest loans are an important motivation for farmers entering contract farming for the first time in Myanmar. Households surveyed were seven times more likely than those in Thailand to report low interest rates as a benefit (Table 7.2), mainly because other loan sources available to poor farmers in Myanmar have prohibitive interest rates.

Fertilizer is routinely provided in rice farming contracts in Myanmar (Table 7.1). Khittayar Hinthar and Gold Delta also offer technical training and seed inputs. Gold Delta's offer comes with the guarantee of a price higher than market value if contracted farmers produce quality rice of a certain yield. In 2011, its first year, Gold Delta's seed offer was only utilized by a small number of farmers who owned more than 4 ha of paddy land. The same contractor also includes the option of cheap access to tractors in their contracts.

Thai sugar contractors offer a comprehensive range of inputs and options (Table 7.1). Contracted farmers receive required material inputs on credit, with some farmers applying only for finance support, which is usually set at US$102/ha (US$1 = THB30,700). Contracted farmers in Thailand were much more likely to have reported benefiting from better access to credit than Cambodian or Lao contract farmers (Table 7.2). Thai contract firms sell fertilizer and cane stock at the market price, offering credit at interest rates equivalent to bank lenders. Large contracted farmers may onward-sell their cane-stock and fertilizer inputs to subcontracted farmers at prices lower than the prevailing market. Alternatively, farmers may purchase inputs with credit obtained from the Bank for Agriculture and Agricultural Cooperatives (BAAC) by showing evidence of their contract (this is only one aspect of the general support that BAAC gives to contract farming as a favored agribusiness model in Thailand).

Many of the Savannakhet farmers we spoke to have little prior experience of industrial sugarcane production, where credit is a standard part of contract provision. Mitr Lao makes a large range of inputs available at cost; many provided through smaller companies who are subcontracted to assist farmers with land development, cultivation, planting, harvest, and cane delivery. These inputs are paid for through long-term credit agreements with the farmers (Table 7.1).

In Luang Namtha, Laos, representatives of Yingmao Sugar Factory schedule crop harvest and arrange for the collection, weighing, customs declaration, and transportation of the cane across the China–Laos border to the mill in Yunnan (Table 7.1). Interest-free credit for seeds, fertilizer, insecticide, and land processing is part of a start-up package, while the provisions of other contracts generally include cash loans and free produce transportation.

Developing new knowledge and techniques

Thai contract farmers were more than 10 times as likely as those in the other study regions to perceive technical support or training as a benefit of contract farming and more than 30 times as likely to perceive a yield increase as one of the benefits (Table 7.2). About half of the farmers in southern Laos and Myanmar received contractor training, compared to 97 percent in Thailand; and exited contract farmers everywhere were one-third as likely to have received technical support as current contracted farmers. As for increases in crop yield, 99 percent of Thai farmers noted this benefit, but in all other regions, the frequency was between 50 percent and 60 percent.

Representatives of the Yingmao factory train new contract farmers in Luang Namtha to prepare their land and plant the cane-stock provided. Representatives visit plantings routinely and farmers ask for technical assistance through their village heads and contract farming liaison committee.

Contracted farmers in Savannakhet are more than twice as likely as those elsewhere to have been challenged by insufficient farm management knowledge (Table 7.2). Mitr Lao has its own model plantation, which in 2010 employed up to 7,000 people per day during the harvest, mostly as day laborers (PEI-Lao PDR 2011). Contracted farmers learn about the establishment of viable sugarcane farms in this plantation and from Mitr Lao's subcontracted companies, but they expressed a desire for more training in production methods and financial planning.

Farmers in Thailand receive support from mill staff on matters such as soil-improvement, pest control, and maximizing the sugar content of cane (i.e. Commercial Cane Sugar, CCS). Companies invite officers from regional agricultural departments to advise and train farmers. Many small farmers who are not under contract also wish to receive more training from the agricultural extension officers, but claim that the latter are often aligned with companies and are neglecting them. One representative of a growers' cooperative said that government should be more involved in educating farmers about contracts; if farmers understood these better, they could more effectively manage their appropriate scale of operation, planting of cane, and finances.

Some contracted farmers in Myanmar have increased both yield and profit by adopting new methods of nursery rice management and transplanting introduced by Gold Delta, despite having to use more inputs (fertilizer, labor). Khittayar Hinthar has trialed new rice cultivars, with varying success for their contracted farmers. In the first year a hybrid rice variety from China was made available to interested farmers, but the experiment did not go well and farmers suffered as a result. In 2011/2012, another new cultivar was trialed with more success.

In Myanmar, contracted farmers would like more training at the village level, on a variety of topics and more often; they also want to learn modern techniques to boost output. Most respondents indicated that household heads always attend the training sessions provided by the Myanmar Agriculture Service, contracting firms, and/or relevant fertilizer and pesticide marketers. In general, this training has focused on agronomy, rather than on livelihood training, financial management, water management, or environmental protection.

In Cambodia, in contrast to AKR, CEDAC does not provide seeds on a continual basis, but instead trains contracted farmers to grow and select high quality seed from their own fields. CEDAC has introduced training on the low-input rice growing method known as the System of Rice Intensification (SRI) (Koma 2011). Both AKR and CEDAC also provide training in compost production, rather than supplying inorganic fertilizers; their training is focused on high quality rice production for the organic export markets. Many contracted farmers report increased yields from SRI compared with their previous low input rice-growing methods and better profits because of the higher prices offered for organic rice.

Markets

Standards and access

By engaging in contract farming, farmers get access to markets that they previously would not have had. Better market access was cited as a benefit of contract farming by a majority (≥86 percent) of farmers in all locations, except Myanmar (46 percent; Table 7.2).

In the past, Cambodian rice was sold to neighboring countries for export processing; however, the country now exports directly, passing on better prices to farmers. CEDAC sells its specialty rice varieties to international niche markets, while the ultimate consumers of AKR's organically grown

Neang Malis rice include those in Hong Kong, East Asia, Europe, and Australia. Cambodia's status as a least developed country exempts the country from paying European Union import tax. However, international buyers have strict quality standards, which puts pressure on farmers.

From 1963 to 2002 rice exports from Myanmar were a state monopoly. However, since this policy was abandoned, and following the formation of the Myanmar Rice Industry Association in 2009, the government has permitted the private sector to engage in exporting rice (Thaung 2011). Initially, contracting firms were motivated in part by export opportunities, yet benefits have been slow to accrue due to volatile international prices, increases in production costs, and inefficiencies in export licensing.

Thailand is the world's third largest producer and second largest exporter of sugar (OECD-FAO 2011). In the 2010/2011 harvest season Thailand produced a record crop from which 9.6 million tonnes of sugar was extracted out of which more than 7 million tonnes were exported (Tapaneeyangkul 2011). The committee structure of the Office of Cane and Sugar Board (see below) provides a centralized forum for juggling the interests of farmers, millers, and domestic consumers in managing interaction with the world sugar market.

In Laos's sugarcane-producing areas, there is an expectation that contracted farmers will supply cane exclusively to the contracting firms. In Savannakhet, Mitr Lao produces refined sugar for the domestic market, and raw sugar (semi-processed) for export and further processing in Thailand. In 2008/2009, 221,000 tonnes of cane were milled; with 200 tonnes of refined sugar for the domestic market and 23,000 tonnes of raw sugar for export (PEI-Lao PDR 2011). In Luang Namtha, cane is transported across the border to the Yingmao factory in Yunnan, for processing and consumption in Chinese domestic markets.

Pricing and quotas

A guaranteed pricing arrangement based on the quantity and quality of produce is a common component of contracts. A quarter of surveyed respondents said they found it a challenge to meet the contracting firm's quality standards; Cambodian farmers were more than three times as likely as others to have this problem (see Table 7.2). The most important contract condition for Cambodian contract farmers is to produce a specified quantity of high quality, pure organic rice with optimum moisture content. Strict methods are needed to meet these standards.

If farmers fall short on quality, however, many can sell their rice to middlemen at a similar price.

Around 2003, when the Myanmar government ceased control over the export of rice, fixed pricing for paddy (threshed, unpolished rice) from farmers was also brought to an end (Kyi 2005). The price of paddy, although regulated, is now determined by supply and demand in domestic and foreign markets. Only 27 percent of Myanmar contracted farmers cited a guaranteed sale price from the contracting firm as a benefit (Table 7.2).

Most contracting firms in Laos are foreign companies (e.g. Thai, Chinese). In both study sites there is no local market, so prices specified in contracts relate to the negotiating power farmers have with contracting firms. In Luang Namtha there is strong competition for land and hence contracted farmers retain some bargaining power. In Savannakhet, Mitr Lao Sugar holds a monopsony, so prices are set by the factory according to market standards. Farmers in this new cane-growing area are wary of the motives of the parent company; Savannakhet sugarcane growers were almost ten times as likely as growers in Thailand to express concern about perceived errors in the measurement of their cane delivered to factory (Table 7.2). The benefit of a guaranteed sale price was cited by only 60 percent of Savannakhet's contracted farmers, whereas in Luang Namtha the frequency was 89 percent, just slightly lower than in Thailand and Cambodia, at 92 percent and 93 percent, respectively (Table 7.2).

Regulations, policies, and laws

Governments in the four countries currently attempt to align policies—not only to promote contract farming, but also to assist subsistence farmers and reduce harmful side-effects of contracts. Many commentators (e.g. Singh 2006b) consider it the government's responsibility to set the boundaries for and monitor the transparency of contracts. However, excessive legalism can stifle achievement of a pragmatic economic balance in contract agreements (Eaton and Shepherd 2001).

Regulations

In Thailand the Cane and Sugar Act of 1984 established: 1) a revenue-sharing agreement which gave growers 70 percent and millers 30 percent of all earnings from sugar sales; 2) a regulating body, the Office of the Cane and Sugar Board (OCSB); 3) a government-supported export company; and 4) a reserve fund for crisis compensation. Among other things, this legislation provides an administrative framework for sugarcane contract farming. The OCSB sets the domestic price for cane; enforces the 70:30 revenue-sharing system and the Commercial Cane Sugar pricing system (based on CCS); and allocates domestic and export sugar quotas to the milling companies. The price-setting mechanism is complex, involving collaborative pre-season estimates, approval by cabinet and post-season adjustment. Implementation is overseen by the OCSB's Sugar and Cane Committee, incorporating representatives from sugar factories, farmers, and government. In 1999, the Thai Department of Internal Trade introduced a standard contract proforma (Singh 2006), which set a precedent for legal and transparent contracts, with specifications of quantity, quality, price, delivery, and payment, but did not offer much in terms of dispute resolution.

In Laos, a company wishing to contract farmers has to pass an investment assessment and approval process in accordance with multiple regulations and laws. These relate to land and forest protection and sustainable agricultural practices—the Agriculture Law (1998), Environmental Protection Law (1999), Regulation on Management of Fertilizers and Use of Insecticides (2000), and Regulation on Environmental and Social Impacts Assessment (2010)—and the promotion of small-scale production and export in border areas, i.e. Instruction on Management of Border Trade (2001). In addition, the Law on Investment Promotion (2009) accords rights, obligations, and incentives to investors and highlights contract farming as a preferred investment mode in agriculture. It includes the One-Stop Service Decree to fast-track exports and imports at border checkpoints, and abolishes the need for licenses. The Instruction on Contract Farming (2002), encourages contractual arrangements to follow the "2 & 3 model," whereby contracted farmers provide two inputs (land and labor) while the contracting firms provide three (capital, technology, and markets).

In Cambodia, the 2011 Sub-decree on Contract Farming was initiated by the Ministry of Agriculture, Forestry and Fisheries, aiming to: strengthen and ensure balanced contracts; expand purchase, processing, and export of the products of contract farming; and facilitate national economic development and poverty reduction (RGoC 2011). At the end of 2012, the Cambodian government issued a regulation on quality standards of milled brown rice for the US and EU markets, further influencing contract farming. It is suggested that the national standard will be aimed at the production, processing, and promotion of brown rice for domestic consumption and export to foreign markets.

To date there are no specific policies or regulations for contract farming in Myanmar. Myanmar's long-standing rice policy emphasized self-sufficiency at the national level, and also within each of the 14 states and regions. The Ministry of Agriculture and Irrigation is in the process of developing a national rice policy that will influence the direction of the export sector. As yet, there are no legislated standards on chemical residues in food items. In March 2011 an Environmental Protection Law was enacted for Myanmar following the reformulation of the Ministry of Forestry as the Ministry of Environmental Conservation and Forestry.

Implications for governance

In our survey, current contracted farmers were more than five times as likely as farmers who were not under contracts to hold the opinion that the overall balance of impacts of contract farming in their communities is positive (Prachvuthy et al. 2013). Across study locations, farmers who exited from contracts did so for a variety of reasons (Table 7.2): they were more than twice as likely to believe that they were constrained in their choice of inputs under contract than were those who remained under contract; much less likely to believe that they had benefited from access to credit and low interest rates; and less than half as likely to perceive they had benefited from technical support and training or access to markets. Even so, from our discussions with farmers, negative opinions voiced about contract farming referred to specific shortcomings rather than to a wholesale rejection of the system, even in the cases of households who had exited from contracts. Therefore, we believe adjusting existing systems is preferable to overhauling or abandoning them.

Current contracted farmers represent only a small proportion of agricultural land holders (e.g. not more than 2.8 percent in the case of Thailand; NSOT 2008). Thus, policies to benefit contracted farmers must address an important principle of equity: the specific needs of contract farmers cannot be considered in isolation from the needs of the wider community. We recommend that contract farming continues to be promoted as a way of broadening opportunities for subsistence farmers in the GMS to participate in the wider economy; reaching a greater and more diverse farmer base should be part of this promotion.

In Laos, the most important factor influencing success for new entrants into contract farming is a management system that includes significant involvement of independent third parties (e.g. the local government contract farming committee in Luang Namtha) as facilitators and for conflict resolution. Evidently, there needs to be a balance of negotiating power between the firms and farmers. Governments can achieve this by continuing reforms towards a more competitive market (important in the rice industry in Myanmar or the sugar industry in Laos and Thailand where milling companies can be local monopsonies) and by encouraging farmer associations and cooperatives with legal authority to act on behalf of their members.

Governments have a responsibility for capacity building in farming communities. Environmental protection, sustainable land use, financial planning, advice on welfare entitlements, and matters of community and farmer health are the proper provenance of government. Governments should be proactive in: training farmers in safe chemical application; monitoring to ensure guidelines are adhered to; and reinforcing compliance laws. The broad issues of environmental degradation, pollution from agricultural practices and response to the risk of climate change are outside the direct commercial interest of contracting firms. On the other hand, training, research and development in specific technologies for a commodity grown under contract is a commercial matter and should be pursued by both firms and farmers. For example, the OCSB of Thailand could coordinate the activities of government departments, contracting firms, and farmer organizations to address issues of soil conservation on sugarcane lands to avoid yield declines resulting from years of monocropping.

In Cambodia, inexperienced entrants into commercial rice growing under contract have been guided by contracting firms in organic production for external markets. Governments can encourage similar endeavors by mediation at the contract development stage. As a matter of principle, wherever a signed document is part of the arrangements, farmers, not only firms, should receive a copy of the signed agreement. This serves both legal and educational functions, and should not be sacrificed even at the expense of one or other party opting out of a formal contract in favor of informal agreements.

Linkages

Vertical: Value chains and hierarchies

Higher prices offered by contracting firms in Cambodia for quality organic rice are dependent on the integrity of the value chain leading to distant niche markets. As a result, the contracting firms' quality standards are so high that 25 percent of Cambodian farmers surveyed said they faced a challenge in meeting them (see Table 7.2). This tempts contracting firms, contracted farmers, or third parties into shortcuts and opportunistic behavior that can thwart fair-trading intentions. To pursue legal redress is not the answer, as it is rarely worth the costs involved (Eaton and Shepherd 2001). Rather, what is required is a culture of awareness of what is at stake for all players when a potentially lucrative market is lost through ill-discipline. Policies contributing to awareness are: voluntary codes of practice (such as Cambodia's Sub-decree on Contract Farming, and Lao PDR's Instruction on Contract Farming); effective training by contracting firms, government extension services and NGOs; and incentives such as contracting firms (e.g. CEDAC) awarding company share offerings to farmers. We recommend share-offerings because they encourage farmers to develop a vested interest in the contracting company and strengthen linkages across contract divides and international borders.

In Thailand, some growers believe that the 70:30 sharing of trading profits between growers and millers respectively should be revised to take into account revenues earned from byproducts, not just from the sale of sugar. In-depth interviews revealed that growers are not always convinced that current price-setting by the OCSB reflects the new realities of financial returns from cogeneration, biofuels, and organic fertilizers processed from wastes. Such issues call for greater transparency and equitable benefit sharing in changing markets.

Horizontal: Cooperatives, associations, and communities

Farmers' associations should be strengthened and included in the negotiation of contractual arrangements. Further, small-scale farmers should be encouraged to form their own networks to better share labor, agricultural machinery, and transportation and to more easily acquire inputs with economies of scale. Networks would also help in negotiations with buyers, both contracting firms, and independent buyers.

Farmers' associations in Thailand in the 1970s represented cane growers in their disputes with millers, paving the way to the political settlement of the Cane and Sugar Act, 1984. Nevertheless, this settlement left some issues unresolved; for example, independent sugarcane farmers are not registered as cane growers under the Act. The OCSB should advocate for milling companies to widen their contract system to include small cane farmers as registered growers. This would stimulate smallholders to greater levels of production, enabling mills to better meet their cane-crushing targets.

In Cambodia, forums involving farmers, contracting firms, government representatives, and NGOs should be organized as annual events (as is the case in Luang Namtha, Laos) to develop the common interests of contracting parties and to resolve problems experienced.

In Myanmar, there has been a call by some farmers for a representative organization (e.g. May Lay 2012). Recently, a 'Farmers Association' was formed under the leadership of the Rice Specialist Companies (contracting firms), based on the justification that, by acquiring a certain amount of rice land with the government's permission, such companies are now 'rice farmers'. Whether these companies are genuine farmers or not and whether the Association really represents the interests of farmers in general is questionable. The issue of land acquisition is ongoing and the government is currently trying to develop a land use policy specifically to deal with land seizures from farmers under the previous administration.

Market failure and recommendations for improvement

In Myanmar, 93 percent of households reported that contract farming had benefited them by providing credit at lower interest rates. There, and in other countries, contracting firms are stepping in to remedy market failures, where banks and other lenders are supplying insufficient or inappropriate financial services to farmers. Access to low interest loans provided by small-scale financial services should be monitored carefully by governments with attention to equity, in the same way they regulate mainstream financial institutions. We support the expansion of contract farming, and would like to see more smallholding farmers benefit from contract farming and associated credit opportunities. However, this objective of including poorer farmers must be economically viable; lenders, private and government, should act responsibly by choosing creditors who are able to repay their loans.

Contracting firms gain the loyalty of contracted farmers by input supply arrangements supported by credit, but lose such loyalty if lending practices are perceived to be exploitative or fail to achieve the results they have been led to expect. In Savannakhet, Laos, inexperienced entrants into commercial agriculture have been burdened with unmanageable debt. Contracting firms must ensure that farmers are adequately trained in both new agricultural production methods and business planning.

Interviewees representing contracting firms in Cambodia have proposed that governments could expand the export of rice to high value markets by providing soft loans during the harvest season to enable the contracting firms to properly process larger quantities of rice. Such value-adding would create more jobs and the byproducts would be used in the livestock industry.

Another example of market failure is the funding of long-term research, for example, plant breeding to develop resilient, high sucrose-producing sugarcane varieties. In Thailand, such breeding efforts have usually been funded by government because markets often fail to reward companies adequately for such long-term risky investment. But there are other ways of funding such research. In September 2012, after a public inquiry into the funding of research in the Australian sugar industry, cane growers and mill representatives voted to raise funds through growers and mills being levied a small amount per ton of cane on a 50:50 basis (Sugar Poll 2012). Countries of the GMS could look more closely at the principle of levying industry as a whole for research that is industry-specific, and which could strengthen the cooperative links between contracting firms and contracted farmers.

Lastly, in Thailand, there is a need to continue efforts toward more effective, less polluting ways of delivering sugarcane to mills. Chetthamrongchai et al. (2001) recommended provision of strategically placed loading stations, a practice which is now in operation in some locations. Small cane farmers benefit from buying stations, especially in remote locations. Increasing these access points could reduce the transaction costs involved in contracting large numbers of smallholders, the main reason why Thai contracting firms have traditionally preferred dealing with large quota-holders who have subcontractees.

Study limitations

In Thailand, the complexity of the contracting pathways was not fully appreciated at the beginning of the project, which meant that the classification of farmers into contracted and non-contracted was questionable; reclassification of some farmers prior to analysis meant the loss of some data, but was preferable to including dubious data. Protocols developed jointly by the team for our literature reviews, household surveys, in-depth interviews, and group discussions could not always be implemented as planned because of logistical and cultural contingencies, particularly in the more remote localities. Despite this, triangulation between different methods of data entry and analysis allowed inconsistencies to be minimized.

Conclusions

Our household surveys suggest that contract farmers are more likely than other farmers to have experienced a perceived increase in profits per hectare in the last five years (Prachvuthy et al. 2013). Such improved profits can be linked to technical and input support, higher or more stable produce prices and sometimes (as in organic rice production in Cambodia) to reduced costs. This finding is in line with a recent meta-study of literature on contract farming (Prowse 2012). However, failed cases are often underreported in the literature, so there tends to be a bias in favor of successful cases. Our research design partly addresses this issue by comparing current contracted farmers with those who have quit contract farming and by removing from analyses some variables that influence whether or not a household can enter into contract.

Many studies have concluded that contract farming provides opportunities for farmers to diversify their agricultural and agribusiness expertise, and thus increase average cash income (Eaton & Shepherd 2001; Prowse 2012). Cai et al. (2008) further suggest that contract farming is a way of empowering farmers "*to move beyond the contract toward independence*". But our study has revealed considerable variation across the GMS in the application of, and perceived benefits, challenges, and impacts from contract farming. Farmers in Thailand are three times as likely as farmers in Cambodia or Myanmar to report increased profits from the switch to contract farming, whereas new contract farmers of sugarcane in southern Laos are by far the least likely to have experienced increased profits (Prachvuthy et al. 2013).

Thus, we have tailored our recommendations for change to the specific circumstances of each of the study areas and their contracting models. Nevertheless, this investigation, by examining contract farming across the boundaries of language, culture, and environmental circumstance in four linked yet diverse countries, has explored what is possible in tackling rural poverty through this important production and marketing mechanism. At its best, contract farming is a way of reducing transaction costs for both contracting firms and contracted farmers. It is an investment in people, involving both vertical linkages (with an exchange of knowledge along supply chains), and horizontal linkages between farmers and their communities.

Note

We wish to thank many people—farmers, companies, government officials and NGO representatives as well as research assistants—who contributed so much to the success of this project. Special thanks go to the SUMERNET team in Bangkok and to our research assistant Ms. Phatcharee Srikuta in Khon Kaen for their tireless work in support of the regional team.

8

Quantifying Carbon Emissions Derived from China's Investment and Trade in the Lower Mekong Countries

Hu Tao and Cao Chunmiao

This chapter examines the environmental impact of China's foreign direct investment (FDI) and trade with the region, specifically through the lens of carbon dioxide emissions. The China-ASEAN Free Trade Area (CAFTA) has become the world's largest free trade area in terms of population, and the third largest in terms of nominal GDP (Sheng L. 2003). China's FDI and trade in the Lower Mekong states—Cambodia, Laos, Myanmar, Thailand, and Vietnam—have increased dramatically since the signing of CAFTA.

According to CAFTA, as of January 1, 2004, the tariff on approximately 600 kinds of products (mainly agricultural) was lowered, and by 2006, the tariff on products manufactured by China and the six ASEAN founding member-states was reduced to zero. In 2003 China and Thailand implemented a zero-tariff agreement on vegetables and fruits, which was part of the early harvest program. Vietnam, Laos, Myanmar, and Cambodia lifted their tariffs on early harvest products between January 2008 and January 2010, in recognition of special treatment for the newest ASEAN member-states. Further agreements were signed with respect to the trade in goods, services, and FDI. CAFTA was formally established on January 1, 2010.

The liberalisation of trade is reflected in patterns of increasing trade and investment by China. China's FDI flows to the Lower Mekong countries soared to US$1.7 billion in 2011, up from US$93 million in 2003 (excluding Myanmar). However, the pattern has been of gradual growth since 2006, and Chinese investment only accounted for 9.1 percent of the total inward investment to the region in 2011 (Fig. 8.1). Only Myanmar has received significantly higher levels of Chinese investment since 2008: US$7.8 billion in 2011, compared to US$258 million in 2003—a staggering increase of 29 times in less than a decade, and particularly growing in 2010 and 2011 (Fig. 8.2).

Fig. 8.1 China's outward FDI flows in Mekong countries, 2003–2011

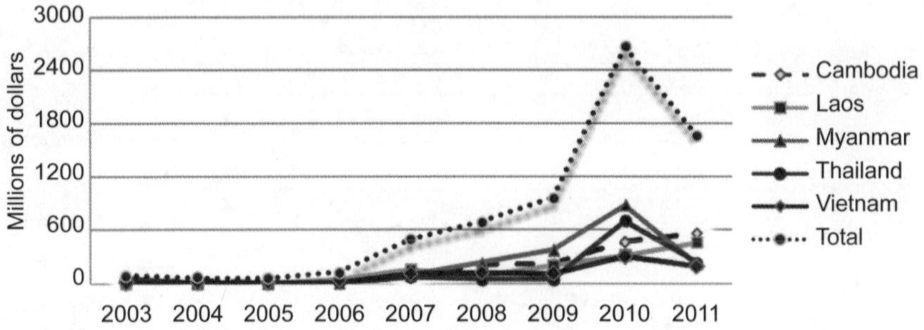

Sources: Ministry of Commerce, PRC (2011). Xinhuanet (2011).

Fig. 8.2 China's outward FDI stocks in Mekong countries, 2003–2011

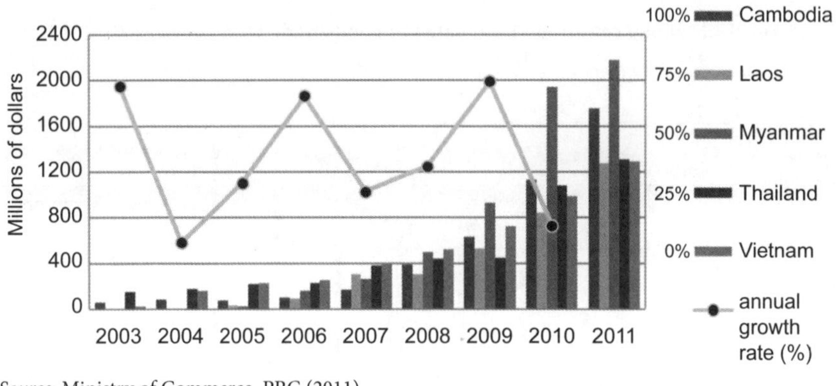

Source: Ministry of Commerce, PRC (2011).

The GMS countries are net goods importers from China, except for Laos in 2010 and 2011 (Fig. 8.3). Net imports have increased to US$27.1 billion in 2011, from US$3.1 billion in 2003. Vietnam's trade deficit with China is large relative to the other countries, and determines the broader trend for the Mekong countries as a whole (Fig. 8.3).

The total value of the goods trade from China has increased rapidly, from US$17.8 billion in 2003 to US$107 billion in 2011, a five-fold increase since 2003 (Fig. 8.4). The corresponding share of the total amount of the goods trade has increased from 9 percent to 16 percent. The amounts for Thailand and Vietnam are larger in comparison to the other Mekong countries (Fig. 8.4).

Fig. 8.3 Mekong countries: Net import vs export trade balance of goods with China, 2003–2011

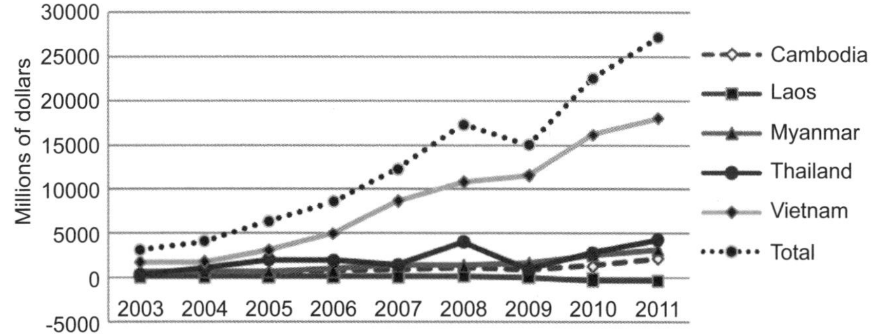

Sources: National Bureau of Statistics, China Statistical Yearbooks, 2004–2012.

Fig. 8.4 Mekong countries: Total value of goods trade with China, 2003–2011

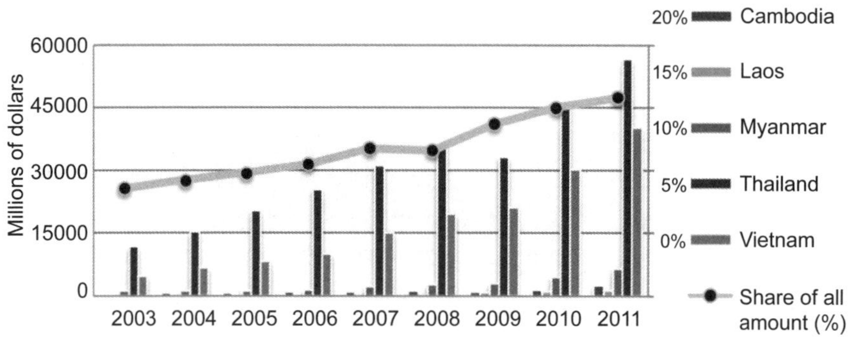

Sources: National Bureau of Statistics, China Statistical Yearbooks, 2004–2012.

Given the region's increasing FDI and trade with China, there could potentially be a corresponding rise in associated environmental impacts, but this has not been quantified. This chapter attempts to calculate the balance of environmental payments in the Lower Mekong countries as a result of China's FDI and trade.

Carbon emissions are an important dimension to measuring environmental impacts. Thus, this chapter aims to quantify the carbon emissions associated with China's FDI and trade in the Mekong region by applying the extended Intergovernmental Panel on Climate Change (IPCC) approach. It treats each country's national economy as a whole rather than by sectors due to a lack of data. Despite its shortcomings, this study helps to fill a major gap in our knowledge about the environmental impacts of China's FDI and trade in the Mekong region.

Estimating emissions derived from FDI and trade

Definition of emissions derived from FDI and trade

Emissions of pollutants are one of the most significant direct environmental impacts of economic activities. Emissions include air pollutants, such as Carbon dioxide (CO_2), Sulfur dioxide (SO_2), Nitrous oxide (NO), Particulate Matter (PM), Mercury (Hg); there are also water pollutants, such as chemical oxides (COD), biochemical oxide (BOD), heavy metals, etc. Here, due to limited data availability, only CO_2 is selected as the key indicator of emissions. The three key features of our estimations of emissions are:

- We define CO_2 emissions derived from FDI as emissions from projects that are *domestic investments*. That means the CO_2 emissions as a result of FDI *have the same emissions status as that derived from domestic investment.*
- We define CO_2 emissions derived from trade as embedded carbon in traded goods, when such emissions are a product of these goods. This can also be termed *virtual carbon* in traded goods.
- Similar to the balance of payment figures in national accounting systems, it is also necessary to have a 'balance of environmental emissions,' here defined as the *sum of CO_2 emissions derived from FDI and trade.*

These definitions and approaches are used in this chapter to calculate the environmental impacts arising from China's direct investment and trade in the Mekong region.

Measurement of CO_2 emissions from FDI and trade

The costs borne by natural resources and the environment is estimated, according to the Trade Expert Group of the Ministry of Environmental Protection of China, by using the balance of trade from the perspective of resources and the environment (Hu Tao et al. 2007). During production and consumption processes, both goods and services will consume resources and emit pollution, so there are embedded resources and emissions in the product, which are called embedded pollutants, such as embedded (virtual) SO_2 and embedded CO_2. We can measure the impact of trade on the environment through tracing these embedded pollutants in the traded goods (Hu Tao et al. 2011).

In terms of the flow of goods, if the value of one country's imports is greater than that of the value of its exports, this means that the country is a net importer, which could decrease its CO_2 emissions by allowing it to reduce domestic production, since imported goods have virtual CO_2 emissions. We consider that a positive net import value means minus CO_2 emissions, while a negative net import value means producing positive CO_2 emissions (Hu Tao et al. 2007, 2011).

The extended IPCC approach can be used to measure CO_2 emissions derived from FDI and trade. This approach has been used in studies estimating the environmental impacts of China's foreign trade (Chunmiao et al. 2011). The approach is summarized below:

- Emissions = activity level × emission factor
- For FDI: directly measuring emissions by investment level with FDI emission factor
- For trade: indirectly measuring emissions by traded goods volume with its embedded CO_2 (i.e. virtual carbon).

Parameters

First, we have to determine the CO_2 emission factors. The unit of measurement used is kg/US$, kg of CO_2 emissions per unit of FDI or trade, which is the same as the indicator of emission intensity. We use two assumptions to determine the CO_2 emission factor for each country.

Assumption 1: The production structure is similar to the imported goods structure, using a per unit of GDP emission factor of CO_2 as trade. That means that virtual CO_2 emissions produced by the unit of value of trade are equal to the CO_2 emissions produced per unit of GDP (kg of CO_2 emissions per unit of value of trade = kg CO_2 emissions per unit of GDP) (See Table 8.1) (Chunmiao et al.. 2011; World Bank 2012).

Table 8.1 CO_2 emission factors of trade (kg per US$ GDP, 2000 prices)

Year	Cambodia	Laos	Myanmar	Thailand	Vietnam
2003	0.6844	0.6597	1.954	1.736	2.057
2004	0.6938	0.6387	2.392	1.780	2.445
2005	0.6518	0.6088	3.188	1.748	2.308
2006	0.6440	0.5965	2.864	1.717	1.708
2007	0.6364	0.5611	2.548	1.586	2.127
2008	0.6660	0.5191	1.823	1.593	2.298
2009	0.6196	0.4578	1.479	1.562	2.418
2010	0.5199	0.4137	1.333	1.457	1.931
2011	0.5197	0.3774	–	1.437	1.693

Assumption 2: The production structure is similar to the FDI structure, hence the use of per unit of GDP emission factor of CO_2 for unit investment value. In addition, unit investment value could produce not only one unit of GDP, but more units of GDP due to the multiplier effect. So we use the FDI contribution as the index indicator to measure the economic development impact (value added) of FDI in the host economy, i.e. the bridge connecting the FDI and CO_2 emissions per unit of GDP, in order to estimate the CO_2 emissions by FDI. That means CO_2 emissions produced per unit of investment amount is equal to FDI's contribution index to GDP multiplied by CO_2 emissions produced per unit of GDP (kg of CO_2 emissions per unit of investment amount = FDI's contribution index to GDP × kg of CO_2 emissions per unit of GDP) (Chunmiao et al. 2011) (See Table 8.2 and Table 8.3).

Table 8.2 GMS countries: Contribution of FDI to GDP (value added)

Year	Cambodia	Laos	Myanmar	Thailand	Vietnam
2003	22.80	11.08	4.37	11.20	2.86
2004	16.69	13.58	5.20	11.32	2.96
2005	6.82	8.88	5.36	9.00	2.84
2006	6.19	1.43	5.52	9.00	2.66
2007	4.09	0.89	2.98	8.96	1.11
2008	5.22	1.36	3.81	13.12	1.00
2009	7.93	1.05	3.43	22.33	1.34
2010	5.90	1.30	5.29	14.40	1.40
2011	5.85	1.30	5.40	18.26	1.75

Notes: FDI contribution index to GDP=GDP*41.1%/FDI (Cambodia, Thailand); FDI contribution index to GDP=GDP*10.5%/FDI (Laos, Myanmar and Vietnam).
Sources: World Investment Report (2012); World Bank, World Development Indicators (2012).

The 2012 *World Investment Report* indicates that "41.1 percent" was the percentage share of the median world value of FDI's contribution index to the value added to the first quartile in 2011. As Cambodia and Thailand belonged to the first quartile, we use the figure of 41.1 percent as the FDI contribution index to GDP. In the same way, 10.5 percent was the percentage share of FDI contribution index median values to the value added in East and Southeast Asia in 2009; we used it as the FDI contribution index to GDP in Laos, Myanmar, and Vietnam. Due to the inadequacy of the data, we assumed that the figure of 41.1 percent in 2011 and 10.5 percent in 2009 stood for FDI's contribution to value added from 2003 to 2011 in these countries.

Table 8.3 GMS countries: CO_2 emissions related to investment (kg per US$ of investment amount, 2000 prices)

Year	Cambodia	Laos	Myanmar	Thailand	Vietnam
2003	15.60	7.311	8.544	19.46	5.890
2004	11.58	8.671	12.43	20.13	7.244
2005	4.445	5.403	17.08	15.73	6.563
2006	3.984	0.8510	15.80	15.46	4.552
2007	2.605	0.4987	7.605	14.22	2.368
2008	3.476	0.7066	6.951	20.91	2.294
2009	4.912	0.4789	5.069	34.89	3.246
2010	3.069	0.5367	7.050	20.97	2.698
2011	3.039	0.4903	–	26.24	2.958

Note: kg of CO_2 emissions per unit of investment amount = FDI's contribution index to GDP ×kg of CO_2 emissions per unit of GDP.

There are three clarifications regarding the parameters above.

- Trade here mainly refers to trade in goods, as trade in services produces relatively fewer CO_2 emissions.
- Our analysis is based on statistical analysis at the national level, not the industry/sector or program levels.
- The time lag between the start of the investment project and CO_2 emissions is ignored.
- All prices in this study are in US$ at base year 2000, unless otherwise specified.

Emissions from China's FDI and trade by country

Cambodia

CO_2 emissions by China's direct investment in Cambodia (Table 8.4) shows that the absolute value of virtual CO_2 emissions by net goods import from China was greater than that of the value of CO_2 emissions from China's FDI in specific years such as 2005, 2006 and 2007. That means CO_2 emissions by China's FDI and trade was negative in these years, and the overall contribution of China's investment and trade was not increasing but decreasing emissions; and vice versa.

Table 8.4 CO_2 emissions: China's investment and trade in Cambodia

Year	CO_2 emissions (investment from China)		Virtual CO_2 emissions (import goods from China)		CO_2 emissions (investment and net goods from China)	
Unit	(kt)	Percentage of total CO_2 emissions (%)	(kt)	Percentage of total CO_2 emissions (%)	(kt)	Percentage of total CO_2 emissions (%)
2003	325	10.4	-175	-5.6	150	4.8
2004	310	8.9	-265	-7.6	45	1.3
2005	20	0.53	-283	-7.6	-263	-7.1
2006	32	0.78	-349	-8.6	-317	-7.8
2007	129	2.9	-406	-9.2	-277	-6.3
2008	486	9.8	-481	-9.7	5	0.10
2009	707	15.3	-359	-7.8	348	7.6
2010	925	22.6	-421	-10.3	504	12.3
2011	1076	24.5	-692	-15.8	383	8.7

Source: Compiled by authors.

Laos

Calculation of CO_2 emissions by China's FDI and trade in Laos (Table 8.5) shows that CO_2 emissions have been increasing. However, the value of CO_2 emissions from China's FDI was much lower, even the highest CO_2 emissions was less than 100 kilotonnes (kt) in 2011, while the corresponding percentage share of total emissions was 7.0 percent. Also, the lowest CO_2 emissions were 4.5 kt in 2003, or 0.33 percent of the country's total emissions. In general, the contribution of CO_2 emissions in Laos resulting from China's FDI was relatively small.

Table 8.5 CO_2 emissions: China's investment and trade in Laos

Year	CO_2 emissions (investment from China)		Virtual CO_2 emissions (import goods from China)		CO_2 emissions (investment and net goods from China)	
Unit	(kt)	Percentage of total CO_2 emissions (%)	(kt)	Percentage of total CO_2 emissions (%)	(kt)	Percentage of total CO_2 emissions (%)
2003	4.5	0.33	-44	-3.2	-39	-2.9
2004	21	1.5	-39	-2.8	-18	-1.3
2005	70	4.9	-30	-2.1	40	2.8
2006	23	1.5	-41	-2.7	-18	-1.2
2007	41	2.7	-27	-1.8	13	0.88
2008	30	2.0	-34	-2.2	-34	-0.26
2009	49	3.4	-0.59	-0.04	48	3.0
2010	77	5.4	22	1.6	99	7.0
2011	99	7.0	58	4.1	157	11.1

Source: Compiled by authors.

The virtual CO_2 emissions by net goods import was negative, as the value of net goods import from China was positive from 2003 to 2009. There was a turning point in 2009, as the value of virtual CO_2 emissions was close to zero. From then, Laos became a net exporting country for China, and produced positive virtual CO_2 emissions that coincided with the increased opening up of its economy to foreign investment.

In addition, the biggest absolute value of the virtual CO_2 emissions was 57.64 kt in 2011, while the smallest was 0.59 kt in 2009. Accordingly, the percentage of CO_2 emissions by net import goods from China was 4.10 percent and (-0.04 percent) respectively.

Compared to CO_2 emissions derived from China's investment and trade, we found that the absolute value of virtual CO_2 emissions by net goods imported from China was greater than that by China's FDI in some years, such as 2003, 2004, 2006 and 2008. This meant that the value of CO_2 emissions produced by China's FDI and trade with Laos was negative, and China's FDI and trade helped to decrease CO_2 emissions.

Further, absolute values were smaller, whether of CO_2 emissions or its percentage share. Exceptionally, both CO_2 emissions and its percentage share increased sharply in 2011, amounting to 157 kt and 11.1 percent, respectively.

Myanmar

The CO_2 emissions from China's FDI and trade in Myanmar (Table 8.6) show a rising trend in the past several years, except in 2006. However, the value of CO_2 emissions from China's FDI was much lower, less than 100 kt until 2007, and then CO_2 emissions increased to 1,215 kt in 2010. The percentage of CO_2 emissions also showed an increasing trend, but a smaller share of the total, less than 2 percent in the first few years. However, this has increased up to 10.2 percent, not a low proportion.

Table 8.6 CO_2 emissions: Investment and net import of goods from China in Myanmar

Year	CO_2 emissions (investment from China)		Virtual CO_2 emissions (goods from China)		CO_2 emissions (investment and net goods from China)	
Unit	(kt)	Percentage of total CO_2 emissions (%)	(kt)	Percentage of total CO_2 emissions (%)	(kt)	Percentage of total CO_2 emissions (%)
2003	–	–	-680	-7.1	-680	-7.1
2004	23	0.20	-793	-7.0	-770	-6.8
2005	75	0.52	-801	-5.4	-726	-4.8
2006	63	0.48	-858	-6.6	-795	-6.1
2007	178	1.4	-854	6.6	-676	-3.7
2008	361	2.8	-542	-4.2	-181	-1.4
2009	407	3.7	-507	-4.6	-100	-0.9
2010	1215	10.2	-658	-5.5	557	4.7
2011	–	–	–	–	–	–

Source: Compiled by authors.

Virtual CO_2 emissions tied to net goods imports was negative as the value of net goods imported from China was positive from 2003 to 2011. This meant that Myanmar was a net importer of Chinese goods. In general, the absolute value of the virtual CO_2 emissions remained stable over a small range from 2003 to 2007, and then decreased to about 550 kt. Similarly, the percentage of total emissions also remained stable over a small range of about 6.5 percent from 2003 to 2007, and then decreased to about 4.5 percent.

In addition, as CO_2 emissions resulting from China's FDI were much lower than that of net goods imported from China except in 2011, the trend of total CO_2 emissions from China's FDI and trade reflected that for trade from 2003 to 2007. In short, CO_2 emissions produced by China's FDI and trade were negative, and China's FDI and trade led to decreasing CO_2 emissions. Although CO_2 emissions from China's FDI were already greater than that of the value of net goods imported from China, this was still a small share of Myanmar's total CO_2 emissions from investment and trade.

Thailand

In Thailand (Table 8.7) CO_2 emissions changed variably, reflecting instability in investments. The biggest value was 10,923 kt in 2010, while the smallest value was 67 kt in 2005. However, the percentage share of CO_2 emissions tied to Chinese FDI was small, even in 2010, just accounting for 4 percent of total emissions in Thailand.

Table 8.7 CO_2 emissions: Investment and net import of goods from China in Thailand

Year	CO_2 emissions (investment from China)		Virtual CO_2 emissions (goods from China)		CO_2 emissions (investment and net import goods from China)	
Unit	(kt)	Percentage of total CO_2 emissions (%)	(kt)	Percentage of total CO_2 emissions (%)	(kt)	Percentage of total CO_2 emissions (%)
2003	1069	0.44	-523	-0.21	546	0.23
2004	439	0.16	-1705	-0.6	-1266	-0.44
2005	67	0.02	-3098	-1.1	-3031	-1.1
2006	207	0.07	-2724	-0.96	-2517	-0.89
2007	888	0.32	-1788	-0.65	-900	-0.33
2008	748	0.26	-4970	-1.8	-4222	-1.5
2009	1340	0.50	-1098	-0.40	242	0.09
2010	10923	4.0	-2995	-1.1	7928	2.9
2011	431	1.6	-4361	-1.6	-50	-0.02

Source: Compiled by authors.

CO_2 emissions as a result of net goods imported were minus, as the value of net goods imported from China was positive. In addition, the absolute value of virtual CO_2 emissions varied widely, from 523 kt in 2003 to 4,361 kt in 2011, and the absolute value of virtual CO_2 emissions in Thailand had an increasing trend on the whole. Even so, the percentage share of virtual CO_2 emissions from China's trade was very small, less than 2 percent.

In Thailand, we found that the absolute value of virtual CO_2 emissions resulting from Chinese imports was greater than that of the value of emissions from Chinese FDI in most years, except for 2003, 2009, and 2010. In addition, there was an irregular pattern of CO_2 emissions. Overall, the combined percentage share of CO_2 emissions from China's trade and FDI remained low, at its peak only 2.9 percent. Hence, imports and direct investment from China were only a small contributor to CO_2 emissions in Thailand.

Vietnam

CO_2 emissions for Vietnam (Table 8.8) have grown slowly. Both CO_2 emissions from China's FDI and its percentage share were smaller; the highest value in 2010 was only US$70 million, or 0.3 percent. This indicates that the contribution of FDI from China to CO_2 emissions was very small.

Table 8.8 CO_2 emissions: Investment and net import of goods from China in Vietnam

Year	CO_2 emissions (investment from China)		Virtual CO_2 emissions (import goods from China)		CO_2 emissions (investment and net import goods from China)	
Unit	(kt)	Percentage of total CO_2 emissions (%)	(kt)	Percentage of total CO_2 emissions (%)	(kt)	Percentage of total CO_2 emissions (%)
2003	66	0.08	-3140	-4.0	-3074	-3.9
2004	100	0.10	-3555	-3.5	-3455	-3.4
2005	103	0.10	-5392	-5.2	-5289	-5.1
2006	140	0.17	-5991	-7.2	-5851	-7.1
2007	171	0.15	-11999	-10.7	-11828	-10.6
2008	147	0.11	-13209	-10.3	-13063	-10.2
2009	183	0.13	-14037	-9.9	-13854	-9.7
2010	370	0.30	-13988	-11.5	-13618	-11.2
2011	208	0.18	-11316	-10.0	-11108	-9.9

Source: Compiled by authors.

CO_2 emissions by net goods imported were negative, as the value of imports from China was positive. This indicates that importing goods from China favored a decrease in CO_2 emissions in Vietnam. In addition, the absolute value of virtual CO_2 emissions had two stages: it was less than 6,000 kt until 2006, and increased to more than 10,000 kt since 2007. The trend was for increasing emissions from 2003 to 2009, decreasing slightly in 2010 and 2011.

Compared to CO_2 emissions from China's FDI and trade, we found that the absolute value of virtual CO_2 emissions by net goods imported from China was much greater than that of the value of CO_2 emissions from investment. So the trend of virtual CO_2 emissions from China's trade stood for the trend of CO_2 emissions from China's FDI and trade.

In addition, as the value of virtual CO_2 emissions from Chinese imports was negative, so China's investment and trade greatly decreased the total CO_2 emissions in Vietnam. Further, Vietnam's trade with China was much larger in comparison with its FDI.

Regional CO_2 emissions from China's FDI and trade

CO_2 emissions derived from China's FDI

Table 8.9 summarizes the combined CO_2 emissions from China's FDI in the Mekong region. The value of CO_2 emissions by China's investment has been changeable, from the lowest value of 335 kt in 2005 to the highest value of 13,510 kt in 2010. There was a big gap of about four times between the two values. However, the percentage of CO_2 emissions from China's FDI was low, with the highest at 3.3 percent of the region's total emissions in 2010. This shows that Chinese investment has had quite a small effect on CO_2 emissions in the Mekong region. Compared to the other countries, Thailand had the greatest volume of CO_2 emissions from China's investment, while Laos had the least.

Table 8.9 CO_2 emissions: China's investment in the Mekong countries

kt	Cambodia	Laos	Myanmar	Thailand	Vietnam	Total domestic emissions	Share of total domestic emissions (%)
2003	325	4.5	–	1069	66	1465	0.4
2004	310	21	23	439	100	893	0.2
2005	20	70	75	67	103	335	0.1
2006	32	23	63	207	140	465	0.1
2007	129	41	178	888	171	1407	0.3
2008	486	30	361	748	147	1772	0.4
2009	707	49	407	1340	183	2686	0.6
2010	925	77	1215	10923	370	13510	3.3
2011	1076	99	–	4311	208	5694	1.4

Source: Compiled by authors.

CO_2 emissions derived from China's trade

Table 8.10 CO_2 emissions: China's trade in the Mekong countries

kt	Cambodia	Laos	Myanmar	Thailand	Vietnam	Total domestic emissions	Share of total domestic emissions (%)
2003	-175	-44	-680	-523	-3140	-4562	-1.3
2004	-265	-39	-793	-1705	-3555	-6357	-1.7
2005	-283	-30	-801	-3098	-5392	-9604	-2.4
2006	-349	-41	-858	-2724	-5991	-9963	-2.6
2007	-406	-27	-854	-1788	-11999	-15074	-3.7
2008	-481	-34	-542	-4970	-13209	-19236	-4.5
2009	-359	-0.59	-507	-1098	-14037	-16002	-3.7
2010	-421	22	-658	-2995	-13988	-18040	-4.4
2011	-692	58	–	-4361	-11316	-16311	-4.1

Source: Compiled by authors.

CO_2 emissions by China's trade in the countries of the Mekong region (Table 8.10) show that goods imports created negative emissions, as the value of net goods import from China was positive. This meant that China's trade had a positive effect on the reduction of CO_2 emissions in the Mekong region, but the effect was limited, as the highest percentage share was 4.4 percent in 2008. Among the five countries, Vietnam had the highest CO_2 emissions as a result of its trade with China, while Laos had the lowest.

Comparing Table 8.9 and 8.10, we found that China's trade in the Mekong region has contributed to a greater proportion of CO_2 emissions than its investment.

Summary: CO_2 emissions derived from China's investment and trade

Table 8.11 Summary: CO_2 emissions derived from China's investment and trade in the Mekong countries

kt	Cambodia	Laos	Myanmar	Thailand	Vietnam	Total domestic emissions	Share of total domestic emissions (%)
2003	150	-39	-680	546	-3074	-3097	-0.9
2004	45	-18	-770	-1266	-3455	-5464	-1.4
2005	-263	40	-726	-3031	-5289	-9269	-2.3
2006	-317	-18	-795	-2517	-5851	-9498	-2.5
2007	-277	13	-676	-900	-11828	-13668	-3.4
2008	5	-34	-181	-4222	-13063	-17495	-4.1
2009	348	48	-100	242	-13854	-13316	-3.1
2010	504	99	557	7928	-13618	-4530	-1.1
2011	383	157	–	-50	-11108	-10618	-2.6

Source: Compiled by authors.

In total, CO_2 emissions resulting from China's FDI and trade were minus (Table 8.11). This indicates that the contribution of CO_2 emissions derived from China's trade is much greater than that of its investment. It also meant that China's FDI and trade had a positive effect on the reduction of CO_2 emissions in the Mekong countries.

However, the percentage share of CO_2 emissions in each country from China's FDI and trade was low, with the highest percentage being 4.1 percent in 2008. This means China's FDI and trade contributed a fairly small percentage of total carbon emissions in the region. Vietnam had the highest percentage of CO_2 emissions from China's FDI and trade, while Laos had the lowest (Table 8.12).

Table 8.12 Summary: Shares of CO_2 emissions derived from China's investment and trade in the Mekong countries

Year	2003	2004	2005	2006	2007	2008	2009	2010	2011
CO_2 emissions by China's investment (kt)	1465	893	335	465	1407	1772	2686	13510	5694
Share of total CO_2 emissions (%)	0.4	0.2	0.1	0.1	0.3	0.4	0.6	3.3	1.4
CO_2 emissions by China's trade (kt)	-4562	-6357	-9604	-9963	-15074	-19236	-16002	-18040	-16311
Share of total CO_2 emissions (%)	-1.3	-1.7	-2	-2.6	-3.7	-4.5	-3.7	-4.4	-4.1
CO_2 emissions by China's investment and trade (kt)	-3097	-5464	-9269	-9498	-13668	-17495	-13316	-4530	-10618
Share of total CO_2 emissions (%)	-0.9	-1.4	-2.3	-2.5	-3.4	-4.1	-3.1	-1.1	-2.6

Source: Compiled by authors.

Discussion

Investment and trade are double-edged. On the one hand, they can lead to the introduction of cleaner and more efficient technologies that improve living standards without corresponding increases in carbon emissions. On the other hand, they could also result in the transfer of inefficient and dirty industries, and their associated emissions, to the destination country.

The key finding of this study was that trade with China has had positive impacts on carbon emissions within the Mekong countries, and that this effect was much larger than the negative impact of FDI on carbon emissions. In contrast, a previous study using panel data for 1970–2006 found that Japanese exports as a percentage of GDP contributed to increasing carbon emissions in ASEAN countries, but that FDI does not (Atici 2012). Likewise, a previous global analysis of panel trade found that trade tends to increase carbon emission burdens in less industrialized countries (Kozul-Wright and Fortunato 2012). Recent research suggests it is important to distinguish processing from normal exports as the former add relatively little emissions (Dietzenbacher et al. 2012). Differences in methods, time periods, and countries all contribute to variance in findings, suggesting the value of further work on how emissions in the Mekong region are influenced by trade and FDI.

Either way FDI and trade policies have the potential to help reduce net carbon emissions. Green policies on FDI within Mekong countries, for instance, might take the form of higher regulatory standards or voluntary schemes that also provide incentives. China's FDI policies could give preferential treatment to companies and banks that invest more responsibly in other countries. Trade policies could also reward or otherwise support firms complying with regulations or exporting higher quality goods with a high standard of environmental compliance. Policies of transnational firms could also be made green to reduce net carbon emissions as part of corporate social responsibility.

Conclusions

China's FDI in the Lower Mekong countries has increased rapidly since 2003, but FDI flows and stocks are still lower than those from other countries such as Japan and the United States. The value of the trade between China and the Mekong countries is much greater than that of the FDI from China. And more importantly, the Mekong countries are net goods importers from China, which not only offset the carbon emissions by China's FDI, but also favor the reduction of carbon emissions in the region.

The calculated results are: CO_2 emissions in this region derived from China's FDI are increasing; CO_2 emissions in this region derived from China's trade have decreased; the total emissions in this region resulting from China's FDI and trade are still decreasing as CO_2 emissions by China's trade are much greater than its FDI. This means China's investment and trade show trends of reducing the carbon emissions in the Mekong region countries. Specifically, in 2011, CO_2 emissions derived from China's investment and trade in the Mekong region was -10.6 Mt, or 2.6 percent of total emissions; in 2008 which was the peak year in the last decade, CO_2 emissions derived from China's investment and trade in the Mekong region was -17.5 Mt, or 4 percent of total emissions. Overall, this is quite a small percentage of the total domestic CO_2 emissions in the countries of the region.

Each country has special characteristics, however. For example, in Vietnam, since CO_2 emissions as a result of trade with China is much greater than emissions resulting from Chinese FDI, emissions by China's investment and trade favor Vietnam's CO_2 emissions reduction. However,

Laos is the opposite story; there were some years when China's investment and trade increased CO_2 emissions in Laos.

This analysis shows that trade and FDI can have positive environmental impacts on the emissions of carbon within a country or region. If green trade and investment policies are put in place, positive environmental impacts on the emissions of carbon derived from trade and investment will arise both in the destination and source countries.

9

Transboundary Fish Trade in the Lower Mekong Basin: Impacts on Fisheries and Rural Employment in Cambodia, Lao PDR, and Thailand

Raphael Glemet, Eleanor Elbert, Hap Navy, and Douangkham Singhanouvong

In the Mekong region, fish are the source of not just food but jobs and income to millions of people living far beyond the waters where they are caught, with an estimated 29.6 million people living and working within 15 km of the Mekong, and 2.1 million in riparian communities within 5 km of the river (ICEM 2010).

Alternative sources of employment are limited for most fisher families, making their livelihoods vulnerable to any decrease in fish stocks. But so far, no studies attempting to quantify employment in the fish trade in the Lower Mekong Basin (LMB) have been undertaken.

The LMB has been estimated to produce 3 million tons of fish and other aquatic animals from wild capture fisheries and aquaculture per year (Hortle and Bush 2003; Baran et al. 2007; Dugan 2008; ICEM 2010), making it one of the world's largest fisheries (Poulsen et al. 2004). The present study examines the fish trade, both domestic and transboundary, in three countries of the LMB: Cambodia (Stung Treng province), Lao PDR (Champassak province), and Thailand (Ubon Ratchatani province). The majority of fish in this area are caught in Cambodia, or on the Southern Lao border with Cambodia, and then shipped north to Laos and Thailand. The trade route from Stung Treng to Laos is the smallest of three routes of fish exportation from Cambodia to Thailand. Two larger routes carrying fish from Lake Tonle Sap—from Kampong Chhnang and Siem Reap provinces, through Poipet, and on to Thailand—can be assumed from the results of this study to sustain even more jobs (FAO 2011). In Cambodia, approximately 374,000 tons of fish were caught, and the total domestic and export trade was valued at 9.3 percent of GDP in 2005 (IMF 2006). The most recent estimate for the volume of the trade in Laos was 167,922 tons of fish, and 13 percent of GDP, in 2009 (UNDP 2009). Most of this income comes from the catch, transportation, and sale of large, wild, migratory riparian "white fish." In addition to the commercial value of the fish, wild fisheries also provide significant non-market values for direct consumption and food security.

Unfortunately, fish stocks are severely threatened by irresponsible fishing methods, such as the use of larger gill nets, illegal fishing, and explosives; pollution from unregulated chemical use; rising demand due to population growth; and future development in the region (MRC 2004: 15, 19). Dams on the Mekong and its tributaries could devastate the populations of these migratory fish, which must have access to both upstream areas during the dry season to spawn and pools created by flooding during the wet season to mature (MRC 2004: 5–6).

Despite its critical economic importance in the region, the fish trade of the LMB has not been researched in detail. The fish trade in the LMB creates income and adds value to the fish at multiple points along its commodity chain. The fisher sells any of the catch he does not keep for household consumption to a trader, who sells it to an exporter, who transports it across the border or to a larger city, where it ends up in the hands of a household consumer or sold to a restaurant, creating additional income for the owner and staff.

This study investigates the trade both quantitatively and qualitatively at each point of the fish commodity chain along one of three major transboundary trade routes through Cambodia, Laos, and Thailand, and attempts to quantify its multiplied effect on livelihoods and employment. Previous studies (Bush 2004 in particular) provide us an excellent entry point into the origins of this trade and how it has been impacted by national economic and trade policies. However, Bush's study, based on surveys conducted in 1999 and 2001, focused on Khong district, Laos. Given the major changes in this region, particularly in terms of economic integration over the last decade, this chapter provides an updated status report of the trade and seeks to provide new insights by pooling together research from all three countries, providing a description of how the trade operates over a wide area.

Methods

The study was conducted in three provinces: the major fishing provinces of Laos (Champassak) and Cambodia (Stung Treng), which contain some of the major fisheries of the LMB, and Ubon Ratchatani, in Thailand, which was at one time a major point of importation of fish from Laos and Cambodia. This study identified that this specific section of the regional trade, from Laos to Thailand, has now virtually ceased due to more competitive prices from the other routes (coming from the Tonle Sap, in Cambodia, to Thailand directly).

Given that this project is investigating the market chain, fish resources, livelihoods, and the legal framework governing this transboundary trade, an interdisciplinary approach was required. Research sought to identify: 1) all the actors directly involved in the trade and their function, inputs, controls, and mechanisms; 2) the employment created by the transboundary fish trade; and 3) benefits of the market chain and transboundary fish trade to fisheries and local people. Analysis of the fish market chain and trade is critical to enabling any investigation of the fish supply chain and distribution channels, in the course of which numerous transactions take place before fish and fish products reach the consumer or export markets. It should be noted that this study is focused on the trade in both capture and aquaculture fish (fresh or iced), because it is difficult to separate these two trade chains, and not on trade in fermented or dried fish, which is also significant, but would require a separate study.

Focus

This study focuses on the key stakeholders who are directly involved in the fish trade, making a first step towards a fuller assessment of all direct and indirect actors. The analysis for this research included:

- An investigation of fishing and trading volume, and prices of fish caught and traded, to determine fish catch numbers from fishers and estimate the size of the trade along each step of this route.
- Analysis of domestic market chains and transboundary fish trade.

- Observation of the functions (activities performed in the fish trade), inputs (materials, capital, and skills necessary to perform job), outputs (amount of fish purchased and/or sold in the trade chain), controls (market situation and rules), and mechanisms (where and how they caught, purchased, and/or sold fish) of all trade actors.
- Analysis of the vulnerabilities of the various stakeholders who invest directly in the market chain and the transboundary fish trade.
- Analysis of policies and regulations that govern the fish trade within each country and between the three countries.

Desk study methodology

To broaden the understanding of this study, a desk study was performed in order to review existing laws, policies, regulations, and legislation related to the trade and exportation of fish, as well as their impact on the trade and export of fish commodities; and overall management and conservation of fisheries resources.

Fieldwork methodology

Fieldwork activities in Cambodia and Laos included:

- Informant interviews using structured questionnaires with key stakeholders and households involved in the trade (fishers, traders, and consumers).
- Focus group discussions with local authorities, officers, and other individual community stakeholders, using semi-structured questionnaires.

During individual and focus group discussions, fishers, traders, and other stakeholders were asked about their awareness of policies and regulations in order to understand how well these are disseminated, and gaps found between policy and regulation in each country and across borders. Assessments of policy and regulatory enforcement were made possible through field observation.

Due to the informal nature of the trade carried out between Laos and Thailand, the research conducted in Thailand relied on detailed, in-depth observation rather than on structured questionnaires. Open interviews were carried out with key informants. Large and small-scale traders, and full-time fishermen, were interviewed by the research team. The research team organized several group discussions as well.

The Lao research team carried out additional interviews with fish sellers and traders in markets in Vientiane, as Vientiane has become a major selling site along the trade chain in recent years, and to verify data collected in Pakse district by cross-referencing it with information from Vientiane.

Sample size and locations

The researchers used their knowledge of the area to select villages indicative of the fishing trade in the region (i.e., villages along the Mekong), rather than randomly selecting a representative sample of villages. Study results therefore apply only to populations in a similar setting, where fishing or fish trading is a primary activity. Researchers interviewed village leaders to identify who in the village was involved in fishing.

Fig. 9.1 Map of study sites

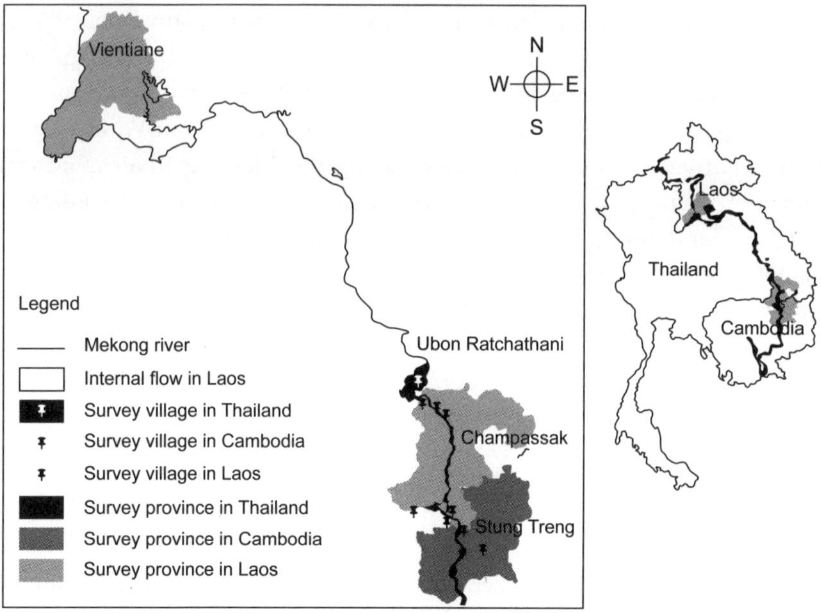

The Cambodian research team surveyed three districts (Siem Bouk, Stung Treng, and Thala Barivat), conducting interviews with 40 small and middle-scale fishermen, 9 farmers, 14 traders, 42 consumers (including fisher and non-fisher consumers), 4 village authority representatives, and held four focus group discussions. The main provinces in Cambodia that export fish to Laos are Kratie and Stung Treng. In the northern part of the country, Stung Treng province is the main trade route that fish from Kratie province—known as one of the most important landing sites for freshwater fish exports—pass through to reach Laos (Rab et al. 2006).

The Lao research team surveyed four districts in two provinces (Xanasomboun, Pakse, Khong districts in Champassak province, and Vientiane capital province), interviewing 40 small-scale fishermen individually and 64 fishermen in four group interviews, as well as 5 farmers, 10 traders, 40 consumers, and 4 key informants. Key informants included local, regional, and national officials and researchers (including one at the Living Aquatic Resources Research Center, LARReC). Researchers studied all the major fish landing sites involved in the fish trade along the Lao–Cambodian border, but were only able to access several fishing villages due to the great quantity and wide geographical range of these villages.

The Thai researcher conducted surveys in Baan Voun Buek village in Khong Jiam district, Ubon Ratchathani province. This village was selected as a pilot site because it is a river-based community and most of the villagers are fishermen. Our study estimates that Baan Voun Buek is representative of nine other villages along the Mekong River in Khong Jiam district that are similarly dependent on fishing because villagers there are landless or have very little land on which to farm. These are: Baan Tapae Tai, Baan Huai Mak Tai, Baan Dan Kao, Baan Dan Mai, Baan Ta Mui, Baan Ta Long, Baan Kum, Baan Huai Pai, and Baan Pak La. Informal interviews were also conducted with restaurant owners and traders in Khong Jiam, where there is substantial demand for Mekong fish from Thai tourists who come especially to sample them.

Data analysis

All data from the field study in Cambodia and Laos was stored in Microsoft Access and analyzed in Microsoft Excel. Descriptive analysis was used to outline the characteristics of the target stakeholders.

In all cases, results were extrapolated from the sample villages, by multiplying the average result by the total fishing population in the area, to represent as accurate a picture of the overall regional fish trade as possible. The total fishing population in the area was identified through interviews with local authorities.

A thorough qualitative analysis of functions, controls, inputs, outputs, and mechanisms; vulnerabilities; and perceptions was carried out for each category of stakeholder in Laos and Cambodia to understand how study participants engage in the fish trade and to what degree their livelihoods depend upon it. The Thai researcher performed a brief data analysis, which he supplemented with long-term experience to contextualize the data. Data gathered on the employment generation of the fish trade was collected and analyzed in Table 9.1 as follows:

- Number of households employed and number of work days per year gathered from stakeholder interviews.
- Number of people employed calculated using number of households employed and average number of people who fish per household.
- Income from fish trade calculated according to the amount of fish traded, top ten fish species sold, and price/kg for each species.
- Percentage of household income calculated by comparing fish income to average rice income per household (rice income for each district divided by the population of the district).

The study of the transboundary fish trade was subject to several limitations. Because of the non-random sample selection, the results are not representative, but indicative. The Thai contribution is descriptive of the trade along the Lao–Thai border; the trade operates informally particularly in this area. The study focused only on key selected actors in the trade chain, omitting indirect beneficiaries of the trade such as ice merchants, truck drivers, mechanics, etc. Underreporting may have occurred, especially as related to questions of money received and volume sold, due to the use of formal questionnaires.

Analysis of the study data was prone to several important limitations and biases. The indicative sample represents a population for which the fish trade is known to be a key source of income. Much qualitative data was acquired in order to thoroughly describe the trade, but was subject to the biases of the study respondents who provided it. And, finally, some population data (including the total population involved in the fish trade) was provided through interviews with key informants and is therefore subject to their biases and errors.

Results

Fish trade between Cambodia, Laos, and Thailand

Most fish captured in northern Cambodia (in Siem Bouk and Thala Borivat districts) or southern Laos (in Champassak province) make their way north to be sold in the large markets of Pakse and Vientiane (Fig. 9.1). Much of this catch used to be sold across the border at Ubon Ratchatani, Thailand, as recently as 2004 (Bush 2004). Our study found that, due to increased demand for fish in growing Vientiane, and a new road that makes the capital easily accessible, a negligible number of fish are now being sold across to Ubon Ratchathani, although there are anecdotal reports of some fish caught in Stung Treng, transported through Laos, and sold in an unrecorded and informal fashion at the Khong Jiam market, in restaurants, and to small traders in Piboon Mangsaharn district in Thailand.

From 2011 to 2012, an average (wet and dry season) of 1,454 kg of fish per day were traded across the southern border from Cambodia to Laos; and an average of over 875 more kg of fish per day were caught in Laos. A total of 2,330 kg of fish per day ended up in Pakse, where approximately 1,730 kg were consumed and 600 to 700 kg per day were transported on to Vientiane. In sum, an average of 1,679 kg of fish moves through the trade chain per day, or an average of 612,835 kg per year (averaged over the wet and dry season) (Fig. 9.2.)

These estimates were established through a single survey taken in 2012, covering the previous year.

Fig. 9.2 Fish trade route and quantities, mid 2011– mid 2012

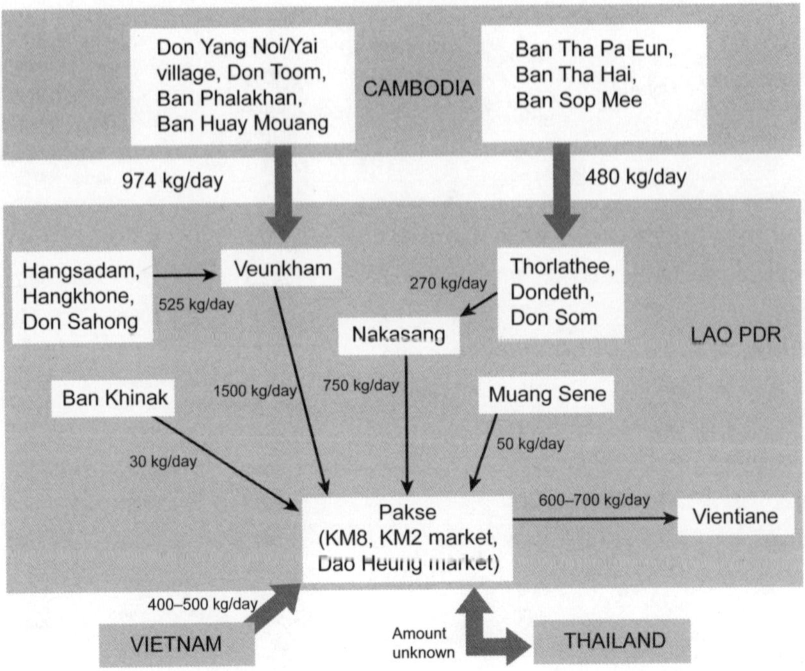

Economic analysis of the fish trade

Key data collected on activity in the fish trade among study populations, income generated by this activity, and basic comparison of this income to other sources in the study area are shown in Table 9.1.

In addition to the many people who rely on fishing for subsistence and/or household consumption, approximately 20,221 people (6,391 in Cambodia, 13,324 in Laos, and 506 in Thailand) rely on the studied fish trade route between Laos, Cambodia, and Thailand to generate a substantial portion of their income.

Table 9.1 Employment generation in the fish trade

Type of Actor	No.of households employed	No. of people employed	Households in village employed (%)	No. of work days/year spent on fishing activities	Income from fish trade (US$)	Household income that comes from fish trade (%)
			Cambodia			
Fishers	2,750	5,500	72.3	174.2	778	23.1
Fish farmers	200	680	5.7	242.3	156	3.22
Traders	87	174	1.4	78.1	806	11.3
Exporters	16	37	0.5	30.2	–	–
			Lao PDR			
Fishers	3,273	13,092	–	227	959	70
Fish farmers	43	172	–	365	2855	87
Traders	30	60	–	365	3305	89
Exporters	–	–	–	–	–	–
			Thailand			
Fishers	–	500	–	365	303–1212	50–100
Fish farmers	–	–	–	–	–	–
Traders	–	2 small 3 medium	–	60 300	181–273 3030–9090	50 –
Exporters	–	1 Thai 3 Lao (nationality)	–	365 60	– 1090–1272	– –

Fishers are defined as actors in the trade chain who sell wild fish caught in the Mekong in addition to fishing for household consumption. Fish farmers rear fish either in flooded paddy fields or in cages, mostly for household consumption. Traders buy fish from fishers and sell to someone else. Cambodian fish exporters transport fish across the border from Cambodia to Laos. Lao fish exporters operating in Thailand transport fish from Thai markets to Laos. Official records, obtained by interviewing border officials, show that no trade occurs between Laos and Thailand. Unrecorded trade does occur, with one known Thai exporter transporting fish from Thailand to Pakse, in Laos. Researchers for this study adopted informal interview methods to attempt to estimate this trade due to its unrecorded nature, but they may still have been unable to account for its total size if respondents were not completely forthcoming, or some traders remained unknown.

Because fish farming (not aquaculture, but rice paddy farming) largely occurs for household consumption in Cambodia, it provides a low percentage of household income: 3.22 percent. In Laos, aquaculture farmers gain a much greater percentage of their income from fish farming: 87 percent. For fishers, though, fishing constitutes a substantial portion of household income: 23.1 percent in Cambodia, 70 percent in Laos, and 50–100 percent in our study village in Thailand. The study results from Cambodia also show that, while fish traders do not receive as high a percentage of their household income from fish trading, only 11 percent, it is also not as time-intensive, requiring only 78 workdays per year by one member of the household to make 11 percent of yearly income for their households. However, these results may also be skewed, as fish trading is a highly specialized occupation. The actual percentage of household income from trading may be much higher, and traders underreported this figure during data collection.

Policy desk study and field observation results

The policy desk review revealed that the governments of all three countries have passed legislation to prevent overfishing, pollution, and control development projects on the river—all major threats to the fisheries. However, the researchers observed that these fisheries laws were incomplete, and have not been strongly enforced. Fishers, traders, and even officials interviewed were not always aware of what the laws are and therefore compliance is still low.

In Laos, a new Fisheries Law, aiming to sustainably manage both capture fisheries and aquaculture, was adopted in 2009. However, an implemention decree has not yet been put in place. As such, the Forestry Law (which includes wildlife and aquatic resources) and other laws with clauses related to fish resource management are still being applied, though with unclear roles and responsibilities.

The Cambodian research team reports that, while Articles 64–69, Chapter 12 of the 2007 Law of Fisheries in Cambodia set out some good regulations and policies for the trade and transportation of fish in the country, the study revealed that these regulations are not well understood by all actors and not well enforced across the board. At a local level some initiatives for fish conservation zones or local management of fisheries exists (Inland Fisheries Research and Development Institute, pers. comm.). Nevertheless, few respondents knew about the existing regulations or local management initiatives, or only knew about them to a limited degree.

The Thai researcher discovered through interviews with restaurant owners and community members that some endangered and critically endangered species such as the Mekong Giant Catfish were being advertised by restaurants to attract customers. The transborder trade of fish species that have been declared endangered or critically endangered by international conventions has not been adequately addressed by the three countries.

Vulnerabilities analysis

The Lao and Cambodian research teams found that fishers normally generated household income through a combination of fishing and farming. Most participants in the study lacked the requisite financial capital and educational background to engage in other activities: 80 percent of people working in the fish trade interviewed by the Cambodian research team stated that there was little access to loans and that, as participants in the fish trade, they were seen as not creditworthy. In addition, the study sites were all in rural areas with few industries except for fishing and farming, so when people needed more income they tended to leave their villages.

The Thailand study site, Baan Voun Buek, provides an in-depth window into how rural Thai villagers are dependent on fishing, and what alternatives to the trade they might have. Baan Voun Buek villagers are of the Kui ethnic group, who migrated from Laos and resettled on the Thai side during the colonial era in the nineteenth century. Villagers in the upland area traditionally carried out swidden agriculture, gathered non-timber forest products, and fished for household consumption. In the 1980s, the government stopped allowing swidden agriculture after the Royal Thai Forestry Department designated the area as the Kaeng Tana National Park. At that point, many villagers became landless (the household surveys taken in the course of this study found that only 6 of 140 households have land titles) and turned to fishing in the Mekong River for their livelihood. Because the villagers are landless, they could not turn to farming as a source of income. Study participants related that during the low season for fishing, many young people migrate to work as wage laborers in Bangkok or other provinces; this is their only alternative source of income.

Analysis of perceptions

Fishers related experiencing unnatural water fluctuations, which they believed was possibly the downstream impact from the construction of large dams on the Upper Mekong and its tributaries. This was a widespread view among fishers interviewed.

The main concern of all stakeholders was the uncertainty of government policies and regulations related to the fisheries sector, specifically for fish trading and fish exportation. Fishers and fish farmers also expressed concern over the impact of environmental changes on the future of their livelihoods, and felt that they are not included in the decision-making processes regarding fisheries management.

Discussion

Context and general discussion

This study reveals that the size of the fish trade in the Cambodian and Lao study areas has been previously underestimated, or has grown in recent years, when compared to a baseline established by the 2001 LARReC study (Phonvisay and Bush 2001) that concluded that only 86,800 kg per year of fish were exported from northern Cambodia to Laos—this present study estimates the size of that trade to be 530,000 kg per year.

The fish trade plays an important, and often undervalued, role in providing income and jobs for vulnerable populations in Cambodia, Laos, and Thailand. Job creation is a primary concern in the region, especially for the developing economies of Cambodia and Laos. An overwhelming proportion of impoverished people in Laos and Cambodia live in rural areas, according to IMF reports on Poverty Reduction Strategies for both countries (IMF 2008, 2006), and the UNDP's Lao Millenium Development Goals progress report. Rural poverty is twice as prevalent as urban poverty in Laos (UNDP 2008: 11), and 90 percent of impoverished people in Cambodia live in rural areas (IMF 2006: 9).

For rural populations, fishing provides a source of income that requires very little start-up capital, no formal education, and no land ownership (something that many fishers lack). Therefore, in addition to providing sustenance and nutrition to the rural poor, the fish trade also plays a crucial role in poverty reduction among these vulnerable communities. Income derived from selling fish provides the rural poor economic access to food and health care, education, shelter, and other basic needs. Fishing is also a part of the cultural identity of populations living on the Mekong, and other trades cannot replace this.

A shift to primarily aquaculture would most likely not provide jobs for the current population of fishers, whom our researchers found lack education and financial capital. Study participants overwhelmingly reported a decrease in fish catches in recent years, a concern echoed in literature on the Mekong fish trade. However, an empirical study of one fishery on the Mekong (Baran and Myschowoda 2008) found that fish stocks had not declined; rather, an increase in the number of fishers had decreased the catch of each. Nevertheless, the anxiety over smaller catches reveals the importance of this source of income and the vulnerability of the populations who rely on the fish trade.

It is beyond this chapter's objectives to make an accurate analysis of how the region's policymakers have valued fisheries and the fish trade. However, a number of studies have pointed out that the capacity of the fish trade to generate employment has been undervalued by policymakers (cf. Friend et al. 2009; Arthur and Friend 2011) in part because fishing is often a secondary activity in terms of time (many fishers spend around half their time farming), but for almost all fishers it is an indispensable source of a substantial percentage of their income. In addition, the fish trade creates relatively high-paying jobs for fish collectors/middlemen, wholesalers, and retailers that do not require formal education.

However, the price of fish depends on capital inputs. If fishers borrow money from fish traders, they then have an obligation to sell their fish to those traders at a lower price than the market (about US$0.2 to 0.5 per kg). These jobs also require experience and good working relationships with officials, and the capacity to store and transport fish. If these jobs were lost, these skills might not translate to new professions that provide as much income.

The transnational nature of the threats to the Mekong's fisheries presents significant challenges to sustainable management of this resource. Researchers observed that the fish trade is not well-regulated or standardized among the three countries, leading to a low level of overall regulation and protection of the fishery as a shared resource, and a lack of a coordinated response to difficulties or environmental threats. Legislation at the individual country level alone is inadequate to address threats to the fisheries: regional cooperation is required.

The research team therefore makes the following key policy recommendations, detailing how best to ensure the continuing abundance of fish in the Lower Mekong Basin, based upon this study.

Policy recommendations

The study results establish the fish trade as a major source of rural employment. A healthy fish population in the Mekong River is vital to maintaining economic prosperity among the people of Laos, Cambodia, and Thailand. Accordingly, the following policy recommendations are designed to counteract the negative effects on the fish trade observed in this study, and fill gaps observed in the policy desk review, in order to maintain healthy fisheries in the Lower Mekong Basin:

Laos, Cambodia, and Thailand (regional recommendations)

- Strengthen and simplify law enforcement through capacity building for the fish trade stakeholders and local officials.
- Promote community fisheries throughout the region, and empower these institutions to protect their resources.
- Consider a transboundary fish trade agreement between the three countries regulating the amount of fish traded per year and per species and implementing improved transboundary control mechanism.
- Raise public awareness on fisheries through the mass media.
- Improve transboundary information sharing.
- Improve multisectoral collaboration within the different ministries in each countries.
- Encourage the creation of a Fisher or Fish Trader Association at national and transboundary levels.
- Include the income generated by the fish trade and the vulnerability of stakeholders in cost–benefit analyses of Mekong River development projects.

Laos

The Lao research team recommends that the following steps be taken by national and provincial government agencies in order to ensure the continuing abundance of fish in the Lower Mekong Basin:

- The Lao research team strongly recommends that the new Fisheries Law be implemented immediately, as their first and foremost policy recommendation.
- Enforce the 2009 Lao Fishery Law through an implementation decree to make it effective.
- Add articles related to the size of fish, amount of fish traded per year per species, fishing gear specification, and number of fishing gear permitted.
- Enforce existing fish conservation zones, and create additional conservation zones monitored by the district officials.
- Involve the Ministry of Industry and Trade and the Ministry of Natural Resources and Environment in overseeing the fish trade, in addition to the Department of Fisheries from the Ministry of Agriculture and Forestry. These ministries should oversee the economic aspects of the fish trade, as the Department of Fisheries focuses largely on ecological issues. Greater coordination between these three ministries will ensure that management decisions made by any one Ministry will be informed by the knowledge and expertise of the other two.

Article 8 of the 2009 Fishery Law of Lao PDR states, "The Government promotes cooperation with different countries [within] the region and with international organizations in the work of fisheries through the exchange of lessons, information, science, technology [and] upgrading of the technical staff [and in] participation in and implementation of those Agreements and International Conventions to which Lao PDR is a party." Accordingly, the results of this study reiterate the need for the Government of Lao PDR to cooperate with the governments of Cambodia and Thailand, and consult these governments and international organizations on issues regarding shared fisheries resources.

Cambodia

The following recommendation is raised through this study, to ensure the continued sustainability of fish stocks and livelihoods:

- Establish fish landing sites at the commune or district level especially at the border (in addition to larger cities) to make the buying and selling of fish operate more smoothly amongst chain actors.

Thailand

The Thai research team recommends the following steps be taken to ensure the sustainability and profitability of the fish trade:

- Conduct comprehensive environmental impact assessments of all development projects on the Mekong River, especially potential hydropower dams.
- Stop the transborder trade of fish species, such as the Mekong Giant Catfish, that have been declared endangered or critically endangered by international conventions, by cooperating with the Lao and Cambodian governments to standardize regulations, penalties, and enforcement of trade laws.

Study limitations and biases

The route examined in the present study represents only the *smallest* of three major fish trade routes in Cambodia, Laos, and Thailand. The implications of threats to the health of the Mekong fisheries are therefore even greater than stated in this study, as many other people rely on the same waterways for income. Policy recommendations were formulated based upon researchers' qualitative observations of fish trade actors' vulnerabilities and analysis of the impact of gaps in existing legislation on actors. Therefore, the results and recommendations of the study are subject to the biases of the providers of data (study interviewees and survey respondents). Nonetheless, the recommendations made in this study are indicative of how the fish trade could be better protected for these two additional routes, as well, and how beneficial a transnational approach is to understanding regulatory and conservation concerns.

Conclusions

The fish trade represents a significant and often irreplaceable source of employment for the rural poor in Laos and Cambodia, and this income source should be factored into decision-making on economic development plans for the Lower Mekong Basin.

More cooperation is needed among the three countries in order to preserve the fish trade. A dialogue should be started among LMB countries to include rural employment on the agenda of transboundary discussions on trade and fish conservation. As this study demonstrates, the trade holds economic value for many people in all three nations, and it is in the best interest of each nation to take a long-term view when making economic decisions and take responsible steps now to preserve fish populations and habitats.

Note

This chapter is the product of research activities supported by SUMERNET, with financial support from SIDA, the DFID through the Climate and Development Knowledge Network, and the Ministry of Foreign Affairs of Finland, through the Mekong Water Dialogue project and with some co-funding of IUCN staff engagement in the MWD project.

The authors would like to thank Mr. Douangkham Singhanouvong, Head of Capture Fisheries Unit, and Khampheng Homsombath, Deputy Biologist of Capture Fisheries Unit, at LARReC, and Dr. Chainarong Sretthachua of Mahasarakam University, who contributed research to this book chapter; Mr. Peter Degen, Chief Technical Advisor of the MRC Fisheries Program; Muanpong Juntopas, research fellow at SEI; and Dr. Chu Thai Hoanh, Principal Researcher-Water Resources at Integrated Water Management Institute for their advice and support. They would also like to thank the Director General and Deputy Director of LARReC, Mr. Lieng Khamsivilay and Dr. Sinthavong Viravong, for their technical advice; the Director and Deputy Director of Champassak Provincial Agriculture and Forestry (PAFO), Mr. Souprany and Mr. Prachit Norraseng; the Head of Livestock and Fisheries Unit, Mr. Khabay; and the District Agriculture and Forestry Offices (DAFPO) of Pakse, Sanasomboun, and Khong districts for their assistance with the field survey.

10

Communicating Water-related Climate Change Risks: Lessons from a Multi-tool and Multi-country Study in the Mekong Region

Ngo Cong Chinh, Yvette Clarke, Nguyen Hung Manh, Louis Lebel, Sakaradhorn Boontaveeyuwat, Seak Sophat, Bach Tan Sinh, and Nguyen Tri Khiem

The potential impacts of climate change on the densely settled, low-lying coastal and delta areas of Cambodia, Thailand, and Vietnam make the Mekong region one of the world's most vulnerable areas to climate change. People living in this flood and disaster-prone region have substantial experience in reducing risk and adapting to climate variability (Berkes 2007; Lebel et al. 2013), and have developed livelihoods and lifestyles that cope with climatic hazards and disasters. While residents in the Mekong region are familiar with current water-related risks and the threat of hazards, many do not recognize how these risks are likely to change, or what effects climate change will have on their future vulnerability to climatic hazards. Improved understanding of changing risks from hydrological hazards such as floods, droughts, storms, and rising sea levels will be essential to ensure that the best adaptation responses are made by individuals and households, as well as to empower the most vulnerable communities to make informed decisions.

It is also necessary to have better tools and processes for communicating climate-related risk and uncertainty and sharing knowledge from different sources to ensure appropriate adaptation decision-making. Communication models need to be tailored to specific places and risks (Few 2003; IDS 2007; ISET 2010; Lebel et al. 2008), and to take into consideration different perceptions among stakeholders about such risks.

There has been a growing body of theoretical and empirical research on climate change communication in recent years (Moser 2010; Nerlich et al. 2010; Rohrmann 2008). Initially, climate change communication focused on disseminating scientific findings about global climate change, such as reports produced by the Inter-Governmental Panel for Climate Change (IPCC), often at high-level conferences, such as the Conference of Parties (COPs) (Moser 2010). There have been an increasing number of studies on public perceptions of climate change in Europe, America, and Australia (Weber 2010; Buys et al. 2012; Hamilton and Keim 2009) as well as how to communicate climate change (Moser 2010; Nerlich et al. 2010). However, effectively communicating the risks and uncertainties of climate change impacts to the public, especially to vulnerable communities in countries where systematic documentation and testing of experiences has been relatively limited, remains a challenge.

Risk communication is the exchange of information between stakeholders about the existence, level, sources, or acceptability of risks and about appropriate and acceptable risk reduction measures (Lebel et al. 2013). Effective risk communication is a social process by which people become informed about hazards, are influenced towards behavioral change, and can participate in decision-making in an informed manner (Rohrmann 2008). The goals of risk communication are to enhance knowledge and understanding, build trust and credibility, encourage dialogue, and influence attitudes, decisions, and behaviors.

In this chapter we describe a process undertaken to create a set of communication models in Vietnam, Cambodia, and Thailand, and the results that emerged from testing these models. Specifically, this study explores how the communication models were developed and tested, which models were most effective, and what factors should be taken into account for improving the communication of climate change risk and uncertainty. Our aim is to contribute to the design of more effective tools and approaches for communicating climate change risks in the Mekong region, and for these findings to contribute to improved climate change risk communication more broadly.

Research questions

The goal of this research was to identify effective ways to improve and share understanding of water-related climate change risks and uncertainties among local stakeholders as a basis for contributing to the sustainable development of communities in the Mekong Delta.

The objectives of this study were to:

- understand how different stakeholders perceived types, levels, and sources of water-related climate change risks and uncertainties;
- develop effective communication models on water-related climate change risks with the participation of local stakeholders in order to promote shared learning and strengthen local adaptation capacity; and
- facilitate sharing good practices and experiences in climate change risk communication and advocate for replication of the communication models to delta communities elsewhere in the Mekong region.

Study location and participants

The potential impacts of climate change on the densely settled, low-lying coastal and delta areas of Cambodia, Thailand, and Vietnam are significant. Floods, droughts, storms, and threats to local water sources present a considerable set of challenges, making the countries of the Lower Mekong region some of the most vulnerable to climate change in the world.

The IPCC (2007) has identified the Mekong Delta—Vietnam's 'rice basket'—as one of three 'extreme' global hotspots in terms of the potential impacts of climate change. Exacerbated rainfall variability will cause floods and droughts to occur more frequently with the region's rice crops being particularly vulnerable, while the Delta's low-lying areas make it susceptible to rising sea levels and saline intrusion. A sea level increase of one meter could result in 39 percent of the Mekong Delta being inundated (MoNRE 2012). These risks threaten the significant economic growth Vietnam has made in recent decades.

Cambodia is one of the most vulnerable countries in Southeast Asia to natural disasters. However, as a post-civil war, developing, predominantly agrarian country, with nearly 80 percent of the population living in rural areas, combined with weak adaptive capacity and poor infrastructure, Cambodia faces particular challenges in terms of climate change. Poor access to, storage, and management of existing water resources, together with erratic monsoon rains and rainfall, have contributed to droughts in Cambodia, whilst approximately 80 percent of the population lives along the Mekong River, which regularly floods. Prey Veng has been identified by Cambodia's Ministry of Environment (MoE) as the province most vulnerable to floods, and the second most vulnerable to droughts (MoE 2006).

In Thailand, floods, droughts, and tropical storms, already regular events, are likely to increase and intensify and have significant impacts on food, particularly rice, production (Marks 2011). These increases will place a greater burden on the Thai government to help those affected by climatic hazards and to respond appropriately. Areas along the Gulf of Thailand are likely to suffer protracted flooding because the tidal range is small and the natural water level is very low, leaving large areas without drainage (Marks 2011). One study predicts that Thailand's surge zone will increase by 33 percent, the third largest in East Asia (Dasgupta et al. 2009).

The study sites in Thailand, Cambodia, and Vietnam (Fig. 10.1) were selected as they are situated in areas frequently affected by climatic hazards such as floods. All are rural areas, and are demographically similar yet are situated in different countries, with distinct characteristics and politics:

- Thoai Son, a rural district of An Giang province in Vietnam's Mekong Delta;
- Peam Chor, a rural district in Prey Veng province, southeastern Cambodia;
- Muang, a sub-district in Samut Sakhon, at the mouth of the Tha Chin Klong River on the Gulf of Thailand.

Fig. 10.1 Location of study sites

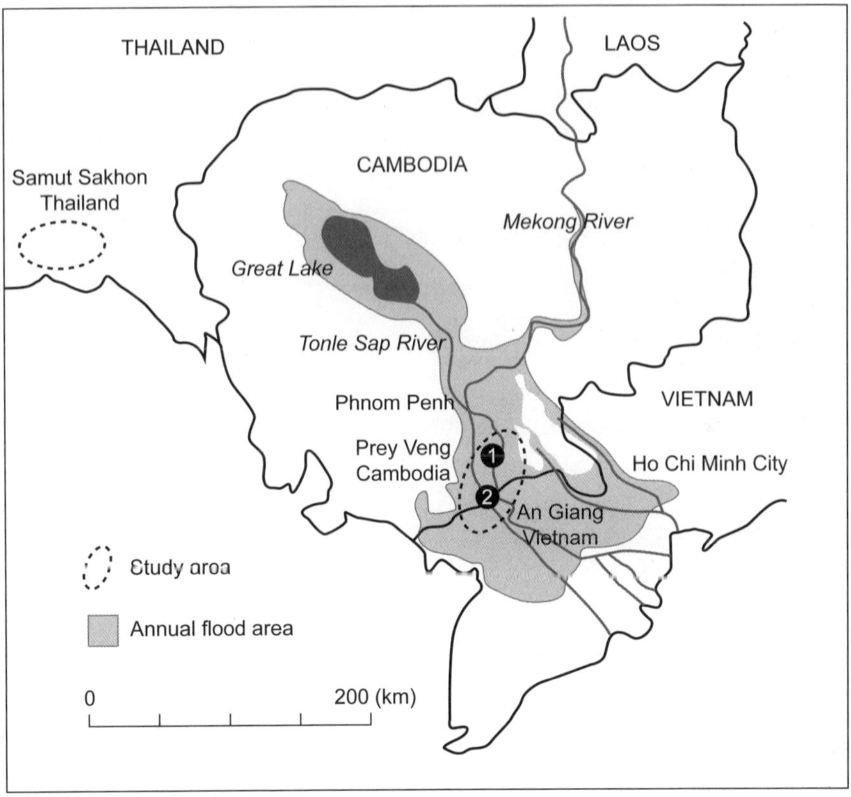

A sample of households was selected from each site, and a baseline KAP (knowledge, attitude and practice) survey was conducted to determine existing perceptions and knowledge of water-related climate change risks. The survey explored household makeup; livelihoods, and income generation; perceptions of local hazards (including floods, droughts, storms, and saline intrusions), and their impact on livelihoods and well-being; disaster and hazard preparedness; knowledge of climate change; personal observations of climate variability; and participation in communication activities.

Half the respondents were female, three-quarters had a primary school education or less, and 20 percent lived in houses of poor or very poor (temporary) quality. There was almost full TV ownership across the sites, and a high number of households had a cell phone and radio (Table 10.1).

Table 10.1 Characteristics of respondents to baseline KAP survey

	Vietnam	Cambodia	Thailand	Total
Number of respondents	213	180	207	600
Gender				
Male	54%	32%	55%	48%
Female	46%	68%	45%	52%
Membership of community groups (individual respondent)				
Credit/Savings group	15%	12%	2%	10%
Farmers group	15%	2%	14%	11%
Mass organisation	2%	4%	1%	3%
Village committee	6%	11%	39%	19%
Women's group	7%	1%	4%	7%
Religious group	4%	7%	11%	4%
Other community group	51%	0%	28%	28%
Education				
Not completed any education	9%	14%	3%	9%
Primary school	58%	62%	78%	66%
Secondary school	25%	21%	10%	18%
High school or tertiary	8%	3%	9%	7%
Vulnerability of household				
Poor or very poor quality	16%	31%	14%	28%
Average quality	40%	57%	8%	17%
Good or very good quality	42%	13%	77%	34%
Protected by dike	99%	0%	17%	42%
Household ownership of communication channels				
Television	93%	93%	100%	95%
Radio	76%	78%	78%	76%
Computer	12%	1%	20%	12%
Computer with Internet	5%	1%	16%	8%
Cell phone	91%	67%	78%	79%

The most common source of information about climate change was from television (TV), with half of those sampled having heard of climate change from this source (Fig. 10.2).

Fig. 10.2 Source of information about climate change

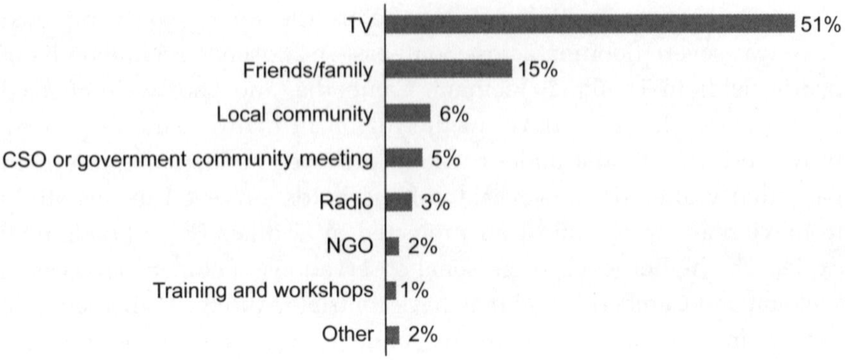

Perceptions of climate-related risk

Understanding the sociological and cultural reasons for variations in the perception of climate change is an essential step to developing effective climate change communication strategies (Weber 2010). The baseline KAP analysis identified several factors that influence perceptions of climate-related risks and climate variability in flood-prone areas of the Mekong region. The most significant factor influencing knowledge, concern, and perception of climate change and climate variability was location. Whilst Thai respondents had more knowledge of climate change, Cambodian and Vietnamese respondents were more likely to have observed changes in local climate and weather patterns (such as temperature, timing of seasons, and rainfall patterns). Vietnamese respondents reported more variability of flood patterns than those in the other countries; however, Cambodian respondents demonstrated much higher levels of concern about the impacts of climate change.

The limitation of the knowledge variable of the term 'climate change,' as opposed to exploring the concept of climate change in less prescribed language, may indicate more formalized awareness-raising about climate change in Thailand compared to Vietnam and Cambodia. The low levels of knowledge of climate change in Vietnam (the lowest of the three countries, at 47 percent) may explain the low levels of concern about its impacts; it is not possible to be concerned about something you are not aware of.

Alternatively, the widespread use of dikes in Vietnam may account for the low levels of concern in that country about the potential impacts of climate change. In 2011, not long before the survey was completed there was severe flooding across Southeast Asia. About 1.5 million ha of paddy fields in Thailand, Vietnam, Cambodia, and Laos were affected by the worst floods to have hit the region in many years. Prey Veng province in Cambodia and An Giang province in Vietnam were both particularly affected; however, the households surveyed in this study from Vietnam were almost all protected by a dike. These floods may explain the higher levels of personal observations of climatic changes in Vietnam and Cambodia, and may have contributed to the higher levels of concern in Cambodia, where the respondents' houses were not protected by dikes. Understanding local experiences such as these, and developing an understanding of local concerns, would assist communicators to shape strategies to effectively engage local communities.

The baseline KAP analysis also indicated that:

- Men were almost 2.5 times more likely to have heard of climate change than women and were 2.5 times more likely to be concerned about it than women.
- Respondents who had completed university or high school were more than three times as likely to have heard of climate change; they were also more likely to think that climate change will impact future generations than those with fewer years of schooling.
- Those whose livelihoods had been most impacted by flooding were more likely to be concerned about climate change than other respondents.
- People who reported having observed changes in the climate were five times more likely to have a lot of or some concern about climate change than those who have not observed any changes; they were more than twice as likely to think that climate change will have an impact on future generations.

Risk communication models

A risk communications working group (RCWG) was formed at each study site by the boundary partners and chaired by An Giang University in

Vietnam, the Royal University of Phnom Penh in Cambodia, and Kasetsart University in Thailand, respectively. RCWG members included researchers from participating institutes and universities as well as representatives from government authorities, local communities, students, and people who have expertise and experience working with the target audiences, communication and media techniques, disaster risk management, and climate change adaptation. The aim of the RCWGs was to get the best knowledge and participation of local informants, build local capacity and encourage the ownership of the models to be developed by the research.

One technical workshop was organized with all researchers and technical advisors to discuss the findings of the baseline survey and develop communication models to be tested. The final design of communication models was made in consultation with the RCWGs at each site to ensure that the models were relevant to the in-country context, while at the same time providing adequate rigor for testing and validation. The final models selected were:

1. Short video
2. SMS messages
3. Talking Farmer.

Content and key messages

The purpose of all the models was to raise the awareness of farmers, citizens, and local authorities about the risks associated with climate change, and to provide them with some recommended actions in the short and long-term to prepare for coping with climate-related risks: In 2011 a serious flood occurred in the three research locations that caused significant damage to households, livelihoods, and local economies. The research team agreed to use floods as a local example of climate change impacts and a common theme for the communication models. The final messages included approximately 65 percent on floods and 35 percent on more general climate change issues. The messages were developed based on the expertise of the members of the RCWG and from findings of the literature review.

Sample messages included:

Climate change

- Climate change is expected to increase the risk of drought because of changes in sea surface temperature, air circulation, and rainfall.
- Changes in ocean dynamics can lead to changes in the migratory patterns of fish and reduce fish stocks, especially in the coastal fishing grounds.
- Extreme weather phenomena such as floods and droughts will become more frequent, increasing the risk of damage to property and infrastructure.

Short-term actions to prepare for floods

- Strengthen house, raise the foundation.
- Reserve food, drinking water, medicines before the flood season.
- Prepare a first aid kit, and safety and rescue equipment.

Long-term actions to prepare for floods

- Support children to develop skills and knowledge to survive during floods.
- Build stronger dikes.
- Plant and protect mangroves, wetland forests, protection forests.

Communication model 1: Short video

Due to the high levels of TV ownership in the test sites, a documentary was raised as a potential communication model. This was deemed unfeasible due to issues with resources, administrative procedures, and time constraints. However, it was still desirable to test the documentary format, so a short video (4'14") was developed and then shown directly to the sampled households.

The video was developed by the Vietnam team and then shared with the teams in Cambodia and Thailand. The video was also shared with experts in risk communication to improve its content, images, audio, and messages, and was tested with the RCWG in An Giang, Vietnam, to fine-tune before translation into Khmer and Thai and piloting.

The video used existing footage from news and documentary coverage of the 2011 floods. It covered the impacts of climate change in the Mekong region, such as extreme weather events; abnormal rain; prolonged heat; abnormal floods at unseasonal times, frequencies, and peaks; whirlwinds and typhoons; salinity intrusions; impacts on people's livelihood and health; and actions individuals can take to respond to these threats.

Communication model 2: SMS

Given that 79 percent of households were found to own a cell phone, it was proposed to explore sending a series of SMSs on flood risks and associated climate change risks to farmers and other citizens.

A bank of messages was developed by the research team in Thailand, and was then improved by the teams in Vietnam and Cambodia. The messages were then tested among the RCWGs to reflect the local context and the capacity of local services to support such SMS messages. As mobile phones are unable to support Khmer text, this method was not tested in Cambodia.

In Vietnam, a final set of 65 messages (43 flood and 22 climate change messages) were sent to 124 households in 6 villages, over 2–3 weeks. Eleven households requested that the messages be stopped after receiving a few. In Thailand, a final set of 80 messages (52 flood and 28 climate change messages) were sent to 104 households in 8 villages over one month. Three messages a day were sent—in the morning, at midday, and in the evening.

Communication model 3: Talking Farmer

The last communication model developed was the 'Talking Farmer', i.e. a two-way approach to communication (face to face) with farmers and citizens. Members of RCWGs were trained by the research team to conduct Talking Farmer visits with farmers and citizens, and then conduct 15-minute visits with selected households. The training included how to engage people, how to use the material to conduct the session, and information on climate-related risks. The materials used during the visit included posters, leaflets, information slips on risks of flood and climate change, warnings, and preparedness. The Talking Farmers encouraged their audience to engage with them and ask questions about what interested or concerned them about climate change.

Results

Reviewing the models

Respondents were asked to what level the tool engaged them. All three models had fairly high levels of self-reported engagement, although this does not indicate comprehension of the content (Fig. 10.3).

Fig. 10.3 Audience engagement with three models

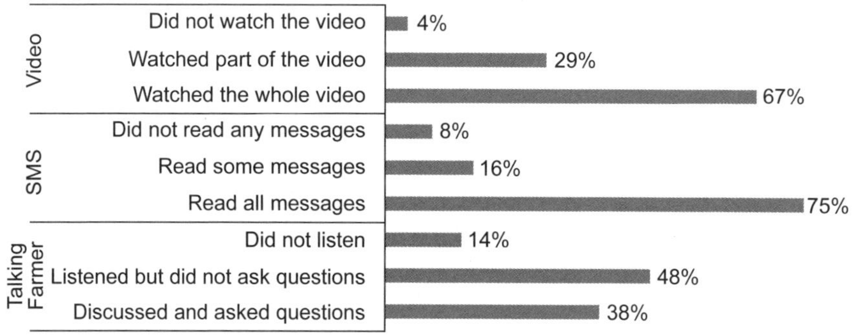

Only 24 percent of respondents who had seen the video reported that the content was interesting, compared to 51 percent of those who participated in a Talking Farmer session (Fig. 10.4). The majority of those who received and read SMS text messages said that some (not all) were interesting. Again, level of interest does not indicate comprehension of the content.

Fig. 10.4 Audience rating of interest of model content

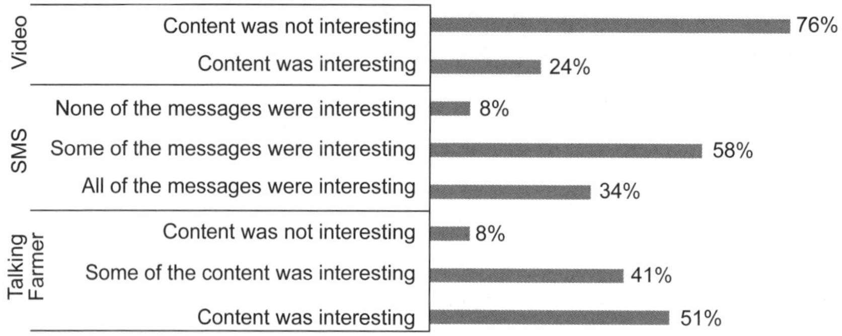

Around half of respondents for all communication models said they shared what they had learnt with a few other people (Fig. 10.5).

Fig. 10.5 Audience sharing of model content

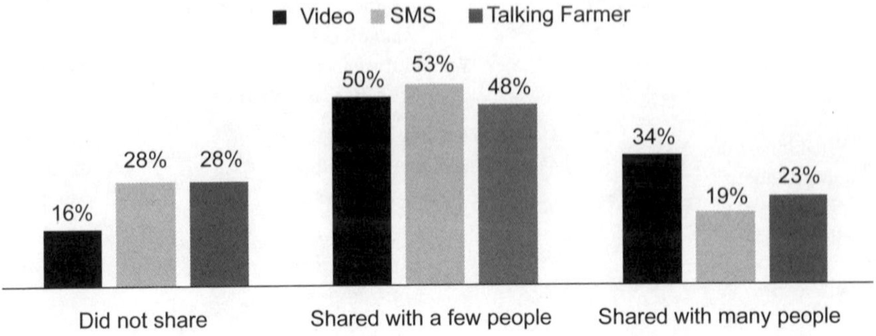

Effects of exposure

A post-KAP survey was conducted with the test households to analyse the effect of the communication models on participants' knowledge on the causes and impacts of climate change. The hypothesis of the study was that the piloted communication models would improve participants' knowledge of the causes and impacts of climate change.

Knowledge index

From the hypothesis and available data, three outcome variables were created: knowledge of the impact of climate change, knowledge of the causes of climate change, and concern about climate change. Each outcome variable was computed as an index where the higher index number indicated more knowledge or concern. The index computed based on the formulas is shown in Table 10.2.

Table 10.2 Creating index for outcome variables

Questions	Answers from respondent	Group	Grading
Knowledge index on the causes of climate change. Index = accumulated points (pt)			
What have you heard about the impacts of climate change?	Nothing		0pt
	Weather change Rainfall change Temperature increase	Changing weather	Give 1pt for each mentioned
	Seasons change		
	Disasters become more intense Disasters become more frequent Sea level rise	Worsening disasters	Give 1pt for each mentioned
	Fish/aquaculture products grow differently Crops grow differently More crop pests Salinization More animal diseases		
	More human diseases	Impact on health	Give 1pt if mentioned
Knowledge index on the causes of climate change. Index = accumulated points (pt)			
What have you heard about the causes of climate change?	Don't know		0pt
	Fossil fuel burning		1pt
	Deforestation		1pt
	Greenhouse gas emissions		1pt
	Methane emissions		1pt
	More cities/development		1pt
Climate change concern index. Index = points for the option selected (pt)			
How much does climate change concern you?	A lot		4pt
	Some		3pt
	A little		2pt
	Not at all		1pt
	Not heard about climate change		0pt

Availability of data

The original research plan was to conduct pre and post-test analysis on the same participants, using seven different combinations of treatments plus a control group, in each site. Difficulties in applying the full combination of treatments in three countries and gathering observations from the same individuals meant we had to instead make use of just the combinations of treatments present in the post-KAP survey in each country.

In Cambodia, people who did not receive any treatment were only able to name one (1.1) of the four impacts of climate change, while those who participated in Talking Farmer were able to name roughly two (1.8). Similarly, in Thailand, people who participated in the SMS and Video section only scored 0.7 on the index of knowledge of climate change impacts (one-third of the group could not name any impacts), while those who participated in all treatments scored 1.1 (on average, each participant was able to name one impact). The mean scores of knowledge of climate change impact for each combination group are shown in Table 10.3.

Table 10.3 Descriptive statistics of knowledge of climate change impact in each communication combination

Country	Communication model	N	Mean	Std deviation
	Video	12	1.00	1.13
	SMS	14	0.71	0.83
	Talking farmer	10	1.50	1.27
Vietnam	Video + SMS	15	1.27	0.96
	Video + Talking farmer	14	0.57	1.09
	SMS + Talking farmer	10	0.50	0.71
	All three models	13	1.23	1.09
	Total	88	0.97	1.04
	None	44	1.16	1.38
	Video	48	1.69	1.27
Cambodia	Talking farmer	44	1.82	1.42
	Video + Talking farmer	44	1.50	1.32
	Total	180	1.54	1.36
	SMS	5	0.80	0.45
	Video + SMS	25	0.76	0.52
Thailand	SMS + Talking farmer	18	1.00	0.69
	All three models	158	1.16	0.79
	Total	207	1.09	0.75

The effect of each communication model was evaluated while controlling for the impact of the other communication models. The comparison design is shown in Table 10.4. The research team used a test to compare the mean of outcome variables (knowledge on climate change impact, knowledge on causes of climate change, and climate change concern) between the combinations in the comparison design. Results from the statistical tests are summarized in Table 10.4.

Table 10.4 Summary of a priori contrast of different combinations of treatments following ANOVA test

Comparison	Difference between means of outcome variables		
	Knowledge of impact of CC	Knowledge of causes of CC	Level of CC concern
Vietnam			
Effect of Video			
SMS vs. Video + SMS (controlling SMS)	0.6	0.0	0.4
TF vs. TF + Video (controlling TF)	-0.9[a]	-0.1	-0.9
Effect of SMS			
Video vs. Video + SMS (controlling Video)	0.3	0.2	0.7
TF vs. TF + SMS (controlling TF)	-1.0[a]	0.1	-0.9
Effect of talking farmer			
Video + SMS vs. All (controlling Video & SMS)	0.0	0.3	0.2
Cambodia			
Effect of Video			
None vs. Video	0.5	0.0	0.8[a]
TF vs. TF + Video (controlling TF)	-0.3	0.0	-0.4
Effect of talking farmer			
Video vs. Video + TF (controlling Video)	-0.2	0.1	-0.2
None vs. TF	0.7 [a]	0.1	0.9[b]
Thailand			
Effect of Video			
SMS vs. SMS + Video (controlling SMS)	0.0	0.1	0.7
SMS + TF vs. All (controlling SMS&TF)	0.2	0.2	0.2
Effect of talking farmer			
Video + SMS vs. All (controlling Video &SMS)	0.4[b]	0.1	0.2
SMS vs. SMS + TF (controlling SMS)	0.2	-0.1	0.6

Notes: [a] Significant P<=0.5; [b] Significant P<=0.1

From these tests, we are able to infer the following:

- The piloted communication models did not improve people's knowledge of the causes of climate change.
- The communication models did not increase or decrease the level of concern about climate change for vulnerable communities in Vietnam and Thailand. In Cambodia, people who received the combination of Video and Talking Farmer

were more concerned about climate change than those did not participate in any section. The mean score of people who did not participate in any section is 1.8 (between "not at all" and "a little" concern), while the mean scores of those who participated in the Video and Talking Farmers were 2.6 and 2.7 respectively (between "a little" and "some" concern).

- The video did not have any effect in Thailand and Cambodia. In Vietnam, people who received the combination of Video and Talking Farmer scored lower in terms of knowledge of the impacts of climate change compared to those who only participated in the Talking Farmer model.
- The SMS model did not have any effect in Thailand. In Vietnam, people who both received SMS and participated in the Talking Farmer segments scored lower on knowledge of climate change impact than those who only participated in the Talking Farmer.
- The Talking Farmer approach had a clear effect in Thailand and Cambodia. People who participated in the Talking Farmer sessions scored significantly higher on knowledge of the impacts of climate change than those who participated in other treatments or did not participate in any at all. Furthermore, those who actively participated in the Talking Farmer scored higher in knowledge of climate change impact (mean of 1.5) than those who just listened (1.07) (Table 10.5).

Table 10.5 Active and non-active participants in Talking Farmer model, following ANOVA test

Level of participation	N	Mean	Std deviation
Not active (just listened, did not ask questions)	205	1.07	.88
Active (asked questions)	128	1.50	1.20

Significant P=0.000

Discussion

Despite some limitations of the analysis as discussed above, there are some meaningful findings to be taken from this study.

The results indicated that the SMS model was not an effective communication tool in Vietnam and Thailand (and completely impractical in Cambodia), and that in fact it decreased retention of information when used in conjunction with the Talking Farmer sessions in Vietnam. Despite the high self-reported incidence of participants having read the messages, it could be that SMS messages were seen as not interesting at best, and a hindrance or annoying at worst. Better targeting of the content of the SMS messages to specific audiences could improve the effectiveness of SMSs to communicate climate risk awareness and information. For example, people with specific livelihoods could sign up for climate change messages specific to their livelihoods, which would target messaging more effectively.

The Short Video also did not have particularly strong results. It could be expected that playing the video on TV would decrease its efficacy even further. Under the test environment, the audience was captive and directly watching the content, whereas when broadcast on TV the audience is likely to be less focused. Furthermore, there is a lot of competition on television from other TV programs that may be more attractive to the audience. Whilst a Short Video may be more convenient than a Talking Farmer, and less irritating than SMSs, its efficacy will be limited if the content is not targeted and if the audience is not paying attention.

The most powerful model tested appears to be the face to face interactive Talking Farmer. As a flexible model, the Talking Farmer can make use of different communication tools such as videos and maps, and encourage questioning and discussion. This two-way, participatory approach is more effective than the one-way approaches of the SMS and video. This becomes particularly clear when considering that those who actively participated in the Talking Farmer sessions scored higher on knowledge of climate change impacts than those who passively listened. The Talking Farmer method could be further improved by tailoring content specifically to audiences, based on livelihoods, vulnerability to disaster, gender, make-up of household, and so on. Nevertheless, it shows potential for an effective approach to communicating climate-related risk. Successful risk communication does not guarantee behavioral change and the adoption of good practices in response to climate change risks. However, the absence of effective risk communication may diminish the chances of success in any climate change adaptation efforts in the Mekong region. We consider this study as a starting point for more sophisticated research into the development and testing of climate change communication models and strategies, which will continue to contribute directly to climate action and disaster risk management plans in this region.

11

Climate Change Impacts on Food Security and Livelihoods: Case Studies from Lao PDR and the Philippines

Linda M. Peñalba, Dulce D. Elazegui, Outhai Soukkhy, Mayo Grace C. Amit, Felino P. Lansigan, and Francis John F. Faderogao

Climate change is a major policy concern in Southeast Asia due to its expected impact on national food security and the livelihoods of the small-scale farmers who form a large percentage of the region's population. Its impacts on agriculture are expected to be multifaceted and varied, however, given the heterogeneity of farming systems across countries and agro-ecological zones (ADP-IFPRI 2009).

Lao PDR (Laos) and the Philippines are among the countries most vulnerable to climate change in Asia. Maplecroft's (2011) climate change vulnerability index ranks the Philippines as sixth among the 170 countries listed; it is classified under the extreme risk category. Another recent report by the Economic and Environment Program for Southeast Asia (EEPSEA) shows that the Philippines as a whole is in the top quartile of the most vulnerable countries because of its concentration of small-farmer populations, its biodiversity hotspots, and its wide exposure to climate hazards such as tropical cyclones, floods, landslides, and droughts (Yusuf and Francisco 2010). An average of 20 typhoons enter the country annually and have done so with increasing intensity since the 1990s (PAGASA 2011). Given this scenario, funding for building local resilience and the capacity to adapt to the effects of climate change would provide the greatest benefits to both people and endangered ecosystems (Hannah et al. 2013). Urgent attention to food security is needed also because of its high population, 92.34 million in 2010 (NSO 2012).

In Laos, the other country in this study, six provinces have been classified in the first quartile of Southeast Asia's most vulnerable regions due mainly to their low adaptive capacity, rather than merely their degree of exposure to extreme climate hazards (Yusuf and Francisco 2010). Low resilience, low productivity, and a largely subsistence-based farming system coupled with high poverty rates place their populations at greater risk from variability in rainfall, which could result in drought or flooding (Lefroy et. al 2010). Climate variability and extreme climatic events, bringing heat and water stress, have already caused significant losses in crop and livestock production (MAF 2010). Between 1966 and 2013, there were 33 recorded floods across Laos, including 10 large floods and 9 droughts (3 classified as severe). In addition, a total of 6 combined flood–drought events were observed. The cost of damage due to flooding was estimated at US$477,085,101 and that due to drought was US$87,262,715 (PAFO 2013).

The situation is expected to worsen because the mean precipitation in the Mekong River Basin is predicted to increase by an average of 4.2 percent per year. Based on World Food Program (WFP) estimates, Savannakhet and Luang Prabang provinces are vulnerable to drought, posing risks to the food security of an estimated 46 percent of the lowland population and 12 percent of agro-pastoralists in the Lao uplands (MAF 2010).

Laos's population growth in the first decade of the twenty-first century has also resulted in an expansion of cultivated land to meet the rising demand for food (ADB 2008, cited in Lefroy et al. 2010). National measures to manage climate change impacts, however, have been limited in scope and scale (Chinvanno et al. 2006). It is only recently that the government has addressed climate change in its latest National Socio-Economic Development Plan (2010–2015). The National Strategy for Climate Change, announced in 2010, includes agriculture and food security as priority areas.

Numerous studies have been conducted to assess climate change impact using various models and tools (Perez et al. 1999; WWF 2002; IPCC 2007). However, studies that foreground the *local* impacts of climate change crucial for policy development and actions are needed. Global climate scenarios must be downscaled to determine how projected climate events will impact local communities with specific social, economic, political, and physical conditions. The downscaled climate projections should be integrated into national development scenarios to inform policymakers about the severity of the potential problems and the need for appropriate actions. Climate change research must therefore continue to improve our understanding of future risks in order to improve decision-making and allocation of limited resources (ADB-IFPRI 2009).

This chapter seeks to do this by presenting a comparative study of projected climate scenarios for two major rice-producing provinces in Laos and the Philippines, respectively, for the years 2020, 2050, and 2080. It will address: the potential impacts of climate change on food security and livelihoods in different agro-ecological zones; climate adaptation measures; and differences in the adaptive capacity of farmers in the two study sites. Finally, the study looks at actions that can be taken by local governments to mitigate or harness the potentially negative impacts of projected climate change.

Methodology

Conceptual framework

Indications of climate change include increasing temperature; gradual changes in precipitation; an increase in the frequency, duration, and intensity of dry spells and droughts; changes in the timing, duration, intensity, and geographic location of rain; an increase in the frequency and intensity of storms and floods; and greater seasonal weather variability and changes in the start/end of rainy seasons (FAO 2008).

Long-term changes in temperature and precipitation patterns are expected to cause changes in production cycles, pest and diseases patterns, and cropping systems, all of which could affect production, prices, incomes, and ultimately, the livelihoods and lives of subsistence farmers. Increased intensity and frequency of storms, altered hydrological cycles, and precipitation variance also have long-term implications on the sustainability of current world agro-ecosystems and future food availability (Edame et al. 2011; Odufuwa et al. 2012). Climate change may affect food systems in several ways. Crops may be directly affected by changes in rainfall (drought or flooding) while warmer or cooler temperatures could lead to changes in the length of the growing season.

It could also lead to changes in markets, food prices, and the supply chain infrastructure which could affect food availability and accessibility. The detrimental effects of climate change in developing countries will be exacerbated by population growth and socioeconomic development, which will entail substantial increases in food supplies (Gregory et al. 2005).

Livelihood opportunities ideally should shield farmers from climate-related risks, but this depends on the households' resources or capacity. Livelihoods refer to the abilities and activities that enable a person or household to survive (FAO 2008). Vulnerability may differ seasonally or temporally across groups or individuals due to differences in their livelihood activities or social standing. The vulnerability of food systems is determined by a combination of the societal capacity to cope with, and/or recover from environmental change, coupled with the degree of exposure to stress. Risks also vary spatially, and impacts on lowland irrigated farms may differ from those on rainfed lowland and upland farms.

Meanwhile, policies, institutions, and processes significantly determine the enabling as well as the constraining environment for the generation, provision, and acquisition of these assets—creating them, determining access, and influencing rates of asset accumulation. Those with more assets are more likely to have greater livelihood options that could help them reduce poverty.

Food security constitutes food availability, accessibility, and utilization. Food availability is determined by the quantity of food that is produced, stored, processed, distributed, and exchanged. Food accessibility refers to the ability to secure food based on resource endowments that an individual requires to acquire food and to food affordability, allocation, and preferences. Food utilization refers to nutritional and societal values and safety, so that food security is, therefore, diminished when food systems are stressed (FAO 2008; Gregory et al. 2005). Central to the achievement of food security is the interplay of demand and supply factors. One important supply determinant is the level and change in yield over time. Given population growth, low production levels means a diminishing food supply over time. As rice production is an important source not only of food, but household income as well, limitations to production can undermine farmers' livelihoods, thus exacerbating food shortages. Climate change can, therefore, affect food production, farmers' livelihoods, and food accessibility.

There are also, in some cases, positive impacts of climate change that may be harnessed through effective climate governance and well-managed climate change adaptation interventions (Odufuwa 2012). Local or national conditions will determine the choice of adaptation interventions. However, most current data and studies of projected climate scenarios have been at the national, regional, and global levels and cannot be used directly in assessing local impacts due to coarse spatial resolution. There is a need to downscale climate data to represent the local-scale surface weather, assess the impacts of climate change at the local level, and help determine more appropriate local responses.

More importantly, relevant government agencies as well as NGOs should recognize the impact of climate change and integrate projected problems into policies and planning. There should also be more collaboration between different stakeholders for meaningful and effective adaptation measures to be taken.

Scope of the study

The study was conducted in two provinces each in the Philippines and Laos. In each province, information was gathered from two municipalities; two barangays/villages in each municipality/district; and three types of rice-growing environments: irrigated lowlands, rainfed lowlands, and rainfed uplands, with varying exposure and risks to climate change, particularly water stress.

In the Philippines, the study focused on two rice-producing provinces on the island of Luzon, Tarlac and Pangasinan, which rank fourth and eighth among 80 provinces, respectively, in terms of national rice production. Both provinces are among the 20 provinces most vulnerable to climate change (Lo 2011). In Laos, the study sites were in Luang Prabang and Savannakhet, two of the top rice-producing provinces that are also very exposed to climate risks. Luang Prabang ranks eighth among seventeen Lao provinces in terms of vulnerability to climate change (Yusuf and Francisco 2010). Most farmers in Luang Prabang subsist on upland rice production where long droughts and increases in temperature have been experienced. Savannakhet has 169,000 ha of paddy fields, 45 percent of which are prone to drought and 25 percent prone to flood. Every year about 15,000 ha are affected by mild flooding, 21,000 ha by moderate flooding, and 35,000 ha by severe flooding (Savannakhet PAFO 2011).

Data collection and sampling

Qualitative and quantitative data were collected from primary sources through a survey of rice farmers, key informant interviews with local government officials, and focus group discussions (FGDs) with farmers' associations, village leaders, and local government officials. Data on livelihoods and crop management practices, and observed climate variability were gathered through farm surveys.

In the Philippines, two municipalities were chosen for each province based on two main criteria: being leading rice-producing areas and having all three rice ecosystems. In each municipality, two barangays/villages were selected. The sample respondents were randomly chosen from a list of small-scale farmers obtained from the Municipal Agriculture Office. In each province, 100 rice farmers distributed proportionately across the three rice agro-ecological zones were interviewed using structured questionnaries (Table 11.1). These smallholders cultivated their own farms, ranging in size from 1–1.4 ha. Data collection was conducted from January to March 2012.

For Laos, two districts were selected from each province and two to three villages in each district were chosen as study areas. However, in Savannakhet, the respondents came only from the dominant rice agro-ecological zones (i.e. irrigated lowlands and rainfed lowlands), which have experienced extensive effects of drought and flooding. A total of 217 respondents (140 in Luang Prabang and 77 in Savannakhet), whose farm size ranged from 0.8–0.9 ha, were interviewed.

Table 11.1 Distribution of sample respondents by province and ecozone, Laos and the Philippines

Countries/Province	Agro-Ecozone					
	Irrigated lowland		Rainfed lowland		Rainfed upland	
	No.	%	No.	%	No.	%
Laos	64	100.00	75	100.00	78	100.00
Luang Prabang	23	35.94	39	52.00	78	100.00
Savannakhet	41	64.06	36	48.00	0	0
Philippines	94	100.00	71	100.00	35	100.00
Tarlac	61	64.89	15	21.13	24	68.57
Pangasinan	33	54.10	56	78.87	11	31.43

To project climate scenarios and estimate their potential impact on rice production, secondary weather, soil, and rice crop data were collected. Data collected included physiochemical properties of soils by horizon and genetic coefficients of rice cultivars. In the Philippines, weather data from 1971 to 2000 were collected from the Philippines Atmospheric, Geophysical and Astronomical Services Administration (PAGASA) while soils data were collected from the Bureau of Soils and Water Management (BSWM). In Laos, weather data from 1982 to 2011 were collected from the Luang Prabang and Savannakhet meteorological stations while data on soils were collected from the National Agriculture and Forestry Research Institute (NAFRI).

Data processing and analysis

Downscaling climate projection

Climate scenarios, which are usually generated using global circulation models (GCM), are useful in characterizing future climate risks and evaluating adaptation options. GCMs are capable of projecting climate changes for the next 100 years. Climate projections derived from GCMs suggest that global temperature extremes, expected to exacerbate due to human activities, will enhance the accumulation of greenhouse gases in the atmosphere.

Downscaling procedures such as the statistical approach and the dynamical downscaling techniques have emerged to enable the derivation of local surface weather variables from large-scale atmospheric predictor variables of GCM outputs. Dynamical procedure uses a limited-area, high-resolution model such as a regional climate model (RCM) to derive small-scale information from GCMs. On the other hand, the statistical technique involves determining quantitative relationships with GCM data as predictors and local climate variables as predictands (Wilby and Dawson 2004; Wilby et al. 2001). Statistical downscaling can generate rapid assessments and low-cost local daily surface weather data. In contrast, dynamical downscaling requires significant, high-end computing resources.

Developing climate change scenarios through climate modeling is largely based on the charting changing greenhouse gas concentrations. These emission scenarios, sometimes referred to as emission pathways, are projected based on development trends or storylines. The

Intergovernmental Panel on Climate Change (IPCC) developed three emission scenarios namely, A2 (high-range), A1B (mid-range), and B2 (low-range). Based on its Special Report on Emission Scenarios (SRES), the different scenarios adopt the following assumptions:

- A1: Very rapid economic growth; population peaks mid-century; social, cultural and economic convergence among regions; market mechanisms dominate. Subdivisions: A1FI—reliance on fossil fuels; A1T—reliance on non-fossil fuels; A1B—a balance across all fuel sources.
- A2: Self-reliance; preservation of local identities; continuously increasing population; economic growth on a regional scale, with fragmented per capita economic growth and technological change.
- B1: Clean and efficient technologies; reduction in material use; global solutions to economic, social, and environmental sustainability; improved equity; population peaks mid-century.
- B2: Local solutions to sustainability; continuously increasing population at a lower rate than in A2; less rapid technological change than in B1 and A1.

Most countries prefer an A2 scenario because if one can adapt to climate change more broadly, it would be easier to adapt to smaller climate changes at the lower end scenarios. While there are a number of future climate scenarios, the choice used in the model for any area depends on plausible socioeconomic trends as well as technological developments and environmental conditions. This study is more concerned with the observed and projected trends rather than distinct absolute values of climate and weather variables in the future. Moreover, SDSM used the available GCM data suited for downscaling site-specific daily weather variables under current and future regional climate forcing. In this study, the Coupled Global Climate Model (CGCM3) with available A2 and A1B scenarios developed by the Canadian Center for Climate Modeling and Analysis was selected. However, downscaling was based only on the A1B scenario as reflective of future emission scenarios for developing countries like Philippines and Laos. In the A1B scenario, climates in the next 30–40 years will be greatly influenced by past emissions, principally due to the long lifetime of carbon dioxide (PAGASA 2011). The IPCC scenario A1B was adopted in the analysis.

It is important to choose a plausible scenario to climate-proof development policies and programs (IPCC SRES 2001; Krittasudthacheewa 2013). In this study, the climate scenario used was based on the most plausible one used in earlier studies in the Philippines and in Laos.

This research used the statistical downscaling technique with SDSM software (version 4.2.7). Only statistical downscaling was used to generate rapid assessments and low-cost local daily surface weather data with reasonable results comparable to dynamical downscaling. The development of statistical relationships between the predictand and predictors is crucial in determining the projected scenario for local climate. It is assumed that the predictor–predictand relationship under current conditions remains valid under future conditions. Combalicer et al. (2010) showed acceptable response and reliability between the observed and generated data using SDSM. The reliability of SDSM can be improved if higher resolution grid boxes of GCMs will be developed. Delfino et al. (2013) have also shown that statistical downscaling of climate projection yielded reasonable estimates of local climate.

Downscaled climate (rainfall and temperature) projections for the provinces of Tarlac and Pangasinan in the Philippines, and Savannakhet and Luang Prabang in Laos were done for three time periods 2020, 2050, and 2080, the years used in most GCMs. The average value of the observed data (1971–2000) for the Philippines and for Laos (1982–2011) are referred to herein as the baseline (baseline climate). The World Meteorological Organization (WMO) defines climatological baseline as the thirty-year 'normal' or average which can be used as a standard reference for climate impact studies.

Analysis of the effects and impacts of climate change on rice productivity

The potential impact of climate change on rice production was also assessed in these three periods using the crop simulation model Decision Support System for Agrotechnology Transfer (DSSAT). This involved the following steps:

1. Characterization of the selected sites in each country in terms of dominant agro-ecological zones considering soil properties, climate, and topography, i.e. irrigated lowland, rainfed lowland, and rainfed upland rice production areas.

2. Evaluation of effects of climate change on rice productivity using process-based crop simulation model e.g., CERES Rice Model (Tsuji et al. 1994).
3. Simulation of rice yields under different climate scenarios of temperature, CO_2 increases and rainfall. Yield levels under different climate scenarios were compared with the corresponding yields under the historical climate condition based on the available weather data for the sites. Simulated yields were also determined using downscaled climate projections for the selected sites. Changes in simulated rice yields for each of the different locations were compared under the baseline condition (i.e., the last 30 years weather data) with those for 2020, 2050, and 2080.

Assessment of vulnerability of rice-based production systems

This involved analysis of impacts of climate change and climate-related hazards such as floods, typhoons, and droughts on the farmers' food security (i.e., food production, availability, and access) as well on their livelihoods. Climate change adaptation measures available and needed in the area were also identified.

The research process, which includes the methodologies for data collection and analysis, are illustrated in Fig. 11.1.

Results

Downscaled climate scenarios

The projected climate scenarios vary depending on location, time period, and weather variable. In Laos, there is statistically significant difference between the baseline temperature and rainfall and the projections centered in 2020, 2050, and 2080 in the two provinces (Figures 11.2 to 11.5). Projected temperatures in Luang Prabang (Fig. 11.2) will be slightly higher than in Savannakhet (Fig. 11.3) while rainfall will be lower in Luang Prabang (Fig. 11.4) than in Savannakhet (Fig. 11.5). The temperature pattern in 2020, 2050, and 2080 will remain the same as in the base year. In Luang Prabang and Savannakhet provinces, the hottest months will still be April and May while the coldest months will be December and January (Figures 11.2 and 11.3). While the GCM climate projections show increased global warming for the region, the statistical downscaling of

Fig. 11.1 The research process

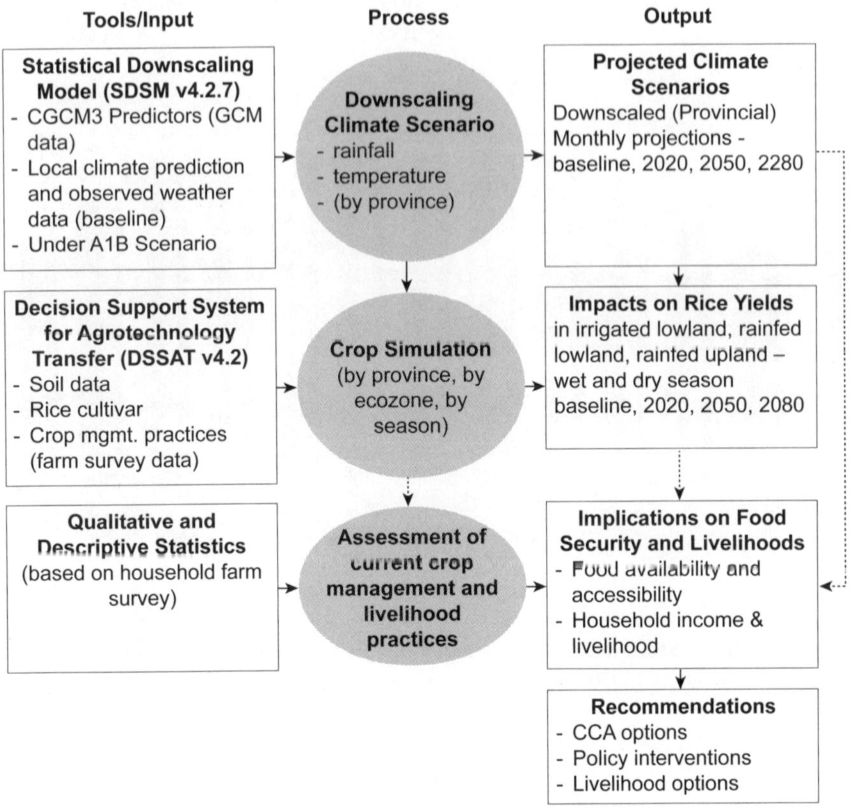

future climate using SDSM yielded lower temperatures. This difference may be attributed to the limited weather data available for the study area in Laos. This is due to the predictor–predictand relationship established based on the available historical data.

Rainfall in Luang Prabang will slightly increase from 12 percent to 19 percent from the base year, while in Savannakhet, it will increase significantly—by 60 percent in 2050 and 2080 (Figures 11.4 and 11.5). On the other hand, rainfall patterns will be the same for the two Lao provinces, but the amount of rainfall will increase slightly in Luang Prabang, particularly during the wet season: the amount of rainfall is expected to increase in 2020 by about 50 mm in June and 75 mm in July. In 2050, an increase of about 50 mm will be expected in May, while in 2080, rainfall will increase from 280 mm to 360 mm in August (Fig. 11.4). In Savannakhet, rainfall will increase significantly during the wet season

Fig. 11.2 Projected monthly mean temperature (⁰C) under medium-range emission scenario (A1B) of Coupled Global Climate Model 3 (CGCM3) for Luang Prabang, Laos, using SDSM

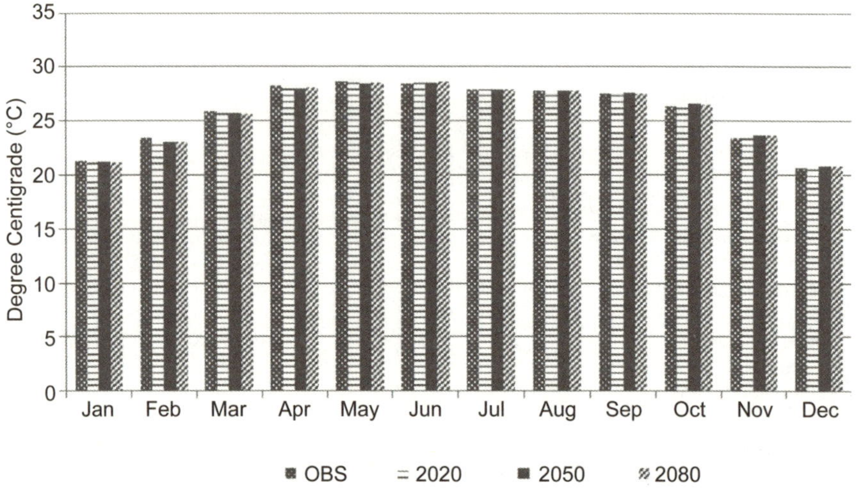

Fig. 11.3 Projected monthly mean temperature (⁰C) under medium-range emission scenario (A1B) of CGCM3 for Savannakhet, Laos, using SDSM

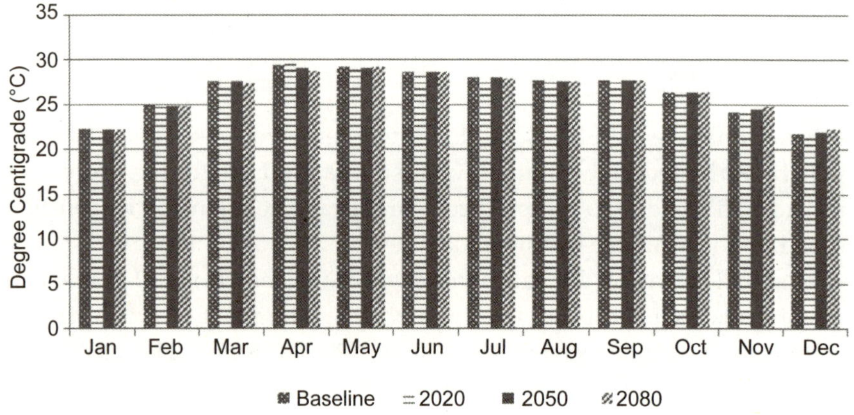

and across periods. It will increase from 200 mm to 325 mm in May, 225 mm to 330 mm in June, 225 mm to 500 mm in July, 350 mm to 450 mm in August, and 200 mm to 325 mm in September (Fig. 11.5).

Fig. 11.4 Projected monthly rainfall (mm) under medium-range emission scenario (A1B) of CGCM3 for Luang Prabang, Laos, using SDSM

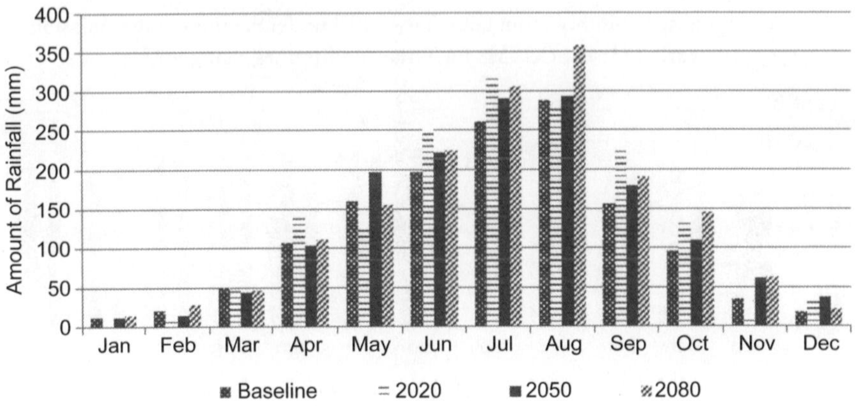

Fig. 11.5 Projected monthly rainfall (mm) under medium-range emission scenario (A1B) of CGCM3 for Savannakhet, Laos, using SDSM

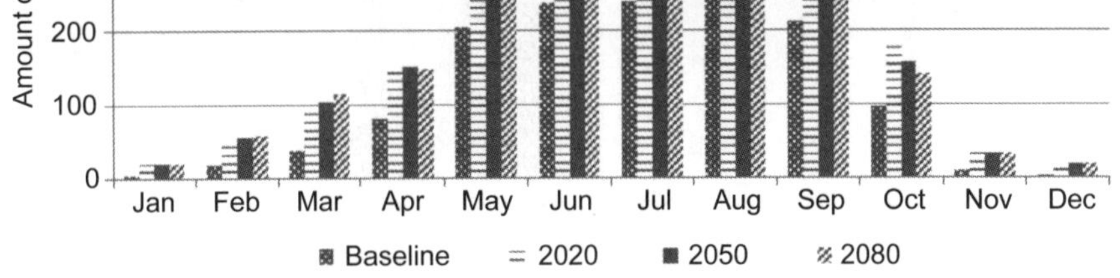

In the Philippine case studies, baseline rainfall and temperature levels in Pangasinan are slightly higher than in Tarlac, but projections revealed that in future the situation will be reversed (Figures 11.6 to 11.9). There will be no significant monthly variations in temperature in Tarlac. The relatively cold and hot months in 2020, 2050, and 2080 are projected to be the same as in the base year. The hottest months will still be April and May while the coldest months will be January and February. The same temperature pattern was projected for Pangasinan. The cold months in this province are December and January while the hot months will also be in April and May.

Fig. 11.6 Projected monthly mean temperature (^{0}C) under medium-range emission scenario (A1B) of CGCM3 for Tarlac, Philippines, using SDSM

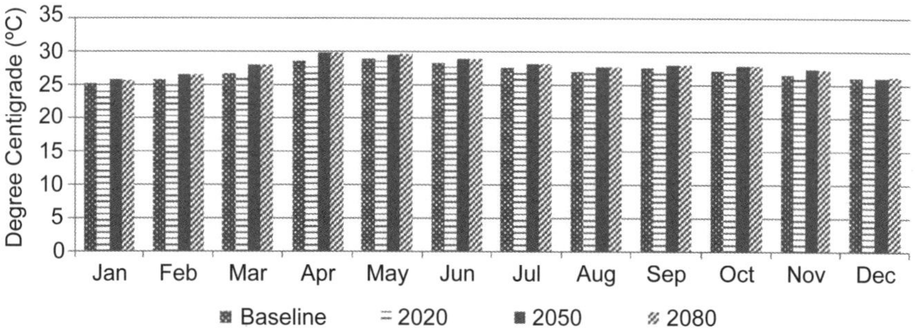

Fig. 11.7 Projected monthly mean temperature (^{0}C) under medium-range emission scenario (A1B) of Coupled Global Climate Model 3 (CGCM3) for Pangasinan, Philippines, using SDSM

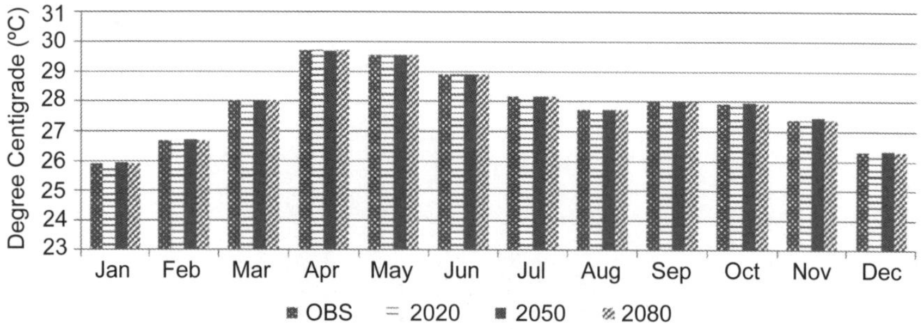

Downscaled climate scenarios show that there will be significant changes in the amount of rainfall in the two Philippine provinces from the baseline to 2020, 2050, and 2080, but the magnitude of change in the two provinces differ. Rainfall in Tarlac during the rainy season will be higher, but it will be drier and hotter during the dry season (Fig. 11.8). In Pangasinan, however, there will be a slight increase in temperature (Fig. 11.7) and rainfall throughout the year (Fig. 11.9).

There are no foreseeable changes in terms of monthly rainfall patterns for the two provinces (Figures 11.8 and 11.9). However, in Tarlac, the amount of rainfall is expected to increase significantly by 200 to 400 mm in the wet season. Rainfall will increase from 175 mm to 300 mm in May, 200 mm to 600 mm in June, 400 mm to 800 mm in July and August, and 300 mm to 525 mm in September. These changes in rainfall might have a serious effect on rice yields since these months coincide with critical periods in the rice production cycle. On the other hand, projected rainfall during the dry season (November to April) will be almost the same during the three periods (Fig. 11.8).

Fig. 11.8 Projected monthly rainfall (mm) under medium-range emission scenario (A1B) of CGCM3 for Tarlac, Philippines, downscaled using SDSM

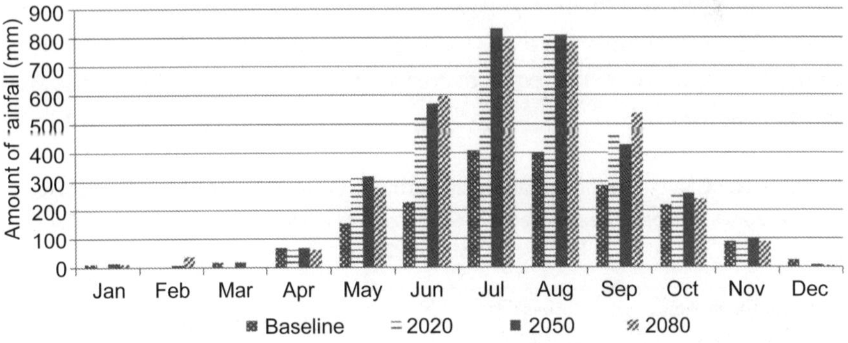

Fig. 11.9 Projected monthly rainfall (mm) under medium-range emission scenario (A1B) of CGCM3 for Pangasinan, Philippines, downscaled using SDSM

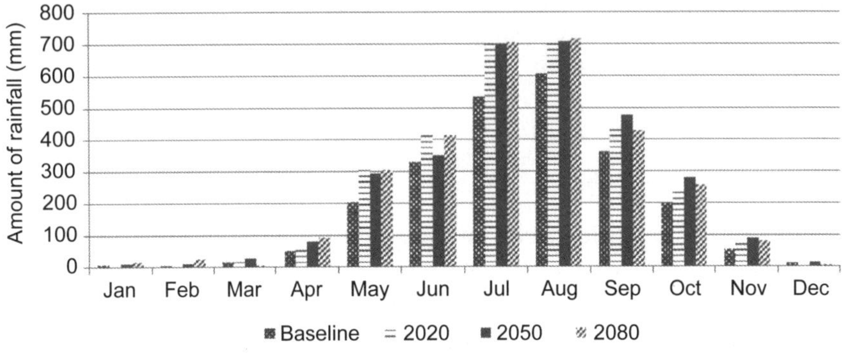

In Pangasinan, the increase in monthly rainfall volume, compared to the base year, during the rainy season was only minimal, ranging from 50 mm to 150 mm. Rainfall in November, the planting time for the second crop, will also increase slightly by 50 mm (Fig. 11.9).

Impact on rice yields

The combined effects of increases in temperature and rainfall on rice production vary depending on the time, location, and planting schedule. There will be a potential increase in rice yields in the irrigated lowlands of Luang Prabang in 2020, 2050, and 2080 during the dry and wet seasons, but the increase in the dry season will be higher than in the wet season (Table 11.2). Rice yields in the rainfed lowlands will increase in the dry season in 2020 and 2050, but it will decrease in 2080. In the wet season, rice yields will slightly increase in 2020, 2050, and 2080. Rice yields in the rainfed uplands are expected to decrease in 2020, 2050, and 2080 compared to the base year.

Table 11.2 Changes in potential yields during dry and wet season in different rice ecozones, Luang Prabang, Laos, baseline and projected climate scenarios 2020, 2050, and 2080

Cropping season/ Ecozone	Baseline (1982–2012) (kg/ha)	Ave. percentage change		
		2020 (%)	2050 (%)	2080 (%)
Dry season				
Irrigated lowland	4270.20	21.61	26.09	35.69
Rainfed lowland	5006.77	15.23	3.07	-2.63
Wet season				
Irrigated lowland	6294.43	5.40	1.64	5.69
Rainfed lowland	6297.37	4.94	5.63	13.66
Rainfed upland	5027.03	-2.20	-1.52	-4.57

In Savannakhet, rice yields in the irrigated lowland will significantly increase during the dry season, but they will significantly decrease during the wet season across time periods (Table 11.3). This could be the effect of the projected increase in rainfall in Savannakhet by about 60 percent in 2050 and 2080. This increase would be beneficial during the dry season, but it will be a problem during the wet season.

Table 11.3 Changes in potential yields during dry and wet season in different rice ecozones, Savanakhet, Laos, baseline and projected climate scenarios 2020, 2050, and 2080

Cropping season/ Ecozone	Baseline (1982–2012) (kg/ha)	Ave. percentage change		
		2020 (%)	2050 (%)	2080 (%)
Dry season				
Irrigated lowland	1450.80	76.54	88.41	83.92
Rainfed lowland	558.4	100.01	129.20	141.86
Wet season				
Irrigated lowland	2193.27	-56.25	-65.83	-72.61
Rainfed lowland	2749.1	-53.93	-63.20	-71.08

In the Philippines study, rice yields will generally decrease in Tarlac, but will increase in Pangasinan over the study period (Table 11.4 and 11.5). In Tarlac, yields in the irrigated lowlands for dry season cropping will potentially decline in 2020, 2050, and 2080. Yields in the rainfed lowland farms will potentially increase in 2020, but will decrease in 2050 and 2080. For wet season cropping, yields in rainfed farms (lowland and upland) will decline through the years, but the decline in yields in upland farms is significantly higher and this downward trend will continue until 2080. In general, yields in all agro-ecosystems will decline significantly by 34 percent to 54 percent in 2080.

Table 11.4 Changes in potential yields during dry and wet season in different rice ecozones, Tarlac, Philippines, baseline and projected climate scenarios 2020, 2050, and 2080

Cropping season/ Ecozone	Baseline (1971–2000) (kg/ha)	Ave. percentage change		
		2020 (%)	2050 (%)	2080 (%)
Dry season				
Irrigated lowland	3268.69	-9.23	-10.68	-11.05
Rainfed lowland	2607.50	6.09	-8.99	-2.76
Wet season				
Irrigated lowland	5710.50	4.16	-1.76	-36.53
Rainfed lowland	6235.65	-0.4	-2.79	-34.3
Rainfed upland	3349.65	-21.54	-32.38	-54.25

Rice yields in Pangasinan are expected to increase in 2020, 2050, and 2080 compared to the baseline, regardless of cropping season and agro-ecosystem (Table 11.5). For dry season cropping, both irrigated lowland and rainfed lowland farms are projected to see increased yields in 2020. Yields will also increase until 2050, but at a lower rate. For the wet season, potential yields in rainfed upland farms are significantly higher compared to potential increases in yields in irrigated lowland and rainfed lowland farms. Yields in rainfed lowland farms will increase in 2020, 2050, and 2080, but the increase will be lower compared to those projected for irrigated lowland and rainfed upland farms.

Table 11.5 Changes in potential yields during dry and wet season in different rice ecozones, Pangasinan, Philippines, baseline and projected climate scenarios 2020, 2050, and 2080

Cropping season/ Ecozone	Baseline (1971–2000) (kg/ha)	Ave. percentage change		
		2020 (%)	2050 (%)	2080 (%)
Dry season				
Irrigated lowland	3531.37	6.23	2.04	9.76
Rainfed lowland	2687.73	3.43	1.85	9.13
Wet season				
Irrigated lowland	7168.33	2.90	7.13	6.58
Rainfed lowland	7768.63	0.68	3.15	5.19
Rainfed upland	2303.33	36.20	46.2	48.26

Impact on food availability and accessibility

Crop simulation results suggest that projected changes in climate will not adversely affect the rice supply in Luang Prabang and Savannakhet in the dry season. In Savannakhet, however, production is threatened during the wet season, which could adversely affect food availability.

The impact of projected climate scenarios on food availability in the Philippines varies according to location, agro-ecosystem, and planting season. In Tarlac, which will be more adversely affected than Pangasinan, the impact on food availability is expected to be more severe in rainfed upland areas, where the probable reduction in rice yield is highest (54 percent). In contrast, it appears that food availability in Pangasinan could benefit from projected climate change, as higher yields are expected in all the three ecozones during the study period.

In general, small rice farmers allocate a sizeable proportion of their produce for home consumption to ensure that food will be available for most part of the year. In many instances, however, stored food runs out before the next harvest comes, particularly in the areas where farmers can plant rice only once a year due to water shortages. This is particularly true for about 50 percent of the rainfed lowland and upland farmers of Tarlac and 80 percent to 90 percent of rainfed lowland and upland farmers of Pangasinan. The situation is expected to improve in Pangasinan since rice yields in rainfed uplands are expected to increase by 36 percent to 48 percent.

In the rainfed uplands in Luang Prabang province, more than half of the respondents (55 percent) reserve their rice produce for home consumption. In contrast, irrigated lowland farmers save 45 percent and rainfed lowland farmers save 36 percent of their harvest for home use.

The projected increase in rice yields in Luang Prabang and Savannakhet are expected to provide additional rice, ranging from 103 to 1,524 kg, which could supply the needs of one to nine individuals annually, given an average per capita consumption of 171 kg. In Tarlac, however, the projected decline in rice production will affect the food supply of about one to nineteen individuals, while the additional rice supply in Pangasinan is expected to make rice available to one to ten individuals per year, with a per capita consumption of 113 kg.

Most farmers in Laos plant for home consumption while in the Philippines, the majority of farmers produce rice primarily to earn income. The significant decline in projected rice yields in Tarlac and Savannakhet will consequently affect farmers' income and their ability to buy food, thereby adversely affecting food accessibility. Reduction in yields will also mean a reduction in the amount of rice produced for home consumption. On the other hand, the projected increase in rice yields in Pangasinan and Luang Prabang will mean greater food accessibility either through the market or from their own harvests.

Based on the reports of farmers, losses in farm income due to recent flooding were as high as 38 percent of household income in the irrigated lowlands and 47 percent in the rainfed lowlands of Savannahket. The losses were lower in Luang Prabang at 6–10 percent of household income. To cope with food shortages, farm households resorted to exchanging or selling household items for food, and borrowing money. Rice farming contributes the highest amount to household income, but farmers also rely on extra income derived from selling byproducts of the rice crop such as livestock feed and rice wine.

In the Philippines, losses due to flooding and typhoons in the rainfed lowland areas accounted for 28 percent and 20 percent, respectively, of the household income. In the irrigated lowlands, such losses accounted for around 15 percent. Losses due to drought are highest for the rainfed areas, accounting for 15 percent of income. Non-farm income complements farm income. During the dry season, income is derived from non-farm activities, especially carpentry, charcoal-making, firewood gathering, and tricycle driving. During the wet season, non-farm activities are limited, and some households have small retail stores. In Pangasinan province, gold panning is a source of employment after the rainy season.

Impact on livelihood

Total household income varies depending on location, ecozone, and the importance of the different income sources. In the Philippines, the contribution of rice farming to total income is highest among irrigated lowland rice farmers in Tarlac (55 percent) and lowest (35 percent) for farmers with rainfed upland rice fields (Table 11.6). Furthermore, the irrigated lowland farmers are also more heavily dependent on farm income sources (60 percent) compared to the rainfed lowland (50 percent) and rainfed upland farmers (38 percent). Hence, while irrigated lowland rice farmers in Tarlac may have the highest total income, they are the most at risk to climate change impacts due to their heavy dependence on farm income sources, which are highly sensitive to climate factors.

Non-farm activities such as employment in public and private sectors and overseas also constitute a major income source for a considerable proportion of rice farmers. About 12 to 20 percent of the Tarlac respondents have family members working overseas, while 20 to 40 percent have family members employed in the private and public sectors. In Pangasinan, the average income of respondents is significantly lower than Tarlac respondents, but remittances from relatives overseas, and private and public sector employment also constitutes 28 to 30 percent of total income. Moreover, they are less dependent (34 percent to 45 percent) on farm income sources compared to their Tarlac counterparts.

Table 11.6 Annual income of sample households by source and ecozone in Tarlac and Pangasinan, Philippines, 2010 (US$)

Province	Source of income	Irrigated lowland	Rainfed lowland	Rainfed upland
No. of respondents in Tarlac		61	15	24
Tarlac	Farm			
	Rice farming	3192	1455	1467
	Other crops	169	175	31
	Livestock/Fishing	100	201	108
	Non-farm			
	Self-employment/business	621	121	1188
	Employment	700	960	904
	Remittance from OFW relatives	840	389	377
	Others (off-farm seasonal waged labor, pension)	215	32	102
	Total HH income	5837	3333	4177
No. of respondents in Pangasinan		33	56	11
Pangasinan	Farm			
	Rice farming	812	413	279
	Other crops	121	271	140
	Livestock/Fishing	70	92	186
	Non-farm			
	Self-employment/business	161	313	66
	Employment	545	416	162
	Remittance from OFW relatives	314	271	286
	Others (off farm seasonal waged labor, pension)	102	126	136
	Total HH income	2125	1902	1255

Note: 1US$ = 42 PhP

As expected, farm income is highly seasonal. In Tarlac, where rice farming contributes 35 percent to 55 percent of total income, farm income in the irrigated lowland farms was highest in March and November, and January and December in rainfed lowland farms, and October and November in the rainfed upland farms. These months coincide with the rice harvests. On the other hand, in Pangasinan, the total income of irrigated lowland farmers is highest in July and August, for rainfed lowland farmers in February and September, and rainfed upland farms in January, March, and October. The levels of income and the magnitude of differences across ecozones are not solely due to rice and climate variables, but may be due to variations in other income sources which are not seasonal.

There is no similar pattern of relations seen in the behavior of non-farm income relative to climate variability as the magnitude remains relatively constant throughout the year. Non-farm income was commonly higher than farm income in the months of April to August, which is the period when rice farmers are just starting their land preparation and planting activities. During the period where farm income is constrained, the importance of non-farm income sources to augment support for farm and household needs is even more acute.

Among the Lao respondents, the irrigated lowland rice farmers from Savannakhet have the highest (43 percent) dependence while Luang Prabang's rainfed upland farmers are the least dependent on rice farming as their income source (Table 11.7). Overall, however, both the Luang Prabang and Savannakhet respondents are greatly sensitive to potential climate change impacts because they derived about 63 percent to 70 percent of their income from farm sources which are highly exposed to climate risks. In Laos, other sources of income for farm households in all ecozones are aquaculture and small business enterprises.

Table 11.7 Annual income of sample households by source and ecozone in Luang Prabang and Savannakhet, Laos, 2010 (US$)

Province	Source of income	Irrigated lowland	Rainfed lowland	Rainfed upland
No. of respondents in Tarlac		42	36	77
Tarlac	Farm			
	Rice farming	970	750	529
	Other crops	245	344	279
	Livestock/Fishing	641	783	714
	Non-farm			
	Business	305	372	449
	Employment	400	450	356
	Others	146	146	99
	Total HH income	2707	2845	2426
No. of respondents in Savannakhet		23	39	0
Pangasinan	Farm			
	Rice farming	1149	941	-
	Other crops	246	273	-
	Livestock/Fishing	480	442	-
	Non-farm			
	Business	387	415	-
	Employed	297	226	-
	Others	131	81	-
	Total HH income	2691	2378	

Note: 1 US$ = 7,995.50 LAK

However, farmers in the two countries have opposing views or perceptions of climate risks. The farmers recognize that there are technical solutions to mitigate impacts of climate events such as the adoption of flood/drought-resistant seed varieties and water management technologies. They would also like to engage in other livelihood sources such as raising livestock through the government's livestock dispersal programs.

Discussion

This comparative study affirms that climate change does affect localities in different ways and that downscaling climate scenarios is important to enable appropriate local adaptation actions. The nature and extent of these impacts seem to vary over time and across ecozones. The projected scenarios showed that climate variability will have an impact on yields, production, income, and livelihoods of small-scale farmers.

Projected temperature changes may vary at particular periods and sites, but rainfall will generally increase in all areas. The magnitude of change could be from minimal (e.g., in Luang Prabang and Pangasinan) to substantially high (e.g. in Savannakhet and Tarlac). The Philippines data suggests more pronounced increases in both temperature and rainfall compared to Laos. This variation is also manifested in terms of the impact on rice yields between study areas within each country as well as across agro-ecosystems. In Luang Prabang, rice yields are expected to generally increase across ecozones and period (by 2 to 36 percent). In Savannakhet, rice yields are also expected to significantly increase across zones in the dry season (by 77 to 142 percent), but significantly decrease in the wet season (54 to 72 percent). For the Philippines case, rice yields in Tarlac may decline by up to 54 percent, while rice production in Pangasinan may increase by about 48 percent.

Several cropping patterns were noted, indicating the attempts of some farmers to adjust the planting season in a manner that will avert risks from changing and unpredictable weather events. There is not enough information to show, however, that this practice is already widely and sustainably adopted by a majority of farmers in the study areas.

Results also show that the projected climate scenarios do not lead to consistent outcomes in terms of rice yield. For example, the combined effect of rainfall and temperature will cause rice yields to decline in Tarlac, but increase in Pangasinan. The positive effect of projected climate change in rice yields in Pangasinan may be attributed to the optimum level of temperature and water available that is conducive to plant growth. On the other hand, in Tarlac, an excessive supply of water during the critical plant growth period may not be conducive to healthy plant growth. In the two Lao provinces, the combined effect of changes in rainfall and temperature differ depending on the cropping season. Rice yields will generally increase during the dry season, but decrease during the wet season.

The climate impact on rice farmers' livelihoods is reflected in their socioeconomic conditions. Rice produced is either sold partially or wholly to generate income or saved for subsistence. As climate adversely affects yield, income and food, rice consumption, which averages 113 to 171 kg per capita, is likewise compromised and may even exacerbate present inadequacies. The substantial number of farmer households across ecozones that presently allocate their produce for home consumption, either wholly or partially, were at risk of food and income insecurity. Farm income is expected to decline as a result of lower yields, thus, the importance of non-farm income livelihood sources. In Laos, although rice farming generates the highest income, farmers' households (except for the rainfed lowland farmers in Luang Prabang) also earn from off-farm activities. In the Philippines, no farmer respondent in either province depends on rice farming alone for his or her livelihood. Important sources of household and family income are off-farm and non-farm employment.

The climate and crop simulation projections approximate future scenarios based on current available data and science-based assumptions. In downscaling climate scenarios using SDSM techniques under a medium-range emission scenario, future uncertainties are recognized, but simulation results can guide researchers and policymakers in planning coping mechanisms. The technology and development policies of both countries could help address this uncertainty. Further research is also needed to:

- determine optimal levels of ecozone resilience to climate variability;
- determine the differential effects of climate change on households across ecozones considering the recognized variability of impact across this area;

- develop a broad range of crop and water management technologies that farmers can choose from; and
- determine the social acceptability of these technologies.

Viewed from the broader context of local development, the current state and future direction of policies and programs in agriculture, land use, and population will have significant implications for the findings. To compensate for farm losses due to climate events, appropriate insurance programs have to be considered. Laos has no existing crop insurance system, but farmers with crops damaged by flood and drought are exempted from paying tax. The Philippines already has a crop insurance system, but usually, this is part of the loan package of formal credit providers such as banks. The insurance covers only damages after the climate event and claims depend on the stage of crop growth. The process is cumbersome to farmers as well as to insurance providers. A weather-based index insurance system is another scheme that could be examined to mitigate climate risks to farmers. This scheme has been implemented in some parts of the country and has to be assessed for wider adoption. In addition, the government also provides advice on adaptation strategies and measures including the types and varieties of crops that may be planted.

In Laos, there have been measures from the Office of the Prime Minister to alleviate the impact of floods and droughts, including a production plan for the wet and dry seasons in 2011. The Provincial Agriculture and Forestry Office (PAFO) and District Agriculture and Forestry Office (DAFO) also provide assistance such as suitable seed varieties and repairing irrigation canals. They also advise farmers on cropping calendars and planting different varieties to cope with extreme climate events, crop disease, and insect pests. Farmers in the irrigated lowlands mostly used improved, short-cropping varieties, whose maturity ranged from 135–140 days. Farmers' households in rainfed lowlands use *Dornumpa,* which is tolerant to drought, or a mixture of traditional and improved varieties. Farmers in the rainfed uplands use varieties tolerant to drought, such as *Pair, Deang-dou, Mak-kheu.*

Agriculture remains a significant driver of the socioeconomies of Laos and the Philippines. In both countries, rice remains an important staple, a primary source of income for many households, and still employs a sizeable proportion of the labor force, although farm households have learned to rely on other means of livelihood. This dependence is particularly prevalent in the rural areas where many people reside, thus complications brought by climate change may compound current vulnerabilities in the sector. Policies and programs that will usher new practices revolving around anticipated climate variability will help avoid such exposure or at least allow for pre-emptive actions to minimize adverse effects. Practices currently in place such as the adjustment of the cropping calendar and the choice of appropriate crops for particular ecosystems need to be studied and disseminated alongside the development of rice farming technologies for more conducive plant management. Such assistance needs to be provided on an equitable basis across the provinces and agro-ecosystems. Incentives for the adoption of alternative and complementary non-farm as well as off-farm sources also needs to be explored to augment and expand support for rice farming income, diversify income sources, and render farm households' more resilient.

As climate change stands to adversely impact production in existing farm areas, the threat to food security is likewise heightened. More diversified livelihood sources within and outside farming need to be sought. This implies the need to open up more opportunities for both alternative and complementary income sources as well as learning new skills to access these opportunities. For instance, employment opportunities in both the public and private sectors were noted for farm household members in all ecozones of Tarlac and Pangasinan. Therefore, continued investment in education and skills development may help improve and sustain non-farm livelihoods for farm households.

The current pattern of an expanding population and that of increasing the built environment in order to service housing, industry, and services, constrains the expansion of agricultural/farming areas. Policies to preserve land for agriculture/rice farming in both countries need to be in place and seriously implemented, given that increased yields alone are not secure under a scenario of climate change. An examination of land suitability for growing alternative crops may also be useful and necessary to complement the strategies for diversifying the income sources proposed above.

Local governments should also provide interventions such as information, education, and communication campaigns to raise farmers' awareness about climate change and variability, and appropriate adaptation options. Communities should be encouraged to participate in climate change adaptation programs.

Conclusions

The downscaled climate scenarios for the provinces studied in Laos and Philippines vary by location over the long term. The combined effect of projected temperature and rainfall changes will also affect rice yields and food availability in different production environments. The impact on livelihood will be greatest for farmers who greatly depend on farm income because of their sensitivity to climate risks. Farmers whose income/ livelihood sources are more varied are more resilient to climate change impacts.

Downscaled climate projections and crop simulation made local policymakers more conscious of the potential risks and the necessary appropriate policy actions. The results reinforced local adaptation planning to cater to communities within specific social, economic, political and physical conditions. The projections could be integrated with broader socioeconomic development planning. Climate change research must also continue in order to improve understanding of the coming uncertainties and allow for more confident decision-making and allocation of limited resources.

Note

We acknowledge the assistance and cooperation of our farmer-respondents and key informants who shared with us their knowledge and valuable time in answering our survey questionnaires. To the municipal and provincial offices of Tarlac and Pangasinan in the Philippines, and the DAFO and PAFO of Luang Prabang and Savannakhet, who are involved in preparing agricultural development and DRRM plans, for sharing with us data and information about the situation and plans of their respective area of responsibility. We also acknowledge the assistance provided by Samantha Geraldine G. De los Santos and Jennylyn P. Jucutan in collecting and analyzing secondary data.

12

Impact of Urbanization on the Hinterlands: Perceptions of Households in the Hinterlands of Khon Kaen, Thailand, and Vang Vieng, Lao PDR

Maniemai Thongyou, Bounthavy Sosamphanh, Thanapauge Chamaratana, Monchai Phongsiri, and Louis Lebel

Urban areas in the Mekong region are expanding annually at a rate of 4.9 percent in Lao PDR, 4.3 percent in Cambodia, 3.6 percent in Vietnam, and 2 percent in Thailand. These rates are around 2.5 times the national population growth averages. In the region as a whole, 31 percent of the population lives in urban areas, and this is expected to exceed 50 percent by 2050 as the trend towards urbanization continues (UN Habitat 2011).

Urbanization is defined as the massive growth of, and migration to, cities and urban areas. Urbanization is driven by a combination of population growth, migration, and the incorporation of rural areas into urban areas (Middleton et al. 2013, this volume). Urbanization can have both positive and negative consequences. Concerns with rapid urbanization extend beyond the challenges faced within urban areas to the impact on peri-urban and urban hinterlands (Hoggart 2005; Simon 2008; Ravetz et al. 2013; Kubes 2013). In these zones of interaction, rural and urban socioeconomies become intertwined, for example, through: migration; daily commuting; commodity chains based on the extraction of resources from rural areas for consumption and processing in urban areas; and investment of income earned from urban areas back into agriculture.

The urban hinterland can be defined as the landscape interface between town and country, or as the transition zone where urban and rural land use interacts. Alternatively, the urban hinterland can be viewed as a landscape type in its own right, one forged from this interaction. The documented impact of urbanization on rural hinterlands include, for example, changes in ecological balance, loss of agricultural land and production (Mekonnen 2012), changes in land use patterns, which creates pressure on water resources (Narain 2010), changes in farming practices, livelihoods and lifestyles, and pollution (Brockherhoff 2000). Urbanization also has many positive impacts on the hinterlands, such as enabling the rural population access to more income-generating and job opportunities, creating a higher demand and bigger markets for agricultural produce, and improving access to better education, health care and other services (Satterthwaite and Tacoli 2003). The impacts of urbanization on the hinterlands of small and medium cities are less well understood than for large cities, and therefore deserve more study (Boyle 2004; Redman and Jones 2004).

In the Mekong region significant urbanization is now taking place around these smaller towns (Kammeier 2003). Hinterlands around smaller urban centers may have distinct features arising from the lack of specialized governance arrangements, and more modest levels of the industrial development typical of the metropolitan regions of mega-cities.

Research objectives

This study focuses on the impact of urbanization on urban hinterlands as perceived by hinterland communities and households, and aims to identify the issues that are of most concern to hinterland populations. The analysis presented in this chapter fills a knowledge gap through an examination of survey data in the hinterlands of two small urban centers in the Mekong region, Khon Kaen city and Vang Vieng town. The results yield a more nuanced understanding of the policy issues that may be of particular relevance to these areas, as well as to various similar constituencies in the region.

Conceptual framework

This research focuses on the perceptions residents have of the impacts of urbanization on the hinterland. Based on the classification of linkage types by Potter and Unwin (1989) and Rondinelli (1985), and the framework for effects of urban–rural linkages of Barkley et al. (1996), we have classified the impacts of urbanization on the hinterlands into five categories: economic, social, political and administrative, ideological, and environmental. In addition, based on our qualitative fieldwork in Khon Kaen and Vang Vieng, we have added a women and youth category. We hypothesize that these perceived impacts may be influenced by the demographic characteristics of the household and household heads, and the linkages of the hinterland households with the urban area in work/migration, farming relations, non-farm relations, and urban service usage (Fig. 12.1).

Fig. 12.1 Conceptual framework

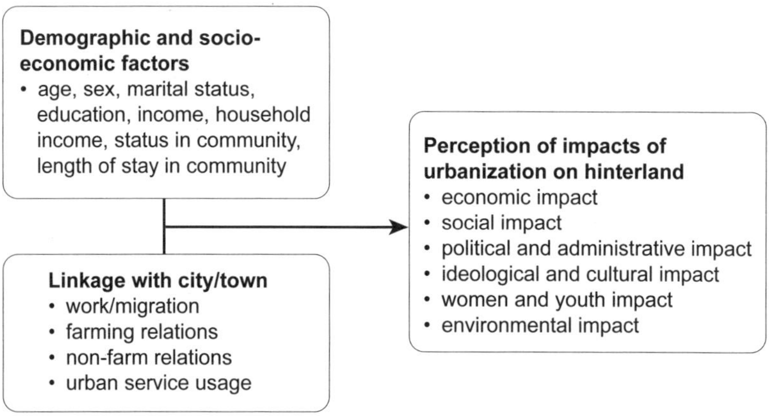

Methods

Khon Kaen city in Thailand and Vang Vieng town in Lao PDR were selected as case studies (Fig. 12.2). These two urban centers are of different sizes, emerged and grew over different periods of time, and are situated in different regimes. But both grew as a result of state policy and support with the expectation that they would develop and that their economies would integrate into world markets. Another shared characteristic is that both grew up from a rural base. Khon Kaen was in the 1950s selected as the growth center for the largest and poorest region of Thailand; it was intended to be a center for the Green Revolution, and for the economic growth and industrial development of northeast Thailand under a modernization model of rural development. Before such planned growth, Khon Kaen was just a tiny rural town, and had no historical township-base like other older towns in the northeast. Khon Kaen grew slowly; despite competitive basic infrastructural investment such as roads, it still could not compete with Bangkok, or even other major secondary cities (Thongyou and Savangnokg 2011). In the 1980s the investments in infrastructure began to pay off—in 2013 Khon Kaen municipality, with a population of 119,585, was Thailand's ninth largest city. It should be noted that there are also a lot more urban residents now living beyond the municipal boundary.

Vang Vieng, on the contrary, grew after the socialist state of Lao PDR (Laos) adopted more open economic strategies after 1986. Vang Vieng, a small town on the main and only road from the capital city of Vientiane to the country's tourist capital of Luang Prabang (Fig. 12.2), slowly grew as another tourist attraction for young backpackers. The Lao government welcomed this new trend of tourism and informally waived certain environmental and particularly social concerns to attract young tourists from the West and also from Asia. At present, although the city of Vang Vieng has a population of only 20,000 persons, its levels of consumption, lifestyles as affected by foreign tourist culture and service sector investors, investment, and its migrant worker population, are making the town more urbanized than many larger towns in Laos.

In order to examine the impact of urbanization on the hinterland, this study used households and hinterland villages of these two localities as units of analysis. The village (*ban* or *mooban* in Thai, and *ban* in Lao) is the smallest administrative unit in both Thailand and Laos. This study used a mixture of quantitative and qualitative research methods. Five main research methods were employed.

First, secondary data were collected and documents reviewed to analyze urbanization in Khon Kaen and Vang Vieng. Second, a community profile was created to identify basic characteristics of hinterland communities and key issues on urbanization impacts. These data were important inputs for the survey questionnaire design. Third, semi-structured interviews were then used to interview key informants from four hinterland communities each in Khon Kaen and Vang Vieng. In addition, semi-structured interviews were conducted with key city and town administrators and planners. In Khon Kaen, interviews were conducted with Khon Kaen Municipality's mayor and two advisory

Fig. 12.2 Study sites in Khon Kaen and Vang Vieng hinterlands

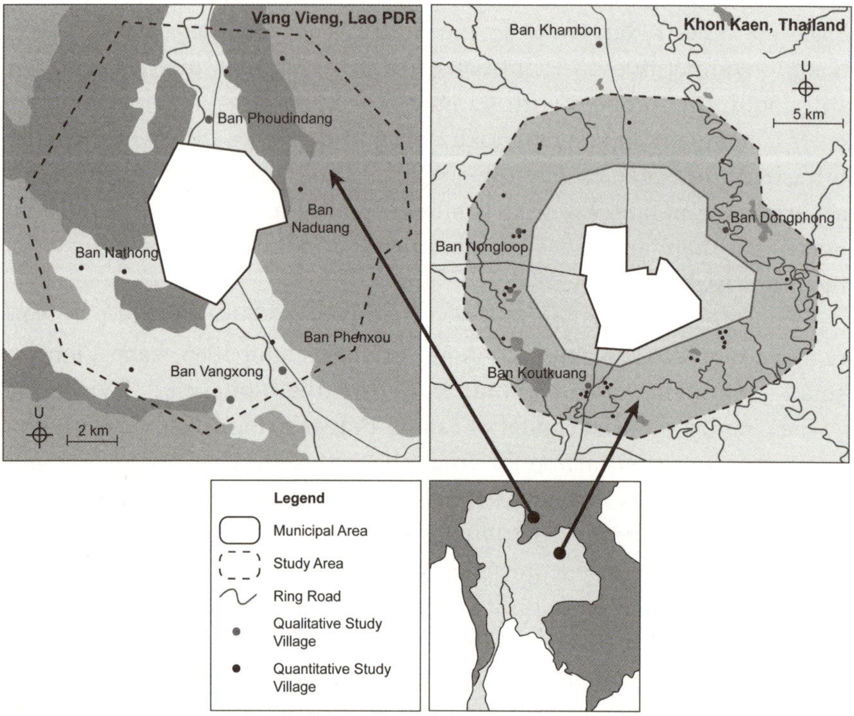

board members, and the mayors of two hinterland's sub-district (*tambon*) municipalities, namely Muangkao and Nonthon, as well as the permanent secretary of Khon Kaen's Provincial Administration Organization. In Vang Vieng, semi-structured interviews were conducted with the acting Vang Vieng district chief and heads or key staff of district offices responsible for administration and public services.

Fourth, quantitative samples of households in the hinterland areas of Khon Kaen and Vang Vieng were taken to identify factors that influence the perceptions of impacts of urbanization in hinterlands. The hinterland of Khon Kaen was identified as those areas lying up to 5 km from a ring road surrounding the city, comprising of 12 sub-districts, 35 villages, and 10,476 households, based on 2011 election data from the Ministry of Interior. A sample size of 409 households was determined by using the F Test of Variance Proportion in Multiple Regression/Correlation Analysis (MCA) (Cohen 1988). A systematic sampling method was used to select sample households from the 2011 election data, from household number 1 to 10,476, yielding 409 households proportional to the size of the villages in 12 sub-districts around Khon Kaen city.

The hinterlands of Vang Vieng were identified as the villages surrounding Viang Vieng town's municipality area. This area contains 12 villages with a population of 1,668 households. A sample size of 330 households was determined by using F Test of Variance Proportion in MCA (Cohen 1988). A systematic sampling method was used to draw sample households proportional to the size of each village's population, by using village household data available at each village chief's office. The structured interviews were completed during November–December 2011 for Khon Kaen, and March 2012 for Vang Vieng. Data from the quantitative survey were analysed using SPSS software. Linear models for perceived overall impact (based on responses to multiple questions and standardized to between -1 and 1) were explored using the General Linear Model function in SPSS. Variables with unusual distributions were transformed or converted into categorical variables prior to inclusion. Only a single representative variable of highly intercorrelated or logically-related variables was included in candidate set of predictors. Only significant variables were retained in final models reported.

Fifth and finally, qualitative methods were used to get a deeper understanding of the impact of urbanization and people's perceptions in specific sites. Four villages in the hinterland of Khon Kaen and five villages in that of Vang Vieng were chosen for in-depth study. Semi-structured interviews, unstructured interviews and group interviews were used to collect data at the household, community, and organization levels.

Khon Kaen city and its hinterlands

Hinterland households

The majority of the sample households in Khon Kaen case were located in the sub-district municipal areas (67 percent of 409 households). The average distance from the households to the city center was 9.5 km. People in the hinterland traveled to the city by bus (41.6 percent), private car or truck (35.7 percent), and motorcycle (22.0 percent). More than half (56.7 percent) spent between 16–30 minutes traveling to the city, 32.5 percent spent less than 15 minutes, and 10.8 percent more than 30 minutes.

With regards to household monthly income, 22.5 percent earned less than 10,000 baht, 38.4 percent between 10,000–20,000 baht, and 39.1

percent more than 20,000 baht. The average income was 23,075 baht. Income disparities among the sample households was high, with the minimum income of 500 baht and the maximum income of 140,000 baht per month. With regards to land holding, 56.0 percent of the sample households owned small areas of farmland of less than 5 rai (2 acres), 21.1 percent owned between 5–10 rai, and 22.7 percent owned more than 10 rai. The average land holding per household was 6.51 rai. Much of the land was used for rice farming, with each household having an average of 5.4 rai of rice fields.

Hinterland households and their village community

Household heads in the hinterlands were asked to talk about their subjective feelings about their relationship with their village community. Some 43.8 percent felt that the relationship was good and very good, 54.3 percent thought it was moderate, while 1.7 percent thought it was bad and 0.2 very bad. Almost half of the households (48.2 percent) had lived in the village for 30 to 60 years, with the average being 42.9 years. Almost half could enjoy the company of close neighbors within the village; 46 percent had more than 20 neighbors whom they described as close, and 54 percent had less than 20 persons.

Linkages between hinterland households and urban centers

Linkages with other urban centers

Households in the hinterlands of Khon Kaen have established linkages with several urban centers, but Khon Kaen was the major one. Other important centers were sub-district towns, which, though smaller in size, were located closer to the villages (Fig. 12.3). These were Ban Thum sub-district town in the west and Tha Phra south of Khon Kaen.

Linkages with Khon Kaen city

Hinterland households had diverse linkages with Khon Kaen city. Up to 55.5 percent had some household members working in the city, while 27.9 percent supplied their agricultural produce directly to the city. Almost one-third (30.1 percent) had non-agricultural economic activities with the city, mainly in small entrepreneurial activities such as running micro-businesses in construction; mechanical, electrical and electronics repair; trade; running dormitories for urban workers; and village–city transportation. In addition, up to 40.1 percent had 1 or 2 household members studying in the city. The city was also the major supplier of medical services, government services such as birth, marriage, death, and land inheritance and transaction registration. Shopping, recreation, and travel services were enjoyed by 90.2, 75.3, and 72.8 percent of all households respectively. Notably, up to 18.3 percent of the households had already sold land to city people during the previous ten years (Fig. 12.5).

Perceived impacts of urbanization on the hinterlands

The impact of urbanization on the hinterlands was analysed by using the mean ranking of each item of the perceived impacts in the seven themes, namely household economy, village economy, village society, politics and administration, ideology and culture, women and youth, and the environment. Each theme contained 5–12 items. The household heads were asked to give their perceptions of the impact of urbanization on each item, with a ranking from 1–5 to reflect whether the situation seemed to be getting much worse (1), had worsened (2), had no impact (3), was getting better (4), or much better (5).

Favorable impacts

Analysis of the mean ranking indicated that the most favorable impact as perceived by hinterland households was of health care services. Other favorable impacts ranking at the top six were related to increased women's rights and empowerment. These included employment opportunities for women in the village, women's participation in family decisions, women's participation in village decisions and women's social status. Other highly favorable impacts were increased educational opportunities, and more diversified employment opportunities. Households in the hinterlands also had positive attitudes towards urban impact on local government, particularly in terms of its investment in infrastructure development and in increasing residents' participation in local government organizations (see Table 12.1).

Unfavorable impacts

Urbanization has both positive and negative impacts. In what follows, we will discuss the unfavorable impacts as perceived by hinterland households. In contrast to the most favored impacts that were centered around issues of women's empowerment, the negative attitudes focused on the problems of youth. Urbanization is perceived to have had the most negative impact on youth: increasing premarital sex and cohabitation, inappropriate dressing, and youth gangs and inappropriate activities, e.g. motorcycle racing and tattooing. Following these top four most negative impacts, the following two items were social problems concerning people of all ages, namely drug addiction and gambling.

Noise pollution was one of the most worrisome environmental concerns. Our interviews revealed several causes of noise pollution, including traffic, particularly motorcycles, pubs and karaoke bars, factories, and construction machines. Others in the top 10 undesirable items caused by urbanization were related to the village and household economy, including a shortage of farmland, rising household expenditure, and debts (see Table 12.2).

Factors influencing perceptions

A Multiple Regression Analysis (MRA) was used to identify the most important factors that influenced the household heads' perceptions about the impact of urbanization. Based on this analysis, the research found that close neighbors in the village (Beta = 0.155), the household head's main occupation as farmer (Beta = 0.152), and the respondent's gender as female (Beta = 0.104) were *factors that influenced perceptions* on the impacts with the $R^2 = 0.090$ (Table 12.3).

Factors that were included in the MRA analysis but were found *not to significantly influence perceptions* were: whether a household had land holdings of more than 10 rai, rice fields, farm or agricultural land; how long they had been resident in the area; whether anyone was studying in the city, working in a city/town in other provinces, was affiliated to a political party, had sold land to city folk in the last 10 years, conducted non-agricultural activities in city markets, sold produce in the city, or had changed the type of agricultural production to serve city markets; and lastly, the age of the household head.

Fig. 12.3 Average proportion of households with linkages to different urban centers, Khon Kaen

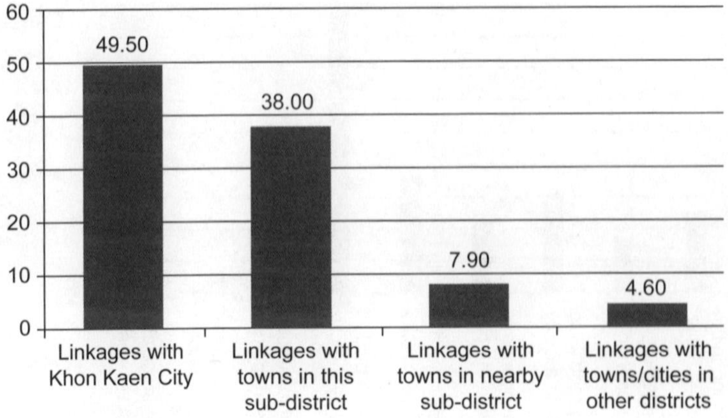

Fig. 12.4 Average proportion of households with linkages to different urban centers, Vang Vieng

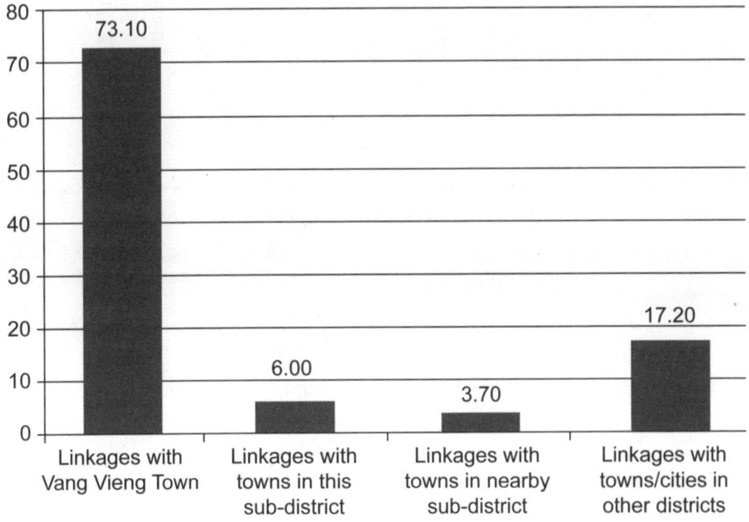

Fig. 12.5 Percentage of households classified by linkages with urban centers

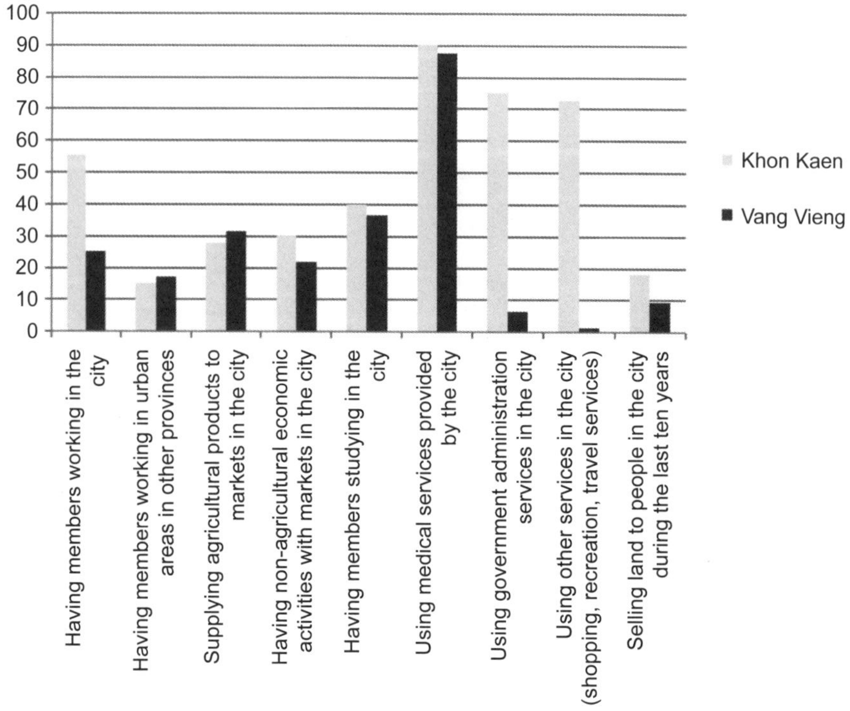

Table 12.1: Top ten mean rankings of most favorable impacts of urbanization in Khon Kaen and Vang Vieng

	Khon Kaen		Vang Vieng	
Rank	Item	Mean	Item	Mean
1	Access to health services	3.86	Education opportunities	4.17
2	Employment opportunity for women in village	3.85	People's participation in local administration (village committee)	4.17
3	Women's participation in family decisions	3.84	Overall quality of life—social	4.13
4	Education opportunities	3.82	Overall standard of living of people in the village—economic	4.05
5	Women's participation in village decisions	3.78	Employment opportunities for villagers	4.05
6	Women's social status	3.77	Overall standard of living of the households—economic	4.00

	Khon Kaen		Vang Vieng	
Rank	Item	Mean	Item	Mean
7	Employment opportunities for villagers	3.72	Vang Vieng Town becoming international tourist attraction creates awareness of the outside world among the villagers	4.00
8	Local government investment in infrastructure development	3.64	Mutual help and support among the villagers	3.97
9	Employment opportunities for household members	3.59	Community solidarity and social cohesion	3.96
10	People's participation in local government organizations	3.58	Women's participation in village decisions	3.95

Table 12.2 Top ten mean rankings of most unfavorable impacts of urbanization in Khon Kaen and Vang Vieng

	Khon Kaen		Vang Vieng	
Rank	Item	Mean	Item	Mean
1	Premarital sexual relationships of village youths	2.10	Quality of air in village	2.21
2	Premarital co-habitation of village youths	2.11	Quality of water resources	2.30
3	Inappropriate dressing of village youths	2.18	Availability of food from natural resources	2.47
4	Youth gangs and inappropriate activities	2.19	Noise pollution	2.60
5	Gambling	2.40	Quantity of solid waste from Vang Vieng town	2.70
6	Drug addiction	2.44	Shortage of farmland	2.74
7	Noise pollution	2.51	Building (roads, houses, offices) creates environmental problems in village, e.g. floods, scenic views	2.79
8	Shortage of farmland	2.54	Quality of soil for agriculture	2.79
9	Household expenditure	2.57	Urban bias policy creates problems, e.g. urban wastes dumping and polluted water leakages	2.83
10	Household debt	2.57	Overall quality of natural resources and environment	2.84

Table 12.3 Factors influencing household heads' perceptions of urbanization in Khon Kaen and Vang Vieng

Independent variables	Khon Kaen			Vang Vieng		
	B	Beta	Sig.	B	Beta	Sig.
Close neighbors in the village	.007	.155	.002	.001	.033	.539
Household head's main occupation is farming	.043	.152	.005	-.004	-.016	.782
Respondent is female	.029	.104	.043	-.025	-.089	.106
Household has farmland	.010	.026	.680	-.076	-.283	.000*
Duration of residence in the village	-.006	-.094	.097	.011	.152	.009*
Household contains members of a political party	-.024	-.034	.500	-.096	-.271	.000*
Household sells agricultural produce in the city markets	.018	.061	.276	.033	.127	.041*
Khon Kaen R2 = 0.090 F = 2.262 +Sig of F = 0.003 n = 409						
Vang Vieng R2 = 0.211 F = 4.680 Sig of F = 0.000 n = 315						

Note: Composite independent and standardized dependent variables are used for the calculations.

Vang Vieng and its hinterlands

Hinterland households

About half of the sample households in Vang Vieng were located outside the municipal area (56 percent of 315 households). The average distance from the households to the city center (Vang Vieng town office) was 5.4 km. Most of the people living in the hinterland of Vang Vieng (82 percent) traveled to the town by motorcycle. Almost half (46.7 percent) spent less than 15 minutes, while 44.1 percent spent between 16–30 minutes, and 9.2 percent spent more than 30 minutes. The longest time required was 2.32 hours—on foot.

The average monthly household income was 2,874,682.1 kip (1 US$ = 7,882 kip). Most of the households (70.8 percent) earned between 1–5 million kip, 18.1 percent less than 1 million kip, and 11.1 percent more than 5 million kip. The average land holding was 8.2 rai per household; 37.1 percent owned less than 5 rai land, 34.3 percent owned between 5–10 rai, and 28.6 percent more than 10 rai. An average of 5.1 rai per household was used for growing rice.

Hinterland households and their village community

Household heads in the hinterland were asked about their subjective feelings about their relationship with the village community. Some 37.1 percent felt that the relationship was good and very good, 61.3 percent thought it was moderate, while 1.6 percent thought it was bad, and none thought it was very bad. A little more than half of the households (53.3 percent) had lived in the village for less than 30 years, the average being only 26.7 years. Even though the settlements in the Vang Vieng hinterland were relatively new, 41.3 percent of the households had more than 20 neighbors whom they felt close to, and 58.7 percent had less than 20 such persons.

Linkages between hinterland households and urban centers

Linkages with urban centers

Geographically Vang Vieng is the major town in the area, especially for the people in the hinterland. The survey data showed that a significant majority of the hinterland households had linkages with Vang Vieng town. Another important linkage was with towns or cities in other districts (17.2 percent) including Vientiane (Fig. 12.4).

Linkages with Vang Vieng town

The hinterland households of Vang Vieng had many types of linkages with the town. The most important type of linkage was in health care (87.6 percent). The second most prevalent linkage was in education. More than one-third of households (36.5 percent) had members studying in the town. Apart from receiving social services, hinterland households also had an economic relationship with the town. One-third of them (31.7 percent) sold their produce in town, but most sold less than 50 percent of their produce. A quarter of the hinterland households (25.1 percent) had some members working in the town, mostly as cleaners and workers in small hotels, guesthouses, restaurants, and bars. Others were construction workers as Vang Vieng was growing rapidly. Another 21.9 percent were involved in other non-agricultural activities, mainly through running small businesses such as food shops and stalls; or providing services such as building; mechanical, electrical and electronics repairs; trading; and transportation. Notably, 9.5 percent had already sold land to people from the town (Fig. 12.5).

Perceptions of impact of urbanization on the hinterlands

The impact of urbanization on the hinterlands as perceived by household heads was analysed by using the mean ranking of each item of the perceived impacts in the seven themes above (see Table 12.1).

Favorable impacts

Analysis of mean rankings indicated that the most favorable impacts were connected to better educational opportunities and participation in local administration, specifically the village committee. Both received very high and equal rankings. Other favorable impacts related to the overall quality of life and standard of living, socially and economically, at both household and village levels. People were also satisfied that Vang Vieng had become an international tourist attraction, which gave them some exposure to the outside world. It is notable that Vang Vieng's hinterland households could still enjoy communal life in their villages. Along with the expansion of the town, local people could enjoy mutual help and community solidarity. In addition, they also appreciated the participation of women in village decisions.

Unfavorable impacts

Urbanization does not bring about positive impacts alone. In what follows, we will discuss the most unfavorable impacts, i.e. those that received negative rankings (see Table 12.2).

The results strongly point out that urbanization was having a negative environmental impact, which was seen as the biggest problem. Nine out of the ten items with the lowest mean rankings were related to environmental impacts. Hinterland households felt that urbanization had worsened air quality. Vang Vieng's urbanization also negatively affected the quality of its water resources and food supply.

Noise pollution and solid waste were identified as other major environmental problems. In addition, new buildings (roads, houses, offices) were seen as creating environmental problems, e.g. creating floods, and destroying scenic views. The overall quality of natural resources and the environment was deteriorating due to rapid urbanization. People were dissatisfied with the urban bias policy which created environmental problems for their village. They also perceived that urbanization has created a shortage of farmland.

Factors influencing perceptions of impacts of urbanization

MRA was used to identify the most important factors that influenced the perceptions of the household head on the impact of urbanization. Based on this analysis, the research found that the major factors were ownership of farmland (Beta = .026), duration of residence in village (Beta = -.094), households that included members of a political party (Beta = -.034), and households who sold produce in city markets (Beta = .061), with the R^2 = 0.211 (Table 12.3).

Factors included in the MRA analysis but found to be not significantly influencing the perceptions were: households with more than 10 rai of land; households with rice land; households with agricultural land; close neighbors in village; members studying in the city; members working in city/town in other provinces; households that had sold land to city people in the last 10 years; households that engaged in non-agricultural activities in the city markets; and lastly, households that had changed their type of farming to supply markets in the city.

Discussion

This study shows that urbanization has had both positive and negative impacts in these two areas. Although the urbanization of Khon Kaen and Vang Vieng differs in many aspects, households in their hinterlands shared many similar perceptions on the impacts of urbanization. Increased opportunities for education, employment and participation in local administration—especially by women—were among the most important positive perceived impacts in both places.

Increased educational opportunities for young people were among the most valued impacts. Young people in the hinterlands had more access to higher education, given convenient transportation to the city, where better schools, vocational colleges and—in the case of Khon Kaen—a university, were located. In Khon Kaen more rural hinterland parents could afford to send their children to college because they had more diversified employment opportunities. In addition, market-oriented agriculture provided incomes that made possible the investment in children's education. As for Vang Vieng, the town's tourism-derived growth not only brought about various job opportunities that demanded higher skills and knowledge, but also engendered a different and broader world view among the local people.

Local people now gave higher value to education—in both countries, parents in the hinterlands most valued giving their children the highest possible education. In Khon Kaen's case, this desire held, despite the concerns about youth misbehavior.

An important underlying reason for positive impacts of urbanization in the hinterland observed in this study were the improved roads, communication, and other infrastructure that have improved access to multiple services in the city. As urbanization proceeds, there has also been a relocation of some activities and services into the hinterland itself, creating local jobs, and new market and business opportunities. The increased linkages between rural and urban livelihoods and economic systems, and the resulting complex spatial outcomes, were coined *desakota* systems by McGee (1991) as a challenge to the idea that urbanization transitions completely transform rural areas into urban ones. The hinterland changes observed in this study, including growing land-use patterns along major roads, also share many of the characteristics of *desakota* systems originally described for large cities in Indonesia (McGee 1991). Since then, studies in various parts of the developing world have often noted the perpetual 'incompleteness' of urbanization in Asia, including Southeast Asia (Yap and Lebel 2009), China (Xie et al. 2006; Leaf 2002), and India (Narain 2009). What these studies collectively show is that the quality of infrastructure, linking peri-urban hinterlands to cities, is a key factor determining negative and positive impacts from urbanization.

The most obvious adverse impacts were on hinterland environments. Here, the differences between the two cities, however, were substantial. Environmental problems in the hinterlands of Vang Vieng were caused by rapid economic and infrastructural development of the town, in particular, and of Laos in general. Many environmental problems were caused by the cement factories being set up near Vang Vieng town, as well as by the quarry blasting and stone grinding industry for the cement factories, which serve not only the town, but also and more importantly, construction in the central and northern regions. All these created smoke, dust, and smells at the mountain sites and along the roads from the mountains to the town, passing through a number of villages. In addition, Vang Vieng's urbanization has negatively affected water quality, especially that of the Xong River which runs through the town. The town's waste water was draining into the river with very limited or no treatment at all. Villagers who lived downstream could no longer use the water from the river for home consumption, due to the deterioration of water quality. People also complained that fish became rare, and some edible freshwater weeds had vanished, due to riparian pollution. This contributed to a reduction of food from natural resources. Noise pollution was caused by traffic, quarry blasting for stones and the cement industry, as well as late-night bars serving young backpackers. Some solid waste was being dumped into the river and landfills. People complained about plastic bags floating in the river after heavy rains. Notably, the survey showed that local (hinterland) people linked their environmental problems to policies biased towards urbanization.

As observed elsewhere, the rapid integration of the urban hinterland economy into cities leads to resource extraction, waste flows, and other environmental problems that can quickly outpace institutional development (Brockherhoff 2000; Narain 2010). Urbanization in the hinterland brings to the fore administrative ambiguities caused by overly decentralized local government structures and highly centralized city administration. The complex organization of space with linear developments along major routes with urban islands in otherwise low-density fields, also makes it more likely that nonessential functions fall through jurisdictional cracks. Studies in the hinterlands of large cities elsewhere have documented stronger negative environmental impacts of urbanization than those reported here—for instance, in Mexico (Aguilar 2008), Brazil (Torres et al. 2007), and Vietnam (Leaf 2002)—but for some similar reasons. In this research, we have advanced knowledge of this process by showing the most important areas of urbanization impacts on the hinterlands, based on the perceptions of the hinterland households themselves. However, the study of the factors which influence these perceptions was limited to some demographic characteristics and some urban–rural linkage factors only, which constrained the explanatory capacity. We therefore suggest that in future research, other factors might be included, for example, civic engagement, social capital, and cultural values. It would also be worthwhile to relate perceptions of impact for some dimensions to independent evidence about actual practices.

Recommendations

Urbanization has created adverse environmental impacts in the hinterland related to water, air, noise, soil, solid waste, and the availability of natural food. Environmental problems seemed to be more prominent in the perceptions of the hinterland households in Vang Vieng, even though the size of the town and its ecological footprint were much smaller than those of Khon Kaen. This suggests that environmental protection should be a high priority, particularly considering the fact that Vang Vieng has been promoted as a nature-based international tourist town. Environmental protection should expand beyond the urban space to cover the hinterland areas.

In Khon Kaen, whose development has given rise to considerable environmental problems in its hinterlands, this is not an easy task. Thailand's decentralization policy has resulted in fragmented planning and budgeting. In addition, the hinterland population are not constituents of the city's mayor. As a result, environmental problems in the hinterland caused by the city are not taken seriously, except in a few cases where there are strong reactions or protests.

This research thus recommends that Khon Kaen and its surrounding hinterland municipalities should work together to plan a more integrated and sustainable development of the 'city region'. Such city planning can overcome the limits to administrative authority. A Khon Kaen city region planning committee should be established with representatives of related municipalities and *tambon* (sub-district) development organizations. In the case of Vang Vieng, the town's administration has broader authority covering the whole district. However, there are often planning conflicts, particularly between industry, tourism, and environment. Therefore, in both cities, representatives from hinterland communities and civil society organizations should be included in town planning.

Conclusion

In this chapter, we have investigated the perceived impacts of urbanization on the hinterlands in two smaller urban centers in Thailand and Laos. We have argued that although these two urban centers differed in terms of their size, level of urbanization, political and economic context and reasons for growth, hinterland households in both cases identified many common concerns about the impacts of urbanization. Increased educational and employment opportunity, as well as better participation, especially by women, in local administration were among the most important positive impacts perceived in both places. As for adverse impacts, the biggest concern among Khon Kaen hinterland households were social problems related to youth behavior. However, the impact of urbanization on the hinterland environment garnered the most negative ratings. Similarly in Vang Vieng, concerns over the environmental impacts of urbanization were explicitly expressed.

Although negative environmental impacts on peri-urban and hinterland areas have been noted elsewhere, this research demonstrates that those who live in the hinterland are very aware of the issues. This point is very important for urban and regional development. Poor management of an urban area and its hinterlands can lead to resource competition, conflicts, and unsustainable development. The juxtaposition of these two different urban centres is not only to stimulate debate, but also to show that their similar problems and the levels of awareness among the affected populations may be a more widespread phenomenon, as small towns and intermediate cities now represent a majority of urban areas globally. Our study also discovered that neither the partially decentralized system of Thailand, nor centralized governance in Laos, can manage this problem. Therefore, future urban, peri-urban and hinterland planning needs to go beyond conventional policy and planning approaches.

Note

We acknowledge the support of the Center for Research on Plurality in the Mekong Region, Faculty of Humanities and Social Sciences, Khon Kaen University; and the Faculty of Social Sciences, National University of Laos.

13

Greenhouse Gas Emissions from Tourism Service Providers in Chiang Mai, Thailand, and Hue, Vietnam

Sivannappan Kumar, Kyoko Kusakabe, Pravakar Pradhan, Pujan Shrestha, Srujana Goteti, Tran Anh Tuan, Ekawit Meteejaroenwong, Trinnawat Suwanprik, and Khanh Linh

Travel and tourism-related activities contributed about 16.3 percent and 10 percent, respectively, to the GDP of Thailand and Vietnam in 2011; this sector also generated about 1.8 million jobs in each of these countries. The total number of jobs in this sector would be more than 3 million and 2.1 million in Thailand and Vietnam, respectively, in 2022 (WTTC 2012a, 2012b).

Tourism, both domestic and international, although making important contributions to these economies, also contributes to greenhouse gas (GHG) emissions (UNEP et al. 2008; WTO and UNEP 2008). It is estimated that tourism-related activities contribute around 5 percent of global carbon dioxide (CO_2) emissions (WTO and UNEP 2008). And, although major cities offer the widest range of tourist attractions, small and medium-sized cities with considerable cultural offerings and heritage sites are increasingly becoming popular destinations. Promoting tourism in such cities will need to take into consideration environmental impact issues (e.g. GHG emissions, local air pollution, waste management, soil degradation) so that they will continue to be attractive destinations.

This study principally focuses on the hitherto little-documented impact of tourism on GHG emissions. Estimating GHG resulting from the products and services of tourism is also a necessary step in order to develop and evaluate strategies toward mitigating such impacts (Becken and Patterson 2006; Filimonau et al. 2013). In a city that also uses its cultural heritage as a tourist attraction, such mitigation strategies can lead to a 'win-win' situation for locals and visitors alike by reducing GHG emissions while creating 'green jobs'—environmentally friendly, decent work and income—and a sustainable tourism industry.

This approach—starting with estimating emissions and then identifying activities that reduce emissions as well as have a positive socioeconomic impact—was followed in the SUMERNET/CDKN-sponsored project on sustainable tourism in Chiang Mai, Thailand, and in Hue, Vietnam. Both cities are historically significant and medium-sized growing cities in the Mekong region where tourism is an important and growing economic sector. They are also mid-sized cities where relatively small initiatives are visible, with municipal authorities that are also keen on climate-compatible tourism development. However, a major stumbling block in achieving sustainable tourism is a lack of knowledge of the tourism sector's impact on GHG emissions. As a first step, this study discusses how tourism service providers (TSPs)[1] and their activities in Chiang Mai and Hue are linked to GHG emissions, and mitigation options that would also help to create green and decent jobs for the local people.

We have used the Bilan Carbon®[2] tool developed by the Agence de l'Environnement et de la Maîtrise de l'Energie (ADEME) to estimate GHG emissions by considering direct and indirect emissions from the TSPs (ADEME 2009; Bader and Bleischwitz 2009). Direct emissions by the tourism sector for this study refer to emissions within the boundary of the city by the TSPs and indirect emissions refer to those produced outside the city (Ahmad and Wyckoff 2003; WBCSD and WRI 2004; Baumert et al. 2005; Forsyth et al. 2008) as shown in Table 13.1. Data was collected from four major types of tourism-related entities (hotels/accommodation, restaurants, tour and travel operators, spas, etc.) in both cities.

Methods

Estimates of GHG emissions by the urban tourism sector were limited to those produced by TSPs within the administrative boundaries of Chiang Mai municipality and Hue city (henceforth Chiang Mai and Hue), respectively (see Fig. 13.1), using the steps listed in Fig. 13.2.

Fig. 13.1 Tourism: GHG emission sources (direct & indirect) within city boundaries and wider zones (based on ADEME 2009, *Bilan Carbone®*)

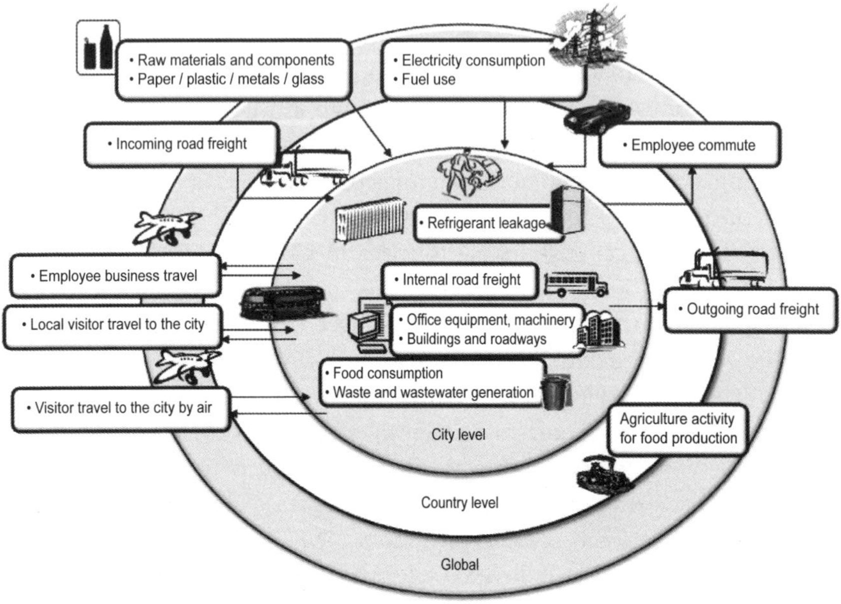

In Chiang Mai, the number of tourist arrivals was 5,545,009 in 2011. The average stay was 3.6 days. The main tourist attractions are: the Ping River, temples, the night bazaar, Sunday Market (Walking Street), spa and massage treatments, and golf.

A total of 686 entities (175 hotels, 321 restaurants, 100 tour and travel companies and 90 other activities) are engaged in the tourism sector in Chiang Mai[3] (S2 in Fig. 2). However, data collection was restricted to 36 hotels, 12 restaurants, 12 tour and travel businesses, and 24 other activities. This amounted to a total of 84 entities, including small, medium, and large hotels, tourism agencies, car rentals, spas, and restaurants (S3 in Fig. 13.2).

Fig. 13.2 Methodology used in this study for GHG emissions estimates

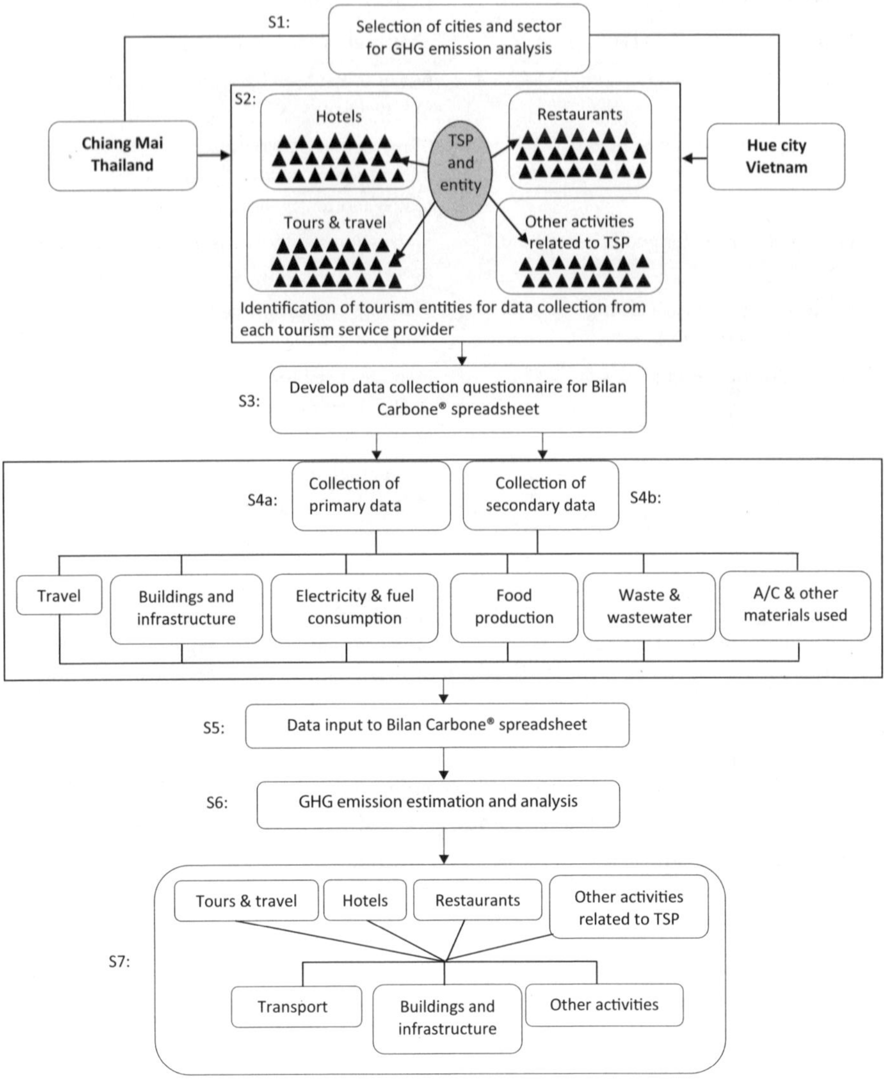

Table 13.1 Direct and indirect emission sources considered for GHG estimates

Direct emissions caused by	Indirect emissions caused by
Cooking using fossil fuels (CO_2)	Air travel by visitors (CO_2)
Own electricity generator used (CO_2)	Bus or coach used by visitors to travel to city (CO_2)
Own car and motorcycle used (CO_2)	Construction materials used for infrastructure, such as buildings, roads, and footpaths (CO_2)
Own vehicle used for goods transport (CO_2)	Electricity used by hotels, restaurants, travel agencies, tours and travels, spas, etc. (CO_2)
Fertilizer used for agriculture in Garden Houses (N)	Materials and products used, such as, glass, paper, metal, plastic, food, etc. (CO_2)
Air conditioner leakages (HCFC & CFC)	Materials used for vehicle manufacture, such as buses, cars and vans used in tours and travel (CO_2)
Waste and wastewater generation at own premises (CH_4)	Incineration of wastes outside the city boundary (CO_2)

Notes: CFC = chlorofluorocarbon; HCFC = hydrochlorofluorocarbon; N = Nitrogen. Direct emissions refer to emissions within the boundary of the city by TSPs. Indirect emissions are the emissions produced outside the boundary of the city. Electricity used by hotels, restaurants, travel agencies, tours, travel, and spas are indirect emissions because the electricity is supplied by government and produced by the cities themselves. If the cities use their own electric generator (or use diesel) to produce electricity then they are direct emissions.

In Hue, the number of tourist arrivals was 1,590,900 (in 2011) with an average stay of two days. The main tourist attractions are: the Perfume River, temples, monuments, and the Citadel.

In Hue, a total of 265 entities (130 hotels, 97 restaurants, 12 tour and travel businesses, and 26 other activities) were found to be engaged in tourism related activities[4] (S2 in Fig. 13.2). The data collected from these entities using a questionnaire was restricted to 18 hotels, 18 restaurants, 6 tour and travel companies, and 8 other activities that accounted to a total of 50 entities (S3 in Fig. 13.2).

The primary and secondary data was input into a Bilan Carbone® spreadsheet. The primary data refers to electricity and fossil fuel consumption by/in the entity; energy used for the production of food, materials and other products consumed by visitors; travel by visitors and employees (amount of fuel used by vehicles, vehicle distance traveled, etc.); fuel combustion for goods transport, waste generation due to the consumption of food and use of other materials; and construction materials used in buildings and related infrastructure (S4a in Fig. 13.2). These were obtained from a survey of hotels, restaurants, travel agencies, spas, and city authorities using a questionnaire.

The secondary data refers to the number of tourists; number of TSPs; number of employees in the tourism sector; emission factors for electricity, fossil fuels, local trains, food and materials used; traveling distance (by air and road); vehicle weights, etc. (S4b in Fig. 13.2), and this data was collected from city authorities, publications[5] and websites.

A Bilan Carbone® spreadsheet was used to calculate the direct, indirect, and total emissions of every entity within a TSP, and the sum was used to estimate the total GHG emissions of that TSP (S5 and S6 in Fig. 13.2).

Individual (entity and TSP) direct and indirect emissions were also available as an outcome of the Bilan Carbone® analysis. These were used to consolidate the results in terms of tourism activity emissions, namely, accommodation (building and infrastructure related), transport (tours and travel related), and others (recreation, etc.).

Assumptions

GHG emissions by TSPs in both cities were estimated by taking into account the following assumptions:

- *Meal consumption at restaurants*: The total emissions by tourists in restaurants was 70 percent in Chiang Mai and 6 percent in Hue and the remaining was from locals (Box 1; equation 1).
- *Fossil fuel used for public transport*: The total emissions by the amount of fossil fuel used for public transportation from tourists was 70 percent in Chiang Mai, 7 percent in Hue, and the remainder was from locals (Box 1; equation 2).
- *Use of infrastructure and services*: The total emissions by city infrastructure and services related to tourists was 28 percent in Chiang Mai, 3 percent in Hue, and the remainder was from locals (Box 1; equation 3).

  Visitors:

- All domestic visitors to Chiang Mai (3,680,192) were assumed to travel from Bangkok to Chiang Mai. Domestic Vietnamese visitors (888,900) were assumed to travel from Hanoi, Ho Chi Minh City, and Danang, to Hue.

- International visitors (1,864,817) to Chiang Mai were assumed to either fly from Bangkok or to Chiang Mai. Similarly, foreign visitors (702,000) flew from Hanoi, Ho Chi Minh City, or Danang, and then travelled to Hue.

Box 13.1 Basis for assumptions

The assumptions used for the analysis are based on the following considerations:

If, no. of tourists in the year 2011 = T; no. of stay days per tourist = S; no. of local people = P; no. of meals consumed per day per visitor at restaurants in the city boundary = F_V; no. of meals consumed per day per local person at restaurants in the city boundary = F_l; no. of days local people visit restaurants per year = D_R; and no. of days local people used fossil fuel for public transportation = D_T, then

The ratio of 'meals consumption' per year by the visitors at the restaurants was calculated by using the following equation, and converted into percentage:

$(T \times S \times F_V) / [(P \times D_R \times F_l) + (T \times S \times F_V)]$
equation 1

The ratio of 'fossil fuel used for public transport' per year by the visitors was calculated by using the following equation, and converted into percentage:

$(T \times S) / [(P \times DT) + (T \times S)]$
equation 2

The ratio of 'infrastructure and services' used per year by the visitors was calculated by using following equation, and converted into percentage:

$(T \times S) / [(P \times 365 \text{ days}) + (T \times S)]$
equation 3

The values for T were obtained from Chiang Mai municipality and Thua Thien Hue Provincial Department of Culture, Sports and Tourism, while values for S and P were obtained from the published literature (Chiang Mai Provincial Office 2011; Hue City Office for Statistics 2011). The values for F_V, F_l, D_R and D_T were based on a survey and from discussions with the city authorities.

City	T	S	P	a F_V	F_l	D_R	b D_T
Chiang Mai municipality	5,545,009	3.6	137,793	2	1	124	62
Hue City	1,590,900	2	339,000	2	1	128	117

Notes:
[a] Some visitors just came for a day in Hue city (had lunch) and went to other cities (for dinner) later.
[b] The majority of local people (in general all Chiang Mai municipality people) used their own vehicle to travel within the municipality.

Analysis of GHG emissions

The results of emissions estimation using Bilan Carbone® (in terms of TSPs) is presented and discussed in terms of transport, buildings and infrastructure, and other activities (S7 in Fig. 13.2).

The '**transport**' sub-sector covers emissions caused by (a) travel by visitors and employees (of the tourism industry) using various modes of transport, (b) goods transport to support the visitors, and (c) transport related to the services provided by the city/municipality.

The visitor's travels were further categorized into travel within the city, and travel from other cities (within the country or abroad). This categorization would help to develop GHG mitigation options at the city level, as the city authorities do not have direct control over the tourists' mode of travel from other cities (within the country or abroad). The GHG emissions were calculated based on the fuel consumption and distance traveled by vehicles between the visitor's cities within the country or abroad.

Goods transport also took into account travel within and outside city boundaries, and the services provided (30 percent) took into consideration the travel made by the personnel of the city authority or municipality for its services to the tourism sector, e.g. waste collection, maintenance of public parks, etc.

The '**buildings and infrastructure**' sub-sector covers all emissions from (a) electricity and fossil fuels used to produce the construction materials used for hotels, restaurants, travel agencies, and other buildings constructed for tourism services, (b) electricity and fossil fuel used to produce materials for infrastructure development, such as roads, parking lots, footpaths, (c) electricity and fuel used in the buildings, and (d) other activities, such as energy (electricity or fuel) used in the production of food for hotels and restaurants, and waste and wastewater generation. Most of these emissions are created by hotels and restaurants. The GHG emissions from wastewater were calculated using standard values on biochemical oxygen demand (BOD)[6] generated by the Thai Pollution Control Department (PCD) (2002) for Chiang Mai, and Hue College of Science (2011) for Hue.

The '**other activities**' category covered GHG emissions from other activities and sources, such as leakage of refrigerants, use of fertilizers and other materials (e.g. paper, glass, plastic, metals, office equipment, materials that make up vehicles, etc.) used by TSPs. Almost all TSPs use air conditioning, refrigerators, and other energy-consuming devices, while fertilizer use is only for agriculture at Hue's 'Garden Houses.'

Results

Both these cities are popular tourist destinations because of their natural beauty and interesting historical sites. The number of domestic and foreign visitors to these cities during 2008–2011 is given in Table 13.2. Chiang Mai had around 3.5 times more visitors as compared to Hue in the year 2011. The majority of visitors stayed on an average for 3.6 and 2 days in Chiang Mai and Hue, respectively (Chiang Mai Provincial Office 2011; Hue Office for Statistics 2011).

Table 13.2 Chiang Mai and Hue: Visitor numbers

No. of Tourists	2008	2009	2010	2011[a]
Chiang Mai				
Thai	3,842,549	3,101,790	3,345,629	3,680,192
Foreigners	1,470,802	1,241,300	1,695,288	1,864,817
Total	5,313,351	4,343,090	5,040,917	5,545,009
Hue City				
Vietnamese	684,714	734,530	844,030	888,900
Foreigners	703,896	561,570	607,600	702,000
Total	1,388,610	1,296,100	1,451,630	1,590,900

Sources: Chiang Mai Provincial Office (2011); Hue City Office for Statistics (2011).
[a] Data for the year 2011 was collected from Chiang Mai municipality and Thua Thien Hue Provincial Department of Culture, Sports and Tourism.

Greenhouse Gas emissions from tourism service providers

The GHG emissions from the TSPs are categorized into three sectors, namely, transport, buildings and infrastructure, and other activities.

Transport sector

1. Visitors

Visitors' travel is separated into two categories to provide insight on direct and indirect GHG emissions at the city level: visitors within the city boundary (factored as direct emissions) and visitors coming from other cities in the same country or abroad (indirect emissions).

Travel by visitors within the city: The GHG emissions are based on the number of visitors (Table 13.2) and days traveling inside the city boundary only. The GHG emissions includes car rentals, travel provided by tourism agencies, and local transport. Chiang Mai city generated 51,070 tons of CO_2 equivalent (tCO_2-eq) GHG whereas Hue emitted around 2,370 tCO_2-eq. Emissions per visitor per day of travel in Chiang Mai is estimated at around 3 kg of CO_2 equivalent ($kgCO_2$-eq). In Hue, it was 1 $kgCO_2$-eq per visitor per day of travel.

Travel by visitors from other cities (within the country) and abroad: international visitors arriving by car, coach, or train within the country were similar to that of travel by domestic visitors. The total GHG emissions from travel by visitors (both local and international) is estimated to be 4,169,300 and 371,000 tCO_2-eq for Chiang Mai and Hue, respectively (Fig. 13.3). Air travel alone contributed 94 percent of each city's emissions. The average emissions by domestic (Thai) and international visitors' travel to Chiang Mai is about 100 $kgCO_2$-eq and 2,000 $kgCO_2$-eq, respectively. Similarly, average emissions by local Vietnamese and international visitors' travel to Hue is 97 $kgCO_2$-eq and 400 $kgCO_2$-eq, respectively.

Fig. 13.3 CO_2 emissions (%) by different modes of visitors' travel to a) Chiang Mai municipality and b) Hue city

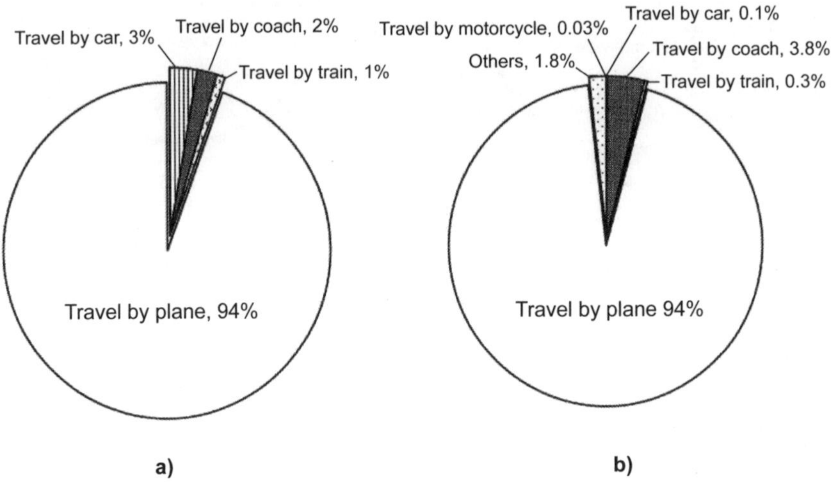

2. Travel by employees

The GHG emissions of employees' traveling between their homes to work, and travel during office hours was also estimated. The home to work (both ways) commute emitted the largest proportion of employees' travel-related emissions. Although the majority of employees used motorcycles to travel to their offices, this contributed to a small portion of the CO_2 emissions in both cities (Fig. 13.4). A few used private vehicles and some also traveled abroad to attend official activities (e.g. meeting travel agencies, tour groups, marketing) particularly in Hue. The total GHG emissions resulting from employees' travel were around 8,360 and 4,320 tCO_2-eq. for Chiang Mai and Hue, respectively, and the per capita emissions from each TSP in each city is shown in Table 13.3.

Fig. 13.4 CO_2 emissions (%) by employees' travel in a) Chiang Mai municipality and b) Hue city

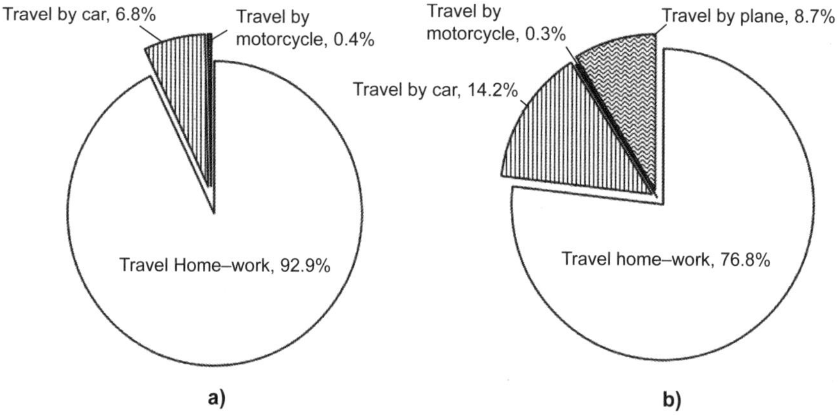

Table 13.3 Per capita emissions of TSP employees in Chiang Mai and Hue

Tourism service provider	Chiang Mai municipality		Hue	
	Employees	Per capita emission ($kgCO_2$-eq)/year	Employees	Per capita emission ($kgCO_2$-eq)/year
Hotels	3,940	1,218	11,545	341
Restaurants	4,213	569	1,195	178
Tours and travel	1,150	814	275	261
Others activities	246	931	244	396

Sources: Chiang Mai Municipality Office (2011); Hue City Office for Statistics (2011).

3. Transport of goods and services (freight)

Some TSPs, such as hotels and restaurants purchased consumable goods and materials from outside the city. The municipality also regularly collects waste and maintains the public parks. The GHG emissions due to goods transport and services provided by the municipality to the tourism sector generated 4,600 tCO_2-eq emissions in Chiang Mai. Hotels alone contributed to 47 percent of total emissions from 'transport goods and services,' while the other two services, namely restaurants and other activities contributed to only 21 percent and 32 percent of emissions, respectively.

The transport of consumable goods to and from hotels and restaurants, regular municipal services, and constructions such the Thuan An resort, Paradise tourist area, Thuy An Ecotourism area, in Hue were used to estimate the emissions related to the transport of goods and services by the city authority. These activities generated around 58,570 tCO_2-eq GHG emissions in Hue. The study shows that hotels alone contributed 61 percent of the total emissions from TSPs, while the other three services, namely restaurants, travel agencies and tours, and other activities, contributed to only 20 percent, 14 percent, and 5 percent of emissions, respectively.

Buildings and infrastructure sector

The buildings of TSPs in Chiang Mai are the largest emitters of GHG with about 48,400 tCO_2-eq. The municipality has also constructed other infrastructure, such as the extension of road networks, footpaths, which contributed to around 5,520 tCO_2-eq emissions.

In Hue, the buildings of TSPs are also the largest source of GHG emissions with about 19,400 tCO_2-eq. In comparison, infrastructure such as roads and footpaths contributed less GHG emissions, only 80 tCO_2-eq/yr.

TSPs also use electricity and fuel (LPG, petrol and diesel), require food, and generate waste and wastewater. These activities cause significant GHG emissions (Fig. 13.5).

Fig. 13.5 Contribution of buildings and infrastructure sub-sector to CO_2 emissions (%) in a) Chiang Mai municipality and b) Hue city

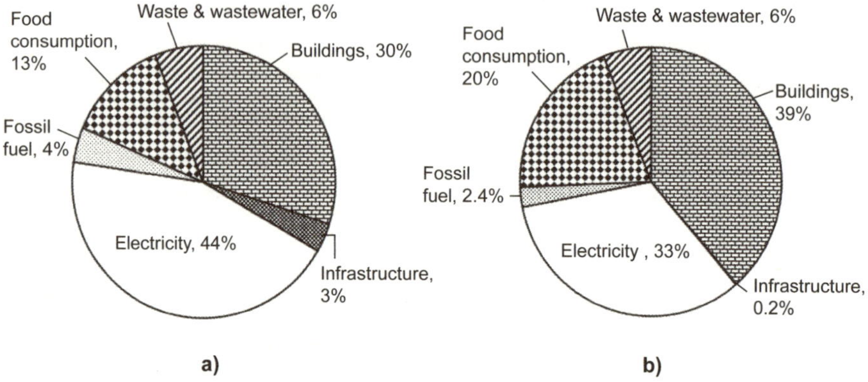

1. Electricity and fuel

Electricity and fuel (LPG, petrol, and diesel) are used in TSPs to run their activities in Chiang Mai and Hue (Fig. 13.5). Electricity is used in almost all TSPs for lighting and air conditioning. The fuels are used mainly for cooking, power generation in case of electricity outages, lawn and tree cutting, etc., at the hotels and restaurants. The GHG emissions from the electricity and fuels used in the TSPs are estimated at about 77,700 tCO_2-eq and about 17,800 tCO_2-eq in Chiang Mai and Hue, respectively. Electricity and fuel consumption per visitor per day are about 4 $kgCO_2$-eq in Chiang Mai and 5.6 $kgCO_2$-eq in Hue.

2. Food production

This accounts for only the emissions during food production processes in farms that consume fossil fuels directly (tractor fuel) or indirectly (manufacture of fertilizers, phytosanitary products, etc.). The emissions from food production depend only on the food consumed by tourists and their duration of stay in the cities (Table 13.2).

Food production emissions are calculated based on the food consumption by visitors, and is estimated at around 20,300 tCO_2-eq emissions in Chiang Mai. The emissions related to food per visitor per day is about 1.2 $kgCO_2$-eq.

Hue generated around 9,920 tCO_2-eq emissions from food production. Emissions related to food per visitor per day is about 3 $kgCO_2$-eq in Hue.

3. Waste and wastewater

Both the cities have a large number of hotels and restaurants to serve visitors, and these hotels and restaurants produce food waste, which also generates large amounts of GHG emissions due to their disposal at landfills. In Chiang Mai, food waste contributed to 83 percent of total waste-related emissions.

In Hue, food wastes contributed to 89 percent of GHG emissions from the total wastes generated in 2011. The GHG emissions from all wastes were 4,860 and 2,170 tCO_2-eq in Chiang Mai and Hue, respectively in 2011.

Besides solid waste generation, hotels and restaurants also use huge amounts of water. Therefore, hotels and restaurants also discharge huge amounts of wastewater that contributes to GHG emissions. The Chiang Mai and Hue TSPs generated around 4,740 and 800 tCO_2-eqs of emissions, respectively, from wastewater in 2011. The GHG emissions related to wastewater per visitor per day in Chiang Mai and Hue were 0.24 $kgCO_2$-eq and 0.25 $kgCO_2$-eq, respectively.

Other activities

Other tourism-related activities which do not come under the above two categories, are discussed below:

Both cities have a hot climate, and air conditioners are used by every TSP, especially hotels and restaurants. The leakage of refrigerants from air conditioning equipment in both cities and the use of nitrogen-containing fertilizers in 'Garden Houses'[7] in Hue were identified as potential/ important GHG emission sources. The common refrigerants used in air conditioning equipment are R22, R134a, and R410a. The GHG emissions by leakage of refrigerants from air conditioning equipment contributes to 5,830 and 2,200 tCO_2-eq/yr in Chiang Mai and Hue, respectively. Only 8 tCO_2-eqs GHG emissions came from the 'Garden Houses' from the use of urea fertilizers.

TSPs utilize various types of materials such as glass, plastic, paper, metals, stationery, IT equipment, vehicles, etc. Most of these items are used in hotels, restaurants, and travel agencies everyday. In Chiang Mai and Hue, these activities emitted around 16,930 tCO_2-eq and 3,690 tCO_2-eq emissions respectively in 2011.

Therefore, the GHG emissions from 'other activities' of the TSPs are estimated to be about 22,760 tCO_2-eq and about 5,900 tCO_2-eq in Chiang

Mai and Hue, respectively. GHG emissions related to 'other activities' per visitor per day were 1 $kgCO_2$-eq in Chiang Mai and 2 $kgCO_2$-eq in Hue.

Discussion

This section analyzes the emissions calculated in the previous section. It also presents the possible opportunities for GHG mitigation and discusses the limitations of the analysis.

GHG emissions by tourism sector

GHG emissions due to tourism in both cities can be summarized as:

- GHG emissions from TSPs in Chiang Mai and Hue were 0.8 and 0.3 tCO_2-eq per capita/yr, respectively.
- The largest proportion of CO_2 emissions from TSPs is, as expected, by the transport sub-sector in 2011 (Fig. 13.6). It contributed to 96 percent of the total emissions in Chiang Mai and to 89 percent in Hue. It is important to note that, out of this, approximately 89 percent and 71 percent of these emissions in Chiang Mai and Hue, respectively, were from air transport alone.
- Emissions from 'buildings and infrastructure' and 'other materials' used were substantially lower than transport emissions (Fig. 13.6).

Fig. 13.6 Contribution of three tourism sub-sectors to total CO_2 emissions (%) in a) Chiang Mai municipality and b) Hue city

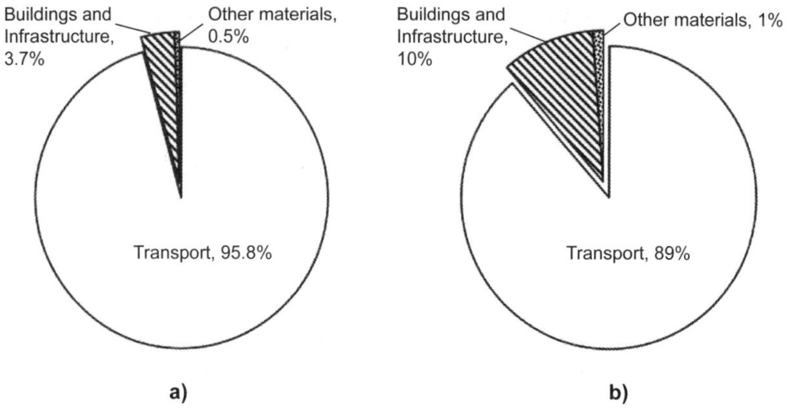

A joint study by the UN World Tourism Organization and the UN Environment Program (UNWTO and UNEP 2008) noted that globally, tourism emitted about 5 percent of total GHG emissions (1,302 $MtCO_2$ equivalent), primarily from transport (75 percent) and accommodation (21 percent, mainly from air-conditioning and heating systems). Air travel alone covers approximately 40 percent of tourist transport emissions. A globally-averaged tourist journey is estimated to generate 250 $kgCO_2$-eq. This study also shows that the tourism GHG emissions in the cities of Chiang Mai (96 percent) and Hue (89 percent) are also clearly dominated by emissions from transport, with emissions from air transport having the major share. The average tourist journey to Chiang Mai and Hue is estimated to generate 800 $kgCO_2$-eq and 300 $kgCO_2$-eq, respectively, and the tourist journey to Chiang Mai causes higher emissions. This is probably due to the establishment of an international airport in Chiang Mai, i.e. 2,000 $kgCO_2$-eq emissions per visitor, as tourists can fly direct to Chiang Mai, and there is no emission sharing with other cities, as in the case of Hue, with its 400 $kgCO_2$-eq emissions per visitor (Table 13.4). In Hue, tourists first have to travel to Ho Chi Minh, Hanoi or Danang, and then travel to Hue. Second, the majority of visitors stayed longer in Chiang Mai (average 3.6 days) as compared to Hue (average 2 days). Chiang Mai received around 3.5 times more visitors as compared to Hue in 2011. Moreover, Chiang Mai has more recreational offerings (spas, massage parlors, the night bazaar/night market, Sunday Walking Street, temples, etc.). Hue has comparatively fewer tourist sites and activities (boating in the Perfume River, temples, handicraft markets, etc.) compared to Chiang Mai.

Yet per capita, GHG emissions from 'buildings and infrastructure', and 'other activities' are higher in Hue than in Chiang Mai (Table 13.4). This is due to the construction of a large number of new hotels, restaurants, as well as more infrastructure in Hue in 2011. It should also be noted that the much higher number of visitors to Chiang Mai reduces the per capita GHG emissions of 'building and infrastructure', and 'other activities' in that city compared to Hue.

Table 13.4 Summary of GHG emissions from TSPs in Chiang Mai and Hue (tCO_2-eq/year)

GHG Emissions								
Transport			Buildings and Infrastructure			Other materials		
Category	Chiang Mai	Hue	Category	Chiang Mai	Hue	Category	Chiang Mai	Hue
Employees	8,362	4,315	Buildings	48,390	19,345	Air conditioning	5,829	2,203
Visitors (local & foreign) travel within the city boundary	51,065	2,368	Infrastructure	5,515	78	Materials	16,927	3,684
Visitors (local & foreign) travel to and from the city boundary	4,169,276	370,934	Electricity & fuel	77,695	17,805	Fertilizer	–	8
			Food	20,256	9,915			
Goods and services	4,601	58,571	Waste	4,854	2,174			
			Wastewater	4,741	800			
Total	4,233,304	4,36,188		161,451	50,117		22,756	5,895
Share	95.8%	89%		3.7%	10%		0.5%	1%
Only air travel (& share)	3,936,530 (89%)	348,951 (71%)						
Total emissions (tCO_2-eq /yr)	**Chiang Mai:** 4,417,511 **Hue:** 492,200							
Emissions per visitor (tCO_2-eq /yr)/visitor/yr)	**Chiang Mai:** 0.8 **Hue:** 0.3							

GHG emissions mitigation options

Earlier studies on GHG emissions of tourism and related activities considered direct GHG emissions from transport, buildings, industry, energy sectors, waste and industrial processes, with little or no mention of indirect GHG emissions (Gillenwater 2008; Schulz 2010; Jun et al. 2011). Similarly, only a few studies dealt with how to reduce GHG emissions in the tourism sector while creating decent jobs for local people (Becken and Patterson 2006; Lebel et al. 2007; Dodman 2011). A comparison of GHG emission distribution for tourism with other cities/countries is difficult because the studies do not consider the same system boundaries, e.g. including emissions of private households, markets, and excluding international flight emissions (Nielsen et al. 2010). For example, at the national level, total GHG emissions from the Swiss tourism sector was estimated to be 2.29 billion kgCO_2-eq (Nielsen et al. 2010). A study by Gossling et al. (2005) showed that the Rocky Mountain National Park in the United States generated a total of 1.1 billion kgCO_2-eq emissions associated with tourist visits. On a smaller scale, Filimonau et al. (2013) estimated the total GHG emissions from a standard holiday package to Algarve, Portugal, to be 627.5 kgCO_2-eq. All these studies indicate that emissions from tourism are the result of large amounts of fossil fuels needed for transport. Therefore, mitigation initiatives in the tourism sector will need to focus on the impact of some particular city or region-specific tourism activities (i.e., particularly those connected with travel, buildings, and infrastructure) if substantial reductions in CO_2 emissions are to be achieved.

The estimated GHG emissions in the tourism sector of Chiang Mai and Hue were discussed by stakeholders[8], and based on these discussions, the feasibility and confirmation of employment-creation possibilities were assessed through local surveys. To ensure that pro-poor and gender concerns were taken as important criteria for selection, three assessment indicators were considered: GHG emission reduction areas; potential employment creation; and stakeholders' interest in pursuing and feasibility of implementation of the strategies/programs. The stakeholders identified measures to reduce emissions by developing 'non-motorized transport' (NMT) in Chiang Mai and promoting 'Garden Houses' in Hue, both of which had the potential to reduce emissions as well as generate jobs, increase income, and provide a much cleaner environment for visitors and locals. It is noted that these measures might not be the best options for reducing GHG emissions per se, but they are decisions by stakeholders to balance both technical and social aspects of the tourism sector.

Non-motorized transport (NMT) in Chiang Mai

NMT is any form of transportation that uses human energy or animal power for personal or goods mobility by methods other than the combustion engine (Adebambo and Yetunde 2010). It includes walking, bicycling, small-wheeled transport, carts, etc., and provides flexible options to mobility. It can also reduce pollution and help to generate income for the poor. The Chiang Mai consultation with stakeholders proposed the Three Kings Monument area near the centre of the Old City in Chiang Mai for developing NMT (around 600 m) that can be connected to the other roads that go around the moat which forms a boundary for the Old City (around 6,000 m). If NMT is promoted in this location, it could replace 535,800–1,339,600 local vehicle (diesel) km travel per year, which could reduce 230–570 tCO_2-eq emissions per year (i.e. reduce 0.6 percent to 1.6 percent of GHG emitted by land-based travel within the city) of the transport sector of Chiang Mai.

For tourists, NMT enhances recreational opportunities by reducing congestion. Facilitating NMT, particularly walking and cycling, will directly reduce the CO_2 which would have otherwise have resulted from their use of other modes of transportation. Cycling and walking are also access modes for public transport and thus their promotion may lead to increased public transport use. Furthermore, promotion of NMT can deliver important co-benefits such as noise and pollution reduction, and improved health (OECD/ITF 2009). On the socio-economic side, the restricted motorized transport will provide opportunities for tourists and others to purchase handicrafts and other goods from local vendors, thereby contributing to the local economy. Moreover, it can provide additional income-generating opportunities for the cyclo drivers, bicycle shops, and related industries. Nevertheless, the city authorities need to develop parking places for motorized and non-motorized transport at different locations to promote NMT and to avoid traffic congestion in the city center.

Garden Houses in Hue

For tourists to Hue, Garden Houses offer serene nature along with a touch of local tradition, and a place to relax within the city limits. More importantly, Garden Houses help to mitigate GHG emissions by acting as carbon sinks as well as lessening the amount of wastes going to landfills by managing household wastes through composting, or as animal feed. At the city scale, such houses can help reduce GHG emissions from freight resulting from importing fruit and vegetables to Hue by producing them for household consumption and sale, and by restricting motorized vehicles around the Garden Houses and promoting cyclo drivers as transport for the visitors.

Thua Thien Hue Provincial Department of Culture, Sports and Tourism estimated that if Hue promotes Garden Houses in the city, it could attract 20–40 percent more visitors. At the same time, if the city authority and Garden House Association encourage visitors to use cyclos or ride bicycles to the Garden Houses, it could replace 127,950–255,900 local vehicle (petrol) km of travel per year, which could help to reduce about 100–200 tCO_2-eq GHG emissions per year (i.e. reduction of 4 percent to 9 percent of GHG emitted by land-based travel within the city) from the city's transport sector. The Garden Houses also help in income generation for the owners and garden workers from the sale of produce such as fruit and vegetables. Besides, restricting motorized vehicles around the Garden Houses could generate additional income for the cyclo drivers and street vendors.

A separate survey conducted in 50 Garden Houses indicates that high development costs (and low returns), low demand, lack of essential skills, and the dominance of mass tourism operators are some of the barriers to promoting Garden Houses as local tourist attractions in Hue. Thus, creating sustainable garden houses requires interventions of a different scale and the involvement of different stakeholders.

Limitations of the study

Only TSPs inside the administrative boundaries of Chiang Mai and Hue were considered for the study. However, many recreational activities aimed at domestic and international visitors such as the Chiang Mai Zoo, shopping malls, night markets, golf courses, temples, karaoke bars, etc., were not included in this study. In the case of Chiang Mai, travel data on goods transport (freight) was not considered for GHG emission analysis. The assumptions are based on discussions with the city authorities. Local emission factors of buildings, infrastructure, waste generation, and some food items were not used, but IPCC (default) emission factors were used for the analysis. The average uncertainties of the results range from +/-18 percent for transport, +/-29 percent for buildings and infrastructure, and +/-37 percent for other materials (see further ADEME 2009).

For the development of recommended strategies, an initial survey was conducted to understand local people's interests in changing their current business or their perceptions about new business opportunities, but it was not possible to estimate the number of jobs that could be created by mitigation projects. Mobilities and businesses newly created through the recommended strategies were only for NMT without actually reducing the use of motorized vehicles as such. A more detailed economic analysis of job creation with detailed employment and income data is needed to determine the extent of newly created jobs and expected job losses.

Conclusion

The tourism sector is one of the most important sectors of the economies of Chiang Mai in Thailand and Hue in Vietnam. Both cities have invested in substantial infrastructural projects to promote tourism. Visitors, local and foreign, travel to these cities because of their good road connections with other big cities, and the establishment of accessible international/domestic airports. The increasing numbers of visitors in both the cities are also generating higher GHG emissions. This study showed that the transport sub-sector generated the largest proportion of CO_2 emissions, equivalent to 96 percent in Chiang Mai and 89 percent in Hue—with 89 percent and 71 percent of these emissions, respectively, attributed to air transport alone. The urban tourism sector in Chiang Mai is more carbon-intensive than Hue mainly because Chiang Mai receives more than thrice the number of visitors as Hue; Chiang Mai also has its own international airport and more recreational activities than Hue. Emissions from buildings and infrastructure and other materials used were low compared to those from transport.

On the basis of this GHG emissions inventory, the cities in consultation with their stakeholders recommended emission reduction strategies, and selected city-specific mitigation options that also support existing government tourism policies. Chiang Mai and Hue have identified strategies to implement NMT and promote 'Garden Houses' respectively. The implementation of these mitigation strategies will not only help to reduce emissions of GHG, but also create more green jobs for local people. The assessment of GHG emissions in this study also enables better understanding of the scope and magnitude of the impact of tourism on climate change for similar cities in the region (e.g. Bali, Phuket, Langkawi) that rely on tourism, and provides a useful and implementable example of developing and designing local strategies for mitigating carbon impacts and increasing employment potential.

Notes

This research for this chapters was funded by SIDA and CDKN through SUMERNET. The authors appreciate the help, support and comments provided by Louis Lebel of USER, Chiang Mai.

1 Tourism service provider (TSP) for this study refers to a group of similar entities. We have used four TSPs for the analysis—hotels, restaurants, tour and travel operators and other activities (including spas, travel agencies, etc.). Entity refers to one (unit) of a hotel, restaurant, travel agency or other institutes/organizations working for the tourism sector in both the cities.

2 Bilan Carbone® is a Microsoft Excel spreadsheet developed by ADEME to calculate GHG emissions. It provides GHG emission results in carbon or carbon dioxide equivalent in kilograms (kgCO_2) or tonnes (tCO_2). See further http://www.terre.tv/?lang=en&vid=1151 or www.ademe.fr.

3 Personal communication with Chiang Mai municipality authorities.

4 Personal communication with Hue city authorities.

5 MONRE (2011); Hue City Office for Statistics (2011); EPPO (2011). The websites consulted were: http://www.1stopchiangmai.com/about_cm/facts/; http://123.242.133.66/tourism/webstorage/download/files/29-20120209045524.pdf; http://thaicarbonlabel.tgo.or.th/filedownlaod/1326646501-12.pdf, .

6 BOD (biochemical oxygen demand) is the amount of dissolved oxygen needed by aerobic organisms in a body of water to break down organic material present in a given water sample at certain temperature over a specific period.

7 Garden Houses are a special style of Vietnamese traditional house. These houses are mostly used as restaurants, and surrounding lands are used for growing vegetables and fruits. See http://www.asianwaytravel.com/blog/hue-garden-houses-model-of-privacy/.

8 Stakeholders refer to representatives from various organizations, such as tour operators, researchers, business associations, media, etc., related to tourism. Some examples of stakeholders in Chiang Mai included the guide association, Provincial Tourism and Sport Office, restaurant clubs, and those in Hue included the Youth Union, Garden House owners, Association of Entrepreneurs, and Hue University.

14

Evaluation of the Impact of Pilot Payments for Forest Environmental Services (PFES) in Lam Dong Province, Vietnam

Bui Duc Tinh, Tran Huu Tuan, and Phong Tran

The forests of Vietnam play a critical role in the economy by providing diverse ecological or environmental services, such as biodiversity conservation, watershed protection, carbon sequestration, and ecotourism necessary for economic development (Chaudhury 2009). There were more than 13.4 million ha of forest in Vietnam in 2008, corresponding to a cover of 38.7 percent. About 34 percent of the population in Vietnam live in mountain/forested areas, most of them from ethnic minority and/or poor communities.

In recent decades, deforestation and forest degradation have occurred throughout the country as a result of illegal logging, poor forestry management, and the continued use of timber and wood for household fuel consumption. More than 1,600 ha of forest is cut down annually in Vietnam (GSO 2009). The most serious deforestation has occurred in the Central Highlands and the Mekong River Delta, where a large proportion of the forests have been cut and the land converted for other uses.

The government has made great efforts to protect and improve Vietnam's diminishing forest resources. There have been several national programs and projects for forest development and protection, such as the replanting of 5 million ha of forest (Program 661), National Program 327 to re-green barren hills, the 'closing of forest gates' policy, control of illegal forest activities, and forest certification. However, these policies have had limited success, and only in some parts of Vietnam, while significant losses of forest cover continue to lead to a loss of biological services (FAO 2007).

In response, in 2008, the government launched a pilot Payment for Forest Environment Service (PFES) scheme in Lam Dong and Son La provinces. Vietnam's Decision 380/QD-TTg is designed to socialize forest protection and development, improve the livelihoods of forest laborers, eradicate hunger, and reduce poverty among the populations of upland forest areas. Decision 380 led to a two-year pilot project in Lam Dong province (2008–10). PFES is considered as a tool for forest protection, carbon sequestration, biodiversity conservation, watershed improvement, local livelihood improvement, and bundled services. Market mechanisms used to protect forest services such as watersheds, biodiversity, and carbon sequestration are still relatively new in the Mekong region, however (Chaudhury 2009). Hence, an evaluation of the performance and impact of the pilot PFES program is very important for learning lessons that can support the adoption of similar schemes throughout the country (and elsewhere in the region). While in general the pilot PFES has had a positive impact on incomes of poor rural households, it may have had some adverse social and economic impacts, and was subject to the same shortcomings of earlier programs. This chapter presents and discusses these positive and negative impacts from the pilot PFES project in Lam Dong province.

Research method

Data collection

A mixture of data collection methods was used in this study, including secondary data collection; key informant interviews (KIIs), focus group discussions (FDGs), and household surveys.

First, information from scientific papers, reports, statistical data, and other published sources related to this study were compiled and reviewed to frame a specific research direction for analyzing PFES in Lam Dong province. This also helped to achieve a better understanding of PFES-like schemes and their implementation as well as their impact on local livelihoods, income, forest protection, ecosystem quality improvement, and carbon sequestration.

Next, secondary data from various sources, including the PFES Fund, Provincial Department of Agriculture and Rural Development (DARD),

Winrock International, hydropower companies, water supply companies, ecotourism operators, and forest management boards to contextualize the study site and relevance of PFES were gathered and analyzed. This included data on the site's socioeconomic and natural conditions, total forest area, area under PFES, number of households involved, the ratio of poor and non-poor households, the ratio of ethnic minority households, PFES organization and implementation, and revenue and costs.

Third, a set of questions was designed for key informant interviews with different stakeholders using forest environment services and officials in charge of PFES implementation in Lam Dong. The 25 key informants were PFES policy implementers, provincial and local officials, PFES Influencers in the Pilot PFES program, members of the PFES Monitoring and Evaluation Boards, and PFES buyers (including water suppliers), hydropower managers, and tourism company staff. Households participating in PFES were also selected. Face to face interviews allowed the team to collect qualitative information on the advantages and difficulties of PFES implementation at different management levels, and the awareness of and responses by PFES buyers and providers.

Fourth, representatives from different stakeholder groups were invited to participate in a series of focus group discussions (FGDs). The agenda included an introduction to explain the discussion purposes and key issues, such as PFES organization, implementation, fee definition and management, monitoring and evaluation, and FES quality evaluation. Eleven FDGs with 116 participants were conducted in Lam Dong province: 1 FDG with provincial stakeholders, 1 FDG with PFES fund managers, 3 FDGs with district stakeholders, 3 FDGs with PFES users, and 3 FDGs with PFES providers.

Finally, a survey was conducted through using questionnaires to collect primary quantitative data on the impact of PFES implementation at the household level and to assess social equity issues. There are two options for collecting such primary data: (1) to sample households 'with and without' PFES participation; (2) to interview the same households at two different times, before and after receiving the PFES. The latter method was selected and applied in this study as it helps to avoid bias due to different demographic characteristics and household resources. This method produced so-called 'panel' data, which is useful for comparable analysis; it also required a smaller budget than the first method as it required a smaller sample size.

Based on the literature review, the questionnaire was developed, pre-tested, and finalized before the survey. Households participating in the PFES program were randomly selected for the survey: 218 questionnaires were completed and cleaned for data entry and analysis.

Analytical framework

The socioeconomic and environmental impacts of the PFES policy on participants were evaluated through quantitative and qualitative data analysis.

Economic impact. The most obvious impact of a PFES program on participants is through the additional income that it provides (Pagiola et al. 2002). The annual household income received from PFES is measured by the total sum of money received for forest protection and management activities: Household income from PFES (VND) = Unit payment (VND/ha) x Forest area managed by household (ha). Additional income generating by employment opportunities and the collection and sale of non-timber forest products (NTFP) such as honey and medicinal herbs were also taken into consideration.

Changes in net profits in pre- and post-PFES periods were calculated as: (Post-income – Post-expense) – (Pre-income – Pre-expenses).

A linear regression model was used to estimate changes in profit during the PFES Pilot study:

Changes in net profit = f (district, ethnicity, education, land area, ratio of labor, ratio of dependent and family labor).

Environmental impact. PFES schemes have proven effective for watershed management, carbon sequestration, biodiversity conservation, maintaining the beauty of landscapes, and bundled services (Mayrand and Paquin 2004). However, these effects were not clearly manifested at the Lam Dong site, hence it was impossible to measure the variables related to the environmental services such as forest cover, biodiversity enrichment, and watershed improvement. It could be that far fewer changes in forest quality can be observed within a mere two-year period. Therefore, we have evaluated the environmental impact by using the following proxy variables:

- *Total labor-days* that households spend per hectare of forest protection or conservation per year before and after participation in the PFES.

- *Numbers of illegal logging cases* per year before and after PFES.
- *Forest areas burnt before and after the application of PFES policy*: this variable implies that forest holders are more responsible for forest fire prevention, thus can improve forest cover area and forest quality. This information was collected from local forest protection agencies and related local staff.
- *Deforestation rate*: this information was collected through household interviews and meetings with local forest managers, participants and non-participants—before and after participation in the PFES scheme—for comparison.
- *Local awareness*: the study evaluates the pilot policy through qualitative indicators such as farmers' and local authorities' knowledge, perception of PFES, training on forest conservation/management, and the awareness of their responsibility for sustainable management and conservation of the forests. These indicators were calculated by using KIIs, FGDs, and household surveys.

Findings and discussions

The study site

Located in the Central Highlands, Lam Dong is the third largest plateau province in the Central Highlands in Vietnam. It is also the highest province lying on a plain of varying elevation, averaging about 1,500 m above sea level. To the north, Lam Dong shares borders with Dak Lak and Dak Nong provinces, to the southeast, with Khanh Hoa, Ninh Thuan, and Binh Thuan provinces, and to the west, with Binh Phuoc and Dong Nai provinces. Lam Dong is considered a forest province with forest coverage accounting for 70 percent of the total area. Given its complex geomorphology and varied elevation, the degree of vegetation cover differs across the province. With 1,179,200 persons living in a total area of 9,776km^2, Lam Dong is the most populated province in the Central Highlands; it comprises 9 districts and 2 cities: Da Lat (the capital), Bao Loc, Lac Duong, Don Duong, Duc Trong, Lam Ha, Bao Lam, Di Linh, Da Huoai, Da The, and Cat Tien.

Pilot payment for forest environmental services

The objectives of Decision 380 were to establish a foundation for the development of a legal framework for a national policy on PFES and to pilot it in Lam Dong province in the South and Son La province in the North. The ultimate objectives are to socialize forest protection and development to improve the livelihoods of forest laborers, eradicate hunger, and reduce poverty for people living in Vietnam's highland and forest regions. This is expected to lay the foundation to enable the government to address other development issues including cultural and social development, and political security in the mountainous areas. Funds for PFES are collected from the fees of VND 20/kw for hydropower, VND 40/m^3 from water supply companies and 1 percent of total revenue generated from a tax on tickets to tourist attractions.

From 2009 to 2010 the pilot PFES program in Lam Dong was implemented successfully in terms of both the number of participants, particularly the poor and ethnic minority groups, and the number and diversity of PFES buyers. A total of 584,396 ha of state-owned forests were allocated to 7,997 households. Of these, 6,328 households belong to ethnic minority groups (more than 79 percent of the total). There is a significant difference in payment rates among catchment areas, for examples, in the Da Nhim and Dai Ninh catchment it is about VND300,000 to 400,000/ha/year while in the Dong Nai catchment it is about VND100,000/ha/year. Participants were told that the PFES payment rate was largely dependent on the number of PFES users in each catchment area. On average, each household participating in the pilot PFES program received about VND10.5 to VND12 million (about US$500–600/household/year at an average exchange rate of US$1: VND20,000 in 2012).

The PFES program has delivered more significant financial benefits than any other program targeting allocation of forest land to households and contracting forest areas for protection such as Government Programs 327 and 661. The viability of the PFES payment mechanism is largely dependent on the size of the forest allocated, the quality of the forest, and the number of PFES users in each forest catchment area. On average, in 2010, PFES providers received about VND300,000 to VND350,000/ha/year in Lam Dong province in comparison with VND100,000/ha/year paid under Program 327 or Program 661. Thus, it can be assumed that the PFES Program has significantly improved the income of these households. This helped to incentivize PFES participants to spend more time and effort on forest protection compared with previous programs.

It is important to note that PFES does not use government budgets for payment—rather, fees are collected from PFES users and paid to households that are contributing to forest protection. The total PFES revenue generated over the two years of the pilot program (2009 and 2010) was VND 98.6 billion (nearly US$5 million).

Fig. 14.1 PFES revenues by its sources from 2009–2010

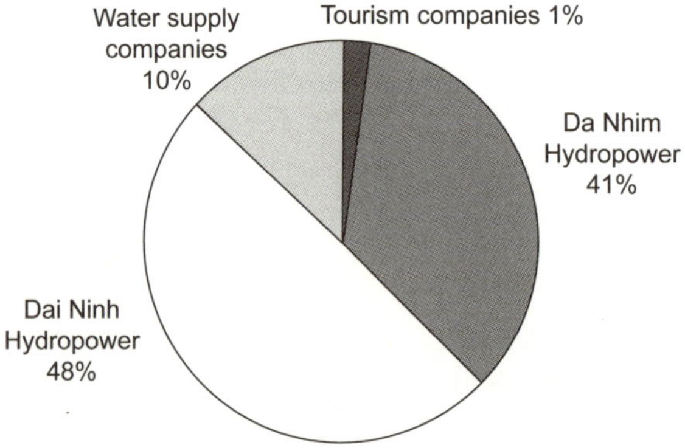

As highlighted in Fig. 14.1, hydropower plants in Lam Dong province (Dai Ninh and Da Nhim) contributed to about 89 percent of total PFES Fund. Water companies such as Saigon Water Corporation (SAWACO) in Ho Chi Minh City and Dong Nai water supply companies paid PFES fees that amounted to about 10 percent of total revenues. Tourism-related companies, with revenue generated from ticketing for forest visits, paid 1 percent of the total FPES fees.

One issue that arose was when key informants stated at the interviews that they were willing to participate in the PFES program, but questioned why fees were only sought from water supply, hydropower, and tourism companies. It was explained to them that the forests thus conserved provide services directly linked and necessary for their businesses, such as biodiversity, carbon sequestration, and environmental conservation. However, this question by the participants indicated that PFES users may not have fully understood the PFES objectives, organization, and implementation mechanisms.

PFES and effects at household level

The PFES scheme contributed to a significant improvement in average household income in the pilot area. Average household revenues in 2010 were VND42.4 million, with a range of values, from VND5 million to VND182 million (Table 14.1). Compared with the pre-PFES period, household incomes as a result of the scheme increased by about 78 percent (or 36 percent per year). This shows the strong economic impact of the PFES program on local households in Lam Dong province. However, it should be noted that this increase had to cover an inflation rate of over 10 percent during an annual economic growth phase of about 7 percent per year in 2008–2010.

Table 14.1 Total income of households in pre-PFES 2008, and post-PFES 2010 (in VND million)

Benefit	Minimum	Maximum	Mean	Std deviation
Total benefit of household gained in 2008	1.0	123.0	24.2	17.1
Total benefit of household gained in 2010	5.0	182.0	42.4	33.5

Source: Household survey in 2011.

Notably, the household income of PFES participants depends on the forest area that they agreed to protect and the payment rate per hectare. With an average of 30 ha of forest allocated to each household for protection, and an average payment rate in 2010 of VND350,000 per ha/year, the average household revenue from PFES was VND10.2 million per year. However, there was a substantial variation in household incomes, ranging from a high value of VND24.8 million/year to a low value of VND5.2 million/year (Table 14.2).

Table 14.2 Average household income from PFES program in 2010

Benefit	Minimum	Maximum	Mean	Std deviation
Ave. forest area allocated to households (ha)	15.0	71.0	29.1	9.9
Ave. household income from PFES program (VND million)	5.2	24.8	10.2	3.5

Source: Household survey in 2011.

Another important point to note is that such household income derived from PFES does not come at zero cost. Costs to households include labor time spent on forest protection, travel, materials, and equipment costs. It is important to estimate these additional costs incurred by households on forest protection to assess whether the payment level and household income from the PFES program provide adequate compensation.

Results show that, on average, each household spent 6.3 labor-days per month for forest protection. The average labor cost in Lam Dong in 2010 was about VND80,000 per day, thus the total cost for labor is about VND6 million/yr. Costs of equipment (farmers need to be equipped with a knife, shoes or boots, a raincoat, flashlight, anti-malarial drugs, etc.) averaged VND 0.6 million/yr. The main cost of materials was for gasoline, as most farmers patrol by motorbike. Gasoline costs averaged about VND15,000/day, so the average costs for gasoline was about VND1.1million/year.

In total, the costs for labor, equipment, and materials is about VND7.7 million per year. It is important to note that farmers use their own household labor for forest protection. Comparing benefits and costs from participation in the PFES program, an average household that received VND10.2 million payment had to spend VND7.7 million/year, so their net benefit was VND2.5 million/yr.

The study also investigated the changes in total profit of households by calculating the difference between Post-PFES benefit in 2010 and Pre-PFES benefit in 2008. On average, there was an increase of VND18.5 million (equivalent US$900) per household.

Changes in household benefits between different ethnic groups also show significant differences. On average, Kinh-headed households gained a substantially higher benefit, about VND34.6 million (US$1,700) compared with about VND12 million (US$600) for households headed by other ethnic groups. Thus, despite the positive effects of the PFES program, it apparently failed in countering social inequities.

Table 14.3 ANOVA of profit changes by ethnicity

Ethnicity of respondents	Mean (million VND)	Std deviation	ANOVA	
			F	Sig.
Kinh	34.6	46.4		
Ethnic minority	12.0	19.6	25.1	1.13E-06
Total	18.5	31.6		

Source: Household survey in 2011.

The study also investigated the profit differentials among four districts in pilot PFES catchments in Lam Dong province and found significant differences, in which households living in the Da Lat catchment gained the most, compared with Duc Trong, Don Duong, and Lac Duong districts. This is partly explained by the fact that a far greater proportion of households in Da Lat district are Kinh, while there is a larger population of non-Kinh in the other districts.

Table 14.4 ANOVA of profit changes by district

Name of district	Mean	Std deviation	ANOVA	
			F	Sig.
Da Lat	33.7	45.8		
Duc Trong	18.4	27.4		
Don Duong	16.4	32.8	4.1	0.008
Lac Duong	11.8	20.9		
Total	18.5	31.6		

Source: Household survey in 2011.

In order to predict the changes in total benefits to households in Pre-PFES and Post-PFES periods with the impact of different predictors, we used a linear regression model because profit changes were normally distributed with a large sample size and an approximately random selection method. We used the functional form:

$$Yi = \beta o + \beta 1X1i + \beta 2\ X2i + \ldots \beta n\ Xni + \varepsilon i$$

Where:

Yi – profit changes of households Pre-PFES and Post-PFES

βi – Coefficient of predictor i

Xi - Predictor i

We then identified the factors that affected profit changes of households, such as characteristics of households and resource availability. The list of predictors is presented in Table 14.5.

Table 14.5 List of predictors used in regression model

Variables	Name of variable	Type of variables	Measure
X1	Ethnicity of respondents	Nominal	1 = Kinh, 2 = Others
X2	District	Nominal	1 – Da Lat, 2 – Duc Trong, 3 – Don Duong, 4 – Lac Duong
X3	No. of years schooling completed	Numeric	No. of years
X4	Distance to PFES forest (km)	Numeric	Km to PFES forest for protection
X5	Total land area	Numeric	Hectares
X6	No. of dependents in household	Numeric	No. of dependents
X7	No. of motors	Numeric	Proxy for poverty by no. of motors
X8	Household credit access	Dummy	Proxy for poverty measure 1 – Yes access 0 – No access

Source: Household survey in 2011.

Table 14.6 Results of linear regression model

Model		Unstandardized coefficients		t	Sig.	Collinearity statistics	
		B	Std Error			Tolerance	VIF
	(Constant)	4.548[a]	12.537	3.63	.0717		
1	Ethnicity of respondents	-7.343[a]	4.285	-1.714	.088	.609	1.643
2	District	0.1280	2.129	.060	.952	.430	2.325
3	No.of years schooling	1.243[b]	.421	2.950	.004	.705	1.418
4	Distance to PFES forest (km)	-.473[c]	.181	-2.609	.010	.888	1.126
5	Total land area	.315[b]	.106	2.973	.003	.897	1.114
6	No. of dependants in household	-2.649	1.880	-1.409	.160	.837	1.195
7	No. of motors	17.731[c]	1.716	10.334	.000	.837	1.194
8	Household credit access	1.578	3.794	.416	.678	.827	1.210

Notes: [a] significant at 10 percent, [b] significant at 5 percent, and [c] significant at 1 percent.
Source: Household survey in 2011.

The results of the regression model highlighted that there were three predictors having negative impacts on profit changes of households, including the ethnicity of respondents; distance to PFES forest; and number of dependants in household (although the impact of number of dependants is not statistically significant) (see Table 14.6). For instance, profit change differences between Kinh-headed and non-Kinh-headed households is -7,343 (significant at alpha 10 percent), which means that Kinh-headed households benefited by VND7.3 million more than non-Kinh-headed households. The analysis on distance to PFES forest, shows that for each increase of 1 km, the profit change of households decreased VND0.473 million. There is a difference, but it is not significant in terms of distance to PFES forests between Kinh and non-Kinh households. The non-Kinh households usually have forests located a little further from their homes than that of the Kinh.

The results from investigating the percentage of households below the poverty line before and after participating in the PFES program according to the national poverty standard issued in 2010 by the Ministry of Labor, Invalids and Social Affairs (i.e., VND300,000/person/month or VND18 million/household/year) showed that 52.7 percent and 26.6 percent of surveyed households were below the poverty line in 2008 and 2010, respectively. After two years of the PFES scheme, there was a reduction in the number of households below the poverty line by about 50 percent, a significant positive effect on poverty reduction.

Table 14.7 Households below the poverty line (%)

	2008	2010
Total no. of households	218	218
No. of households below the poverty line	115	58
Percentage of households below the poverty line	52.7	26.6

Source: Household survey in 2011.

PFES and environmental protection

This study confirmed that the pilot PFES program has generated significant changes in forest management at the household level. Households involved in the PFES program have formed groups to protect the forests allocated to them. Each group operated as a community-based forest management unit. Some groups invested in building a watchtower in order to protect their forests. Households involved in PFES stated that before the program, they annually witnessed many forest fires in their area, but now households involved in PFES spend about 7–10 hours/ per day on their shift on forest protection, including fire spotting and prevention. The number of illegal logging cases has also been reduced significantly in the pilot PFES areas, as stated by FGDs conducted in four districts in 2011.

The study also found that the area of forest invaded/encroached upon for other use has been reduced. Participants at FGDs stated that before PFES implementation households living near forests often encroached forest areas for food cultivation. Some households even used slash and burn techniques. But when households were grouped together to protect the forest, individual households could not conduct such ad hoc cultivation because the group was against such activities. Another positive effect was that there was now regular and consistent forest protection by households and communities which organized 24-hour forest protection shifts.

One incentive was that the income from PFES was more stable than that from illegal forest cultivation. Payment rates largely depend on their achievements in forest protection, however. Households explained that PFES income is reduced if households do not protect their forest well. If there were any forest fires, encroachments or illegal logging in their area, the PFES payment rate was reduced by the forest management board. Hence, the PFES program is considered as an economic instrument for forest management. It has been relatively more effective in getting households involved in forest management compared with earlier government efforts such as Program 327.

Representatives of the PFES Management Board state that the quality of forests has also improved as a result of the PFES program. However, this claim is difficult to evaluate given that the observation period was only over two years. It is unlikely that forest quality improvements can be detected in such a short period, and this report is likely based on wishful thinking. A common concern of PFES users such as SAWACO and Dong Nai water supply companies, Da Nhim and Dai Ninh hydropower plants, and Ho Than Tho Tourism Company was the quality of forest services and responsibility of PFES providers in maintaining it. Their main concerns were how the water quality of their catchment area was to be managed and evaluated, when water quality may be annually affected by many factors, both human-made and natural. These users asked who would take responsibility if water quality worsened. PFES users in this study strongly recommended that it was important to monitor and evaluate PFES services and define the rights and tasks of PFES providers that would maintain the sustainability of the program. Put differently, PFES users are willing to participate in the program as they see it as being good for forest protection and the environment. But users would be more willing to accept the fees if the program's design recognizes the rights and responsibilities of both users and providers. At present, the Pilot PFES program has not designed a monitoring and evaluation mechanism for providers.

Notably, one of the most important contributions of the scheme has been raising stakeholders' awareness of the need for environmental and forest management. The study finds that there has been a significant change in understanding forest and environmental issues by both PFES users and providers (water companies, hydropower plants, tourism companies, and households). All have recognized the importance of forests and their positive correlation with their livelihoods and businesses. Some stated that now that they are aware of how forests were important for their business and their lives, they were willing to contribute to forest development and protection.

However, the study also found that, there are still many challenges to further and larger-scale PFES implementation. The main issue is that PFES users have still not clearly understood the framework of the program, which constrains their willingness to participate in it.

Users strongly recommended that it is important for the PFES program to establish a committee to monitor/control the quality of the forest and water in the catchment area under Pilot PFES and define rights and responsibility of the service providers. PFES providers (households) living in different sub-catchments strongly recommended that it is important to design a mechanism in which the PFES payment rate is defined based on the quality of forests.

This study highlights the fact that the scheme in Lam Dong province was not a full-fledged PFES scheme—i.e. one where there is a voluntary, conditional agreement between at least one 'seller' and at least one 'buyer' over a well-defined environmental service (Wunder 2007). The PFES users in Lam Dong defined in Decision 380/TTg were not voluntary participants. Nevertheless it is important to recognize signs of success from the Pilot PFES program in Lam Dong, such as its positive effects on local livelihoods and poverty reduction. PFES has also contributed to raising awareness about environmental issues and the important role of forests for different stakeholders. Additionally, PFES significantly contributed to forest services improvement and forest protection. This is in line with Mills and Porras (2002) and Richards and Jankins (2007), who concluded that payments for watershed environmental services can be for water quality, flood prevention, water flow regulation, soil salinization control, aquatic habitats maintenance, and local livelihood improvement.

Conclusions and policy implications for PFES

The analysis above shows clear evidence regarding PFES and its impacts, that despite the variance in payments to households in different catchments, PFES has significantly benefited participating households. Income derived from PFES accounted for about 32 percent of household income in 2010; the program reduced the number of households below the poverty line by about 50 percent. However, Kinh-headed households have benefited disproportionately, meaning that the pilot program exacerbated social inequities between Kinh and non-Kinh minorities. It should also be recognized that a full environmental impact assessment of PFES after two years of pilot implementation is impossible as it is hard to measure small improvements in forest and water quality. However, our study affirms that the adoption of PFES in Lam Dong province has shown strong positive effects on environmental services and forest quality, particularly in terms of raising different stakeholders' awareness about current environmental issues and the important role of forests in the economy.

15

Integrating Community-based Participatory Carbon Measurement and Monitoring with Satellite Remote Sensing and GIS in REDD+ MRV systems

Jay H. Samek, Usa Kinhom, David L. Skole, Pornchai Uttaruk, Teerawong Laosuwan, Phung Van Khoa, Sithong Thongmanivong, Chetphong Butthep, Do Xuan Lan, and Nguyen Xuan Giap

Forests play a critical role in mitigating climate change though the sequestration and storage of carbon in perennial woody biomass and soils. Globally, about one-third of anthropogenic emissions of atmospheric CO_2 since 1750 were from land use changes, primarily from deforestation; this figure is about 20 percent from land use changes for the 1990s (IPCC 2007). Recent estimates derived from satellite remote sensing conclude that gross loss of forests accounted for 7 percent to 14 percent of total global anthropogenic CO_2 emissions between 2000 to 2005, with forest loss in South and Southeast Asia accounting for 32 percent of the total deforestation emissions (Latin America and sub-Saharan Africa accounted for 54 percent and 14 percent, respectively) (Harris et al. 2012). It should be noted, though, that these estimates are not without controversy (Zarin 2012).

Climate mitigation opportunities in biotic systems that actively sequester atmospheric CO_2 include afforestation/reforestation (A/R), agroforestry, natural and assisted-natural regeneration (e.g. abandoned fallow), and improved forest management practices (e.g. reduced impact logging and longer rotation periods). Reducing Emissions from Deforestation and Forest Degradation (REDD) also mitigates climate change by changing the amount of CO_2 emissions to the atmosphere, over time, from the existing forest carbon stocks. REDD-Plus (REDD+) twins emission reductions with carbon sequestration activities that include social and ecological co-benefits (Angelsen et al. 2009).

Implementing forest carbon emission reduction and sequestration projects must show that carbon mitigation is real and permanent and it must do so in a cost-effective manner. Hence, accurate measurement and monitoring of carbon stock changes in biomass must use robust, scientific methods. Reporting and verification procedures must follow accepted protocols. Combined, these form the elements of a Measurement, Reporting and Verification (MRV) system. The IPCC's *Guidelines for National Greenhouse Gas Inventories Volume 4 Agriculture, Forestry and Other Land Use* (AFOLU; IPCC 2006) and the preceding *Good Practice Guidance for Land Use, Land-Use Change and Forestry* (GPG-LULUCF) (IPCC 2003) provide important basic methods for measuring biomass and carbon stock changes.

Further, the *GOFC-GOLD REDD Sourcebook* (GOFC-GOLD 2010) gives guidance for implementing measurement and monitoring protocols for REDD implementation, and includes the use of remote sensing satellite data and GIS tools. To generate accurate estimates of carbon stocks and carbon stock changes, remote sensing tools and techniques for measuring forest carbon need to be integrated with ground-based forest and biomass inventories. Where available, National Forest Inventory (NFI) data can be used. In fact such data should be used for national reporting or for national REDD+ implementation. NFI data, however, do not always exist and may be incomplete for a specific project area, for example, NFI data are often collected in a specific national sampling design that may not match the project-level stratification of the landscape.

For project level REDD+ implementation, even if NFI data exists, including community-based measurement and monitoring will enable and can facilitate community co-benefits, such as training and knowledge transfer in the valuation of forest ecosystem benefits and forest management skills. Supplementing NFI data with community-based measurements should become routine for REDD+ project implementation. While there may be some doubts about the accuracy of community-based measurements, Mukama et al. (2012) report on the accuracy of community data collection for forest carbon in three Tanzanian villages. Their study shows that 90, 75, and 85 percent of participants, respectively (for each village), were able to accurately identify vegetation type; 85, 75, and 8 percent, respectively, were able to take accurate basal area measurements using a relascope; 100 percent in each community were able to establish a permanent plot on the ground; and 95, 85, and 95 percent were able to measure a tree's DBH (Diameter at Breast Height, in cm) using a caliper. Furthermore, with a robust training and capacity-building program, strong linkages between communities and responsible government agencies and/or NGOs, and a protocol that includes the verification of some subset of the community-based measurements, the risk of inaccurate measurements can be mitigated.

While there are basic protocols and guides for measuring biotic carbon stocks (IPCC 2003; IPCC 2006), for using satellite remote sensing and GIS tools and technologies for monitoring changes in forest cover and carbon (GOFC-GOLD 2010), and for community-based measuring and monitoring of forest carbon stocks (KTGAL 2009; Skutsch et. al 2009) *none of these integrate all for use in a Measurement, Reporting and Verification (MRV) system.*[1] We demonstrate this capability and a set of online tools to support the integration of using community-based measurements with remote sensing and GIS for reporting biotic carbon in potential REDD+ implementation landscape projects. Many governments are undertaking the development of measurement systems for national REDD+ implementation and our findings could perhaps facilitate their efforts.

Governments implementing REDD+ activities are aware that the inclusion of local communities is essential. Community involvement in measuring and monitoring forest carbon can empower them as co-managers of REDD+ areas, together with local government agencies. Our research team has piloted an approach in three project areas, one each in Lao PDR (Laos), Thailand, and Vietnam, to integrate community-based participatory carbon measurement and monitoring with satellite remote sensing and GIS to support a REDD+ MRV system. In this chapter we report the findings of the three pilot activities and demonstrate the method of integrating community-based measurements with satellite remote sensing analyses. While there are common elements to engaging communities in REDD+ forest measurements there is no 'one-size-fits-all' recipe. We report a number of key elements for working with communities in REDD+ interventions that we believe will prove useful to others, including policymakers, as they continue to develop REDD+ in their respective countries.

Methods

The project methods followed a series of tasks in each of the three country case study areas which provide comparative analysis and insight into our research focus, that is, the integration of community-based carbon measurements with remote sensing and GIS for a REDD+ MRV system. The tasks included the following activities: (1) identification of the case study areas in Laos, Thailand, and Vietnam; (2) establishing project linkages to boundary partners in each of the three countries; (3) developing a survey instrument; (4) conducting survey questionnaires in the communities; (5) training and capacity-building with communities and boundary partners; biomass data collection by the communities; (6) satellite remote sensing and GIS data processing; (7) online Forest Carbon MRV development and data ingest to the MRV system; and (8) outreach.

The research focused on local communities in the three case study areas, one in each of the three participating Mekong region countries.[2] These communities are:

1. Laos: Ban Kouy, Ban Napor and Ban Vangma villages, Sangthong district, Vientiane prefecture. This area is located approximately 80 km northeast of Vientiane. The villages are located in and around the Faculty of Forestry, Training Model Forest (TMF), National University of Laos (NUOL), which covers an area of 4,600 ha. Land tenure and resource rights are established under the Lao government's Land Use Planning and Land Allocation programs, which began in the 1990s. These two programs are designed to transfer rights and responsibilities over land and forest resources use and management to villages and local communities.

2. In Thailand we are working with 31 villages in Mahasarakham province. These villages are located in Borabue, Na Chueak, and WapiPathum districts and surround the Kok Pak Kud–Pong Dang Forest, a conservation forest of 336 ha under the Royal Forest Department but co-managed by communities under limited use guidelines. In addition to this large tract of forest, the same communities are involved in managing 20 additional forest parcels,

which include sacred or traditional conservation forests (Pa Don Pu Ta) and public, community forests.

3. Vietnam: Na Muc village, Van Minh commune and Tu Dooc village, Lang San Commune, Na Ri district, Bac Kan province. These villages are located close to Kim Hy Forest Reserve, an 18,555 ha protected area established in 1997. Tu Dooc village is in the Reserves' official buffer zone. Both communities have 'Red Book' certificates (tenure rights) for managing community forests: Na Muc village—118.3 ha and Tu Dooc village—45.1 ha. The area is located about 200 km north of Hanoi.

The project leaders (Faculty of Forestry, NUOL; Mahasarakham University and the National Research Council of Thailand; and Vietnam Forestry University, VFU) established project linkages with the following boundary partners: the Department of Forestry under the Ministry of Agriculture and Forestry (Laos); the Ministry of Natural Resources and Environment, and the Agricultural and Land Reform Office (Thailand); and the Forest Science Institute of Vietnam (FSIV), Vietnam Administration of Forestry, Ministry of Agriculture and Rural Development, and ICRAF-VN (national office in Vietnam of World Agroforestry Centre, an international agricultural research center). The partnerships with local level agencies working with communities were integral to the success of the project. Agency personnel provided direct linkages to local communities without which it would have been very difficult to implement any of the project activities.

The project team developed a survey instrument designed to assess a number of elements at the community level: awareness and understanding of climate change; knowledge of climate change impacts; uses and benefits of a forest; knowledge and practice of forest management (both traditional/customary practices and governmental/civil or common law regulations); capacity to take tree measurements; knowledge of carbon credits; and knowledge of climate change mitigation in forests. The surveys were initially composed in English and then translated into Thai, Lao, and Vietnamese. Surveys were conducted first in Thailand (September 2011), then in Vietnam (March 2012), and Laos (November 2012).

Communities participated in training for establishing biomass plots for measuring carbon stock in forested areas (see Table 15.1). The training also covered the following topics: introduction to the SUMERNET project, climate change, and climate change impacts; the role of forests and trees in mitigating climate change; the role of remote sensing in forest mapping and carbon monitoring; and the potential and tangible benefits of forests to local people. Trainings were conducted in the three case study areas together with the local-level boundary partners (Thailand in January 2012, Vietnam in March 2012, Laos in May 2012). Following training, communities collected data in the forest areas. In Vietnam, community members partnered with ICRAF-VN, VFU, and FSIV personnel in establishing the plots and recording the data. In Thailand, community leaders assigned individuals from the communities to organize the data collection in the 21 tracts of forest with 56 field plots. In addition, a number of schoolchildren participated in the training and data collection. Data collection by the communities in Laos was completed in December 2012. For the analysis, however, we use data from the sample plots established in Sangthong district in 2011 by the Faculty of Forestry staff at NUOL. Expert foresters revisited a number of community-based biomass plots to validate the accuracy of their measurements.

Table 15.1 Participants at community training

	Laos	Thailand	Vietnam
Dates of training	Jan & May 2012	Jan 2012	Mar 2012
Training organizations	FoF, NUOL	ALRO, NRCT, MNRE, and Mahasarakham	FSIV and VFU
No. of participants	30 adults	86 adults, 28 students	27 adults
Origin	3 villages, 1 district	31 villages, 3 districts	2 villages, 1 district

There are two steps in processing the community-collected field inventory data from the plots. The first step is ingesting the data into the online Forest Carbon MRV database, which calculates the forest carbon stock at the plot, parcel, and project area levels. The second step is to develop landscape-level carbon maps by integrating the field plot measurements with satellite remote sensing data. Figure 15.1 outlines the processing method.

Fig. 15.1 Field-level and remote sensing data processing method

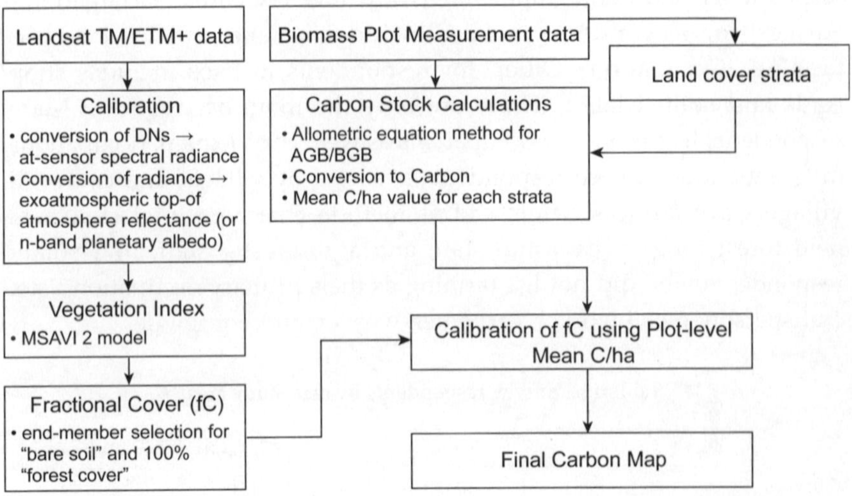

Outreach has been an ongoing part of the overall project design. This includes meetings and workshops with local communities, project boundary partners, and participation in workshops and meetings in the region. The project has also developed and maintained two websites: one in English: http://www.goes.msu.edu/sumernet/ and one in Thai http://www.science.msu.ac.th/sumernet. The online Forest Carbon MRV system itself is an outreach platform.

Results

The project results are described in detail below in the following sub-sections: (1) community questionnaire surveys, (2) carbon stock assessments and mapping results, and (3) outreach. Surveys and training as well as the carbon assessments and mapping are complete. Outreach is ongoing although the project has been completed.

Surveys

The survey results provide a certain level of measurement to assess the knowledge base of the local communities in the project areas. Here we report the summary of the results, highlighting the key elements of community knowledge in the three areas.

Table 15.2 shows the number of surveys distributed, number of respondents, and demographic breakdown for each area. 'Occupation—Farmer' for respondents in Laos includes upland and lowland rice farming. 'Occupation—Other' for respondents in Laos includes shop/trade and skilled labor. These are only the main occupations. Many respondents list secondary occupations as well. For Vietnam, occupational differentiation between respondents is sorted by villager and cadre. All villagers are farmers, while cadres include chairmen, vice-chairmen, field forest rangers, cadastral staff, and agroforestry staff. In Thailand respondents who did not list farming as their primary occupation listed craftsperson, metal worker, carpenter, weaver, and housewife.

Table 15.2 Survey respondents by case study area

	Laos	Thailand	Vietnam
Surveys	190	121	49
Male	133	97	37
Female	57	24	12
Occupation: Farmer	171	115	37
Occupation: Other	19	6	12

One aim of the survey was to measure each community's knowledge of climate change, and their understanding of climate change impacts (see Fig. 15.2). The questionnaire did not ask directly, 'Do you understand climate change?' but was designed to discuss climate change with a series of leading and investigative questions. Through the discussion the interviewer assessed the interviewee's knowledge base on a 'yes', 'no' and 'some' three-level scale; and for the Thai survey those who were classified as 'yes' then were categorized further with basic, moderate, and advanced knowledge classes. A large majority of the respondents from Thailand and Vietnam showed they had good knowledge of climate change (56 percent and 85 percent respectively). In Laos only 30 percent showed a good knowledge of climate change. One explanation for the high percentage in Vietnam could be a result of the work that ICRAF-VN has been doing in the two villages for a number of years now, which is not the case in the other two pilot sites. However, we have no way of confirming this conclusion, beyond it being a reasonable assumption. The disparity in knowledge between Laos and Thailand parallels some of the basic development trends of these two countries with respect to basic infrastructure, education, and economic progress. But even in Thailand of those who indicated that they have knowledge of climate change 73 percent (n=50) indicate only a basic understanding, 19 percent (n=13) moderate understanding, and 8 percent advanced understanding (n=5) (See Fig. 15.3).

Fig. 15.2 Survey respondents' knowledge of climate change

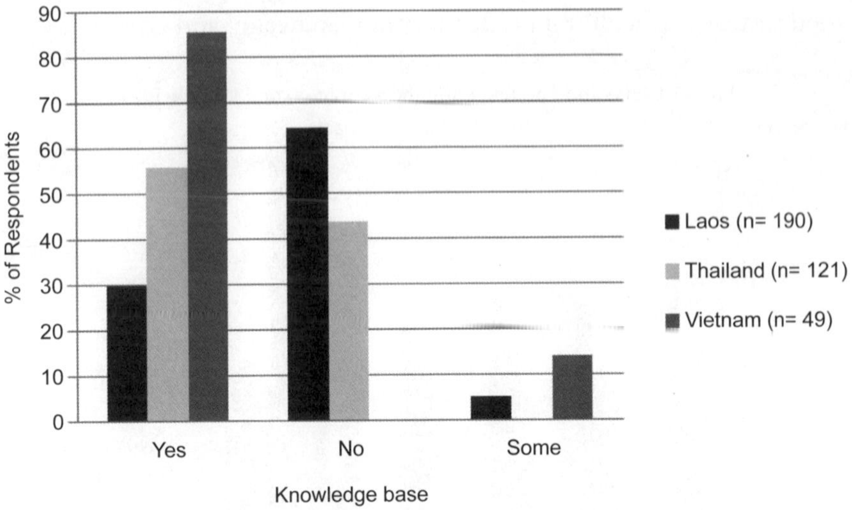

Fig. 15.3 Level of understanding about climate change—Thai respondents

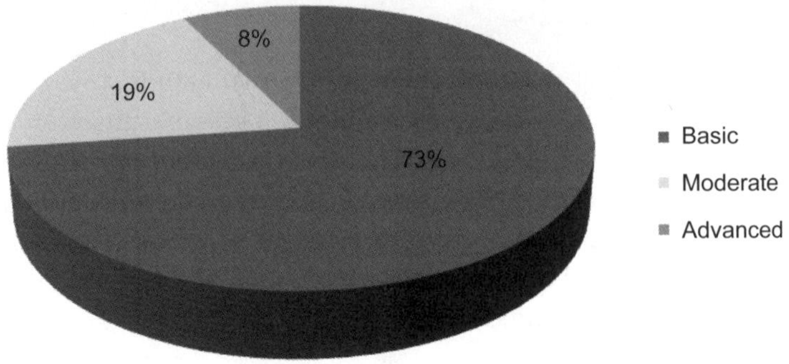

The survey also asked respondents if they were aware of or had knowledge of the impacts of climate change. The results of the survey for the 121 Thai respondents and the 190 Lao respondents are presented in Fig. 15.4. There were a greater percentage of respondents with no awareness of climate change impacts across the six themes in Laos. However, in all areas there were some Lao respondents who did have knowledge of such impacts. In the Thai case study area we grouped respondents who indicated 'no' with the respondents who did not answer either 'yes' or 'no'. A greater percentage of Thai respondents indicated no awareness about the impacts of climate change on non-timber forest products (NTFPs) and society. However, a greater percentage of Thai respondents were knowledgeable about the impact of climate change on food security, agricultural production, human health, and economics.

Fig. 15.4 Thai and Lao respondents' awareness and knowledge of climate change impacts

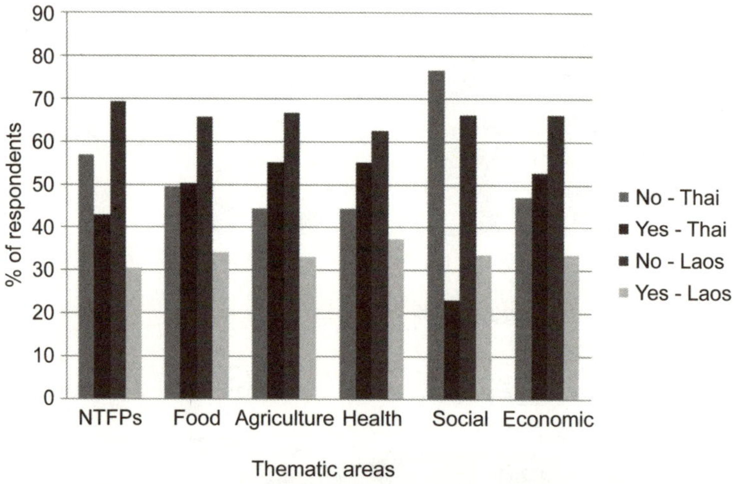

A larger portion of respondents in the Vietnam study area indicated they had little or no clear understanding of the human causes of climate change, the role of deforestation and forest degradation, REDD+, the 'carbon market', and climate change impacts on food security, or on health problems. A greater percentage of respondents, however, acknowledged some understanding of climate change in general and also that there are broad impacts from climate change.

Carbon stock and mapping

Following the survey and training, communities established biomass plots and recorded tree biometric parameters used to calculate carbon stocks.

In the Thai study area 56 sample plots were established in 21 different forest areas and measurements were taken on 1,733 individual trees. The communities organized themselves in field crew teams to collect the data. Team members from Mahasarakham University, who are experts in forest measurement, were able to accompany community members to the field and assist in the data collection. The project site includes two types of forest areas: the first is called 'Pa Don Pu Ta' which is a Sacred Forest or Traditional Conservation area; the second are 'Public' or 'Community' Forest areas. Table 15.3 shows the nine Community Forest areas and the twelve Sacred/Traditional Conservation Forest areas, their size (in rai and ha), and the number of plots established.

Table 15.3 Forest areas for data collection in Mahasarakham, Thailand

Forest Type	Village or Forest Names	Area (rai)	Area (ha)	No. of sample plots
Pa Don Pu Ta (Sacred forests or Traditional conservation areas)	Ban Nong Kham Moo 5,6,7,10 Ban Sabang Moo 8	20	3.2	3
	Ban Kok Sung, Moo 5	8	1.28	2
	Ban Nam Kleang, Moo 8 Ban Weang Chai, Moo 11	33	5.28	5
	Ban Nong Poa, Moo 2 Ban Na Nuea, Moo 11 Ban Na Klang, Moo 12	0.25	0.04	1
	Ban Nong Ruea, Moo 4	0.25	0.04	1
	Ban Pa Rang Naa, Moo 6	0.25	0.04	1
	Ban Pa Rang Naa, Moo 6 Ban Talatmung Moo 1 Ban Talatnoua Moo 13	11	1.76	2
	Ban Sala Moo 8 Ban Sala, Moo 20	5	0.8	2
	Ban Non Thong, Moo 9 Ban Non Tham, Moo 17 Ban Non Sa ard, Moo 13	4	0.64	2
	Ban Pa Kung Hna, Moo 10 Ban Na Charoen, Moo 18	1	0.16	2
	Ban Huai Loa, Moo 7 Ban Huai Lao Nuea, Moo 15 Ban Huai Mai, Moo 19	1	0.16	1
	Ban Non Chaisri, Moo 17	2	0.32	2

Forest Type	Village or Forest Names	Area (rai)	Area (ha)	No. of sample plots
Community or Public Forests	Pa Ba Kan Tan	131	20.96	2
	Pa Kok Hua Lon	16	2.56	1
	Pa Cha Ban Non Thong and Ban Sala – Cemetery Forest	47	7.52	1
	Pa Kok Khum Moung	200	32	2
	Pa Khum Pak Kud	160	25.6	2
	Pa Kok Hin Kong	87.84	14.054	1
	Pa Kok Kalong Pong Dang	2197	351.52	21
	Pa Soon Patibuttham	12	1.92	1
	Pa Ban Na Klang	28.7	4.592	1
Total		2965.29	474.446	56

Note: "Rai" is the unit of area measurement in Thailand for land. One rai is equal to 6.25 ha.

In the Vietnam study area 40 sample plots were established in forest parcels owned and managed by 25 different community members. An FSIV forester from the team accompanied community members to the plots and supervised the data collection. Five community members from the two villages conducted all of the biomass measurements with some oversight by a project member from FSIV. Thirty-two plots were established in evergreen broadleaf forest areas, six plots in *Mangletia glauca* plantations and two plots in *Acacia mangium* plantations. Not all of the plots were located inside the community forest boundary. A total of 1,486 trees in the evergreen broadleaf forest areas, 190 trees in *Mangletia glauca* plantation plots and 51 trees in *Acacia mangium* plantation plots were measured (total all trees = 1,727). We use only the 1,486 trees in the 32 evergreen broadleaf forest area plots to estimate the carbon in the two community forest areas (Na Muc = 118 ha and To Doc = 45 ha). The data collected in the plantation areas do not match the forest type in the community forest area and therefore were excluded from the carbon assessment. Table 15.4 shows the forest owners, area in ha and number of plots established; some own more than one parcel of forest.

Table 15.4 Forest areas for data collection in Na Ri, Vietnam

Forest owner	Area (ha)	No. of plots
Ban Thi An	3.37	1
Cong dong thon ban	119.5	12
Hoang Van Chang	4.79	1
Hoang Van Sang	0.92	1
Hoang Van Soong	2.53	1
Hua Van Cam	10.82	1
Hua van Soong	3.48	1
Luc Quan Luc	2.83	1
Luc Van Bang	1.28	1
Luc Van Bay	13.54	1
Luc Van Cao	2.76	1
Luc Van Dung	0.67	1
Luc Van Hau	5.53	1
Luc van Hoai	0.36	1
Luc Van Hue	4.21	3
Luc Van Hung	1.03	1
Luc Van Huy	3.26	1
Luc Van La	5.85	3
Luc Van Liem	1.82	1
Luc Van Luyen	3.1	1
Luc Van Ninh	2.35	1
Luc Van Son	4.74	1
Luc Van Than	3.51	1
Tran Van Sinh	3.51	1
Name not recorded	3.5	1
TOTAL	119.5	40

Data collected by the Faculty of Forestry staff at NUOL in Sangthong district in 2011 were used in the assessment. The data set consists of 23 nested plots with data collected in 25-m and 10-m radius plots. A total of 711 trees were measured: 332 small-diameter trees (5–20 cm DBH) in the 10-m radius plots and 379 large-diameter trees (>20 cm DBH) in the 25-m radius plots. All of these plots were established in the mixed deciduous forest areas in around Sangthong district.

Because the community groups in both Thailand and Vietnam had supervision from forestry experts in establishing the plots and recording the field data, no assessment of accuracy was felt to be necessary. In Laos, forestry experts from the Faculty of Forestry and NUOL took the measurements used in our analyses.

We mapped the plot data in a GIS using the GPS points taken in the field together with other geographic information and attribute data (see http://www.goes.msu.edu/sumernet/docs/Sumernet_REDDProject_Final_Report.pdf.)

Carbon stock data was calculated using the online Forest MRV tool at http://mrv.carbon2markets.org. This is a secure Internet-enabled project management system with a suite of tools for organizing project documents, spatial data, and tree level inventory data. The system has a carbon stock calculator tool that can estimate project and parcel (or strata) level carbon stocks from plot level field inventory data using allometric equations. These carbon stock estimates are Tier 3 (in IPCC terminology) carbon values. We use the Tropical Moist Forest allometric equation (Eq. 1) developed by Brown (1997) and a default root to shoot ratio of .20 for below-ground biomass (carbon) calculations.

Tropical Moist Forest Allometric Equation Eq. 1

$AGB = 42.69 - 12.8(DBH) + 1.242(DBH^2)$

Where:

AGB = Above Ground Biomass in kilograms of dry matter (kg DM), and

DBH = Diameter at Breast Height in centimeters (cm)

Table 15.5 Carbon stock estimates for pilot area in Na Ri district, Bac Kan province, Vietnam (above) and for pilot area Sangthong district, Vientiane prefecture, Lao PDR (below)

CARBON STOCKS BY PARCEL												
Parcel Descriptors		Carbon Density					Carbon Stocks					
Values below in ha		Values below in tC/ha					Values below in tC					
ID	Area	AGB	BGB	SOC	Litter	Deadwood	AGB	BGB	SOC	LITTER	Deadwood	Total
Na Muc Community Forest	118.00	728.7	145.7	0.0	0.0	0.0	85,986	17,197	0	0	0	103,183
To Doc Community Forest	45.00	826.4	165.3	0.0	0.0	0.0	37,190	7,438	0	0	0	44,628
Project Totals	163.00						123,176	24,635	0	0	0	147,811
Deciduous Forest	3,628.0	93.3	18.7	0.0	0.0	0.0	338,574	67,715	0	0	0	406,289
Project Totals	3,628.0						338,574	67,715	0	0	0	406,289

CARBON STOCKS BY PARCEL												
Parcel Descriptors		Carbon Density					Carbon Stocks					
Values below in ha		Values below in tC/ha					Values below in tC					
ID	Area	AGB	BGB	SOC	Litter	Deadwood	AGB	BGB	SOC	LITTER	Deadwood	Total
R-MA-11	0.17	226.0	45.2	0.0	0.0	0.0	38	8	0	0	0	46
R-MA-10	0.55	481.4	96.3	0.0	0.0	0.0	265	53	0	0	0	318
R-MA-12a	1.28	108.5	21.7	0.0	0.0	0.0	139	28	0	0	0	167
R-MA-12b	1.58	112.7	22.5	0.0	0.0	0.0	178	36	0	0	0	214
R-MA-7	0.62	118.7	23.7	0.0	0.0	0.0	74	15	0	0	0	88
R-MA-9	0.66	51.1	10.2	0.0	0.0	0.0	34	7	0	0	0	40
R-MA-8	1.53	206.6	41.3	0.0	0.0	0.0	316	63	0	0	0	379
R-MA-16	4.41	158.1	31.6	0.0	0.0	0.0	697	139	0	0	0	837
R-MA-17	1.07	158.9	31.8	0.0	0.0	0.0	170	34	0	0	0	204
R-MA-19	0.10	99.7	19.9	0.0	0.0	0.0	10	2	0	0	0	12
R-MA-18	3.99	198.5	39.7	0.0	0.0	0.0	792	158	0	0	0	950
R-MA-6	0.22	76.2	15.2	0.0	0.0	0.0	17	3	0	0	0	20
R-MA-15	14.13	40.3	8.1	0.0	0.0	0.0	569	114	0	0	0	684
R-MA-20	3.06	98.7	19.7	0.0	0.0	0.0	302	60	0	0	0	362
R-MA-5	8.11	36.2	7.2	0.0	0.0	0.0	294	58	0	0	0	352
R-MA-13	32.92	56.4	11.3	0.0	0.0	0.0	1,857	372	0	0	0	2,229
R-MA-21	212.94	41.3	8.3	0.0	0.0	0.0	8,794	1,767	0	0	0	10,562

CARBON STOCKS BY PARCEL												
Parcel Descriptors *Values below in ha*		Carbon Density *Values below in tC/ha*					Carbon Stocks *Values below in tC*					
ID	Area	AGB	BGB	SOC	Litter	Deadwood	AGB	BGB	SOC	LITTER	Deadwood	Total
R-MA-2	3.81	57.5	11.5	0.0	0.0	0.0	219	44	0	0	0	263
R-MA-3	5.20	68.9	13.8	0.0	0.0	0.0	358	72	0	0	0	430
R-MA-14	34.04	72.1	14.4	0.0	0.0	0.0	2,454	490	0	0	0	2,944
R-MA-1	20.25	52.7	10.5	0.0	0.0	0.0	1,067	213	0	0	0	1,280
R-MA-4	4.62	30.2	6.0	0.0	0.0	0.0	140	28	0	0	0	167
Project Totals	355.26						18,784	3,764				22,548

The community-based field plot measurement data is integrated with the remote sensing. The method uses medium resolution (30 m) multi-spectral Landsat TM and ETM+ satellite data. These are raster data. These data are calibrated by converting the processed data digital numbers or DNs to the at-sensor-radiance values and then to exoatmospheric top-of-atmosphere reflectance values. These conversions are done for the visible and near-infrared bands of the data (bands 1–5 and 7). A vegetation index (VI) data set is created using the Modified Soil Adjusted Vegetation Index 2 (MSAVI2) model (Eq. 2). The MSAVI 2 data are the input data to a linear un-mixing model (Eq. 3) that uses two pure pixel end-members—soil and closed canopy forest.

Modified Soil Adjusted Vegetation Index 2 Eq. 2

$2-8 * NIR - RED)/2$

Where:

NIR = Landsat near infrared spectrum band; 0.76 – 0.90 μm (Band 4), and

RED = Landsat visible red spectrum band; 0.63 – 0.69 μm (Band 3)

Linear un-mixing model – vegetation continuous fields (VCF) soil Eq. 3

Where:

VI = vegetation Index; MSAVI 2 pixel value

VI_{soil} = Pure pixel end-member for soil value

VI_{forest} = Pure pixel end-member for closed canopy forest

The output data set from this model is a vegetation continuous field's (VCF) data set. Pixel values range from 0 to 100 and represent the fraction of the pixel that is vegetation. In the forest areas these values approximate tree canopy density. Using a threshold approach we determine through visual interpretation the VCF cut-off point for forest and non-forest areas. This product is a forest fractional cover data set (fC). A calibration coefficient is determined through linear regression of the field plot canopy density values, if they have been collected, with the mean fC value of a 3x3 cluster of pixels geographically co-located where the field canopy data are collected. The first order fC data are then calibrated to the canopy density coefficient. If a land use land cover map delineating forest types and other land uses is available, the fC are further stratified masking out the non-forest areas. Biomass data from the field plots in each stratum or from other sources (e.g. literature, IPCC default) are then assigned to each strata and down-calibrated by the fC value to create a final carbon map.[3]

Discussion

The focus of the project was to implement a community measurement protocol for establishing biomass plots that could be integrated with satellite remote sensing and an online MRV system in support of REDD+ interventions. We have been able to successfully demonstrate this as the results section above shows. An additional outcome as a result of the implementation of this project in the three countries is that we have been able to identify a number of key elements that appear to us, at least, as important in determining success in developing effective community participation in measurement for REDD+ interventions.

We list here eleven important elements that appear to us to affect and impact (1) community involvement, and capacity in REDD+ implementation activities, and (2) integrating community-based field measurements with GIS and remote sensing for a REDD+ MRV. The first seven are important elements for ensuring community participation in REDD+ (i.e. safeguard). These elements are:

1. *Community leadership and organization*: project areas that have an organized community group with strong leadership are able to mobilize community efforts; project areas with less formal and perhaps less experienced community groups appear less effective in participating in REDD+ activities. We witnessed very strong community organization and leadership in the Thai pilot study area.

They were able to mobilize large groups for training and include people from a large geographic area.

An early assessment of community organization strengths and experience will be useful in determining if community organizing and strengthening should be included in the REDD+ set of activities.

2. *Knowledge of climate change*: Local people in even the most remote areas seem to have some knowledge of climate change. The depth of their knowledge and accuracy of their knowledge, however, is not uniform. The result of the pre-training surveys supports this assertion. Basic understanding of the greenhouse effect, carbon cycle, drivers of climate change and the like is useful as a context for why measuring and monitoring is important for REDD+ implementation.

   Training modules for community leaders and community participants in local languages facilitate knowledge transfer and help to ensure success in community involvement as they provide a rationale for why REDD+ projects would seek community support in measuring and monitoring forest carbon.

3. *Linkages to local government agencies and others (universities, NGOs, etc.)*: The strength of a community's relationship with local government agencies in managing natural resources is dependent on many factors—leadership and commitment, cultural and social norms, common goals in managing and using a natural resource, and access and infrastructure. Of the three pilot areas the Thai example proved most effective. The communities already had benefited from strong linkages with the provincial Agriculture and Land Reform Office (ALRO) in Mahasarakham with the implementation of other rural development and environmental sustainable projects. Layering in the ecosystem service that carbon plays in forest areas was not a difficult task for the ALRO staff or for the local community members to grasp and support. In Vietnam, as well, the two community forest groups have strong linkages not only to district level forestry officials but also to ICRAF-Vietnam who are supporting a PES activity with them.

   In projects where linkages between communities and local agencies are weak, or do not exist, community participation in REDD+ implementation is less likely to be successful. In order to identify areas where sub-national REDD+ projects are likely to succeed, an

assessment of the strength of the links between the community and local agencies should be made.

4. *Commitment and capacity of local agencies to support and integrate community participation and facilitate capacity building*: In addition to simply having linkages between communities and local agencies that are part of implementing a REDD+ project, there needs to be knowledgeable and committed agency personnel in place. The local and provincial agency staff, university researchers, or NGO agency personnel must have expert knowledge on REDD+ and climate change science, as well as knowledge of natural resource management and the ability to work effectively with the community. Again, the pilot site in Thailand serves as a good example. One result of this project was a new Memorandum of Understanding between the national ALRO and the Faculty of Science at Mahasarakham University to replicate community training in other ALRO communities in neighboring provinces.

   Training and capacity building, knowledge transfer for local agency personnel should be a part of the overall REDD+ readiness prior to project implementation. University researchers and NGO staff who may be part of a REDD+ activity should also be carefully vetted to ensure they have appropriate background, knowledge and skills.

5. *Tenure and use rights well defined*: REDD+ interventions that target deforestation and forest degradation which may result from land and resource pressures from local people, must address rural livelihoods (Lawlor et al. 2010; Springate-Baginski and Wollenberg 2010). Project sites where there are well-defined land tenure and access rights that include communities in forest land and resource management, or co-management with local agencies, show greater success in also establishing community participation in REDD+ measurement and monitoring (Sunderlin et al. 2009; Naughton-Treves and Day 2012). In the Vietnam project area, one determining factor in identifying these particular two communities was the fact that they both had legal tenure rights to the forest parcel under the Red Book/Titling system. In such project areas, carbon can be viewed very clearly as a co-benefit, a public environmental service, which a forest provides in addition to the many important local benefits that communities benefit from (e.g. NTFP, soil nutrients, regulated water flow, microclimate conditions, etc.).

Identifying land tenure and resource use rights for a project area will help determine the likelihood of successfully integrating community participation in forest carbon measurement and monitoring activities.

6. *Clear, transparent, and effective laws regarding forest management, ownership, and use rights*: Forest Laws need to be clear and enforceable. They should be well defined and transparent to local communities, in particular regarding use and occupation rights, for example, the EU Forest Law Enforcement, Governance and Trade (FLEGT) program. There were clear differences between the three countries in our pilot area regarding governance and forest laws.

   Project areas where Forest Laws are well defined and clearly understood by local communities will have more chance of success in REDD+ implementation. In forests where use and occupation rights are restricted there is less chance that communities will participate in measurement and monitoring.

7. *Support local knowledge integrated with new technologies*: Local knowledge of sustainable resource use and management is often remarkably 'scientific'. Trial and error and observations are certainly part of traditional local knowledge. Technological or modern scientific approaches should not always supersede the methods, knowledge, and practices of local people in managing natural resources.

   Local people have effective tools for measuring and monitoring forest that can be utilized. Simple means of calculating tree heights, ground slopes, etc. do not always require hi-tech tools. The effective computation of carbon from these measurements requires some advanced scientific knowledge. The integration of local knowledge, tools, and methods with those of modern science should be promoted in REDD+ implementation to build stakeholder participation with communities.

The last four elements affect the capabilities of integrating the community measurements at the plot level with landscape scale carbon assessments using geospatial tools and technologies and supporting an MRV system for a REDD+ project:

8. *National committees with clear implementation plans—data custodian, stewardship and standards*: National REDD+ committees as part of the

REDD+ readiness are developing implementation plans under such programs as the UN Collaborative Programme on Reducing Emissions from Deforestation and Forest Degradation in Developing Countries (UN-REDD), Forest Investment Programme (FIP), and Forest Carbon Partnership Facility (FCPF), that define data stewardship, custodian and access rights, and standards. Such plans may differ from country to country based on national needs and requirements.

The use of online MRV systems will need to consider the National REDD+ implementation strategy and plans, in particular, for data standards, management, and access. MRV tools must be flexible enough to support different needs as they are defined by each REDD+ implementing country.

9. *Data management tools*: Forest biometric data collection can be done using simple or hi-tech tools. The diameter at breast height of a tree can be measured with a DBH tape or a simple tailor's tape and the measurement converted to diameter. GPS devices are now common in university labs and with local agencies in the field. The management of data measurements, however, must be systematic and uniform across biomass plots within a project location. Common printouts of spreadsheets, or data field forms in data loggers are important. Tools to manage all project data are also very important.

   Ideally, with web-based services becoming more and more common, data management can be developed as a set of tools for REDD+ implementation that include basic description and management, document and file management, plot level data management and carbon stock calculations, geo-spatial data management, and even emissions reporting.

10. *Measurement and monitoring for REDD+ can effectively combine local, community data collection with expert analysis using remote sensing and GIS*: National-level forest measurements often are conducted under National Forest Inventory (NFI) programs in which permanent sample plots are established and repeat measurement are taken every few years. Unfortunately, not all countries have adequate NFI programs in place. Forest monitoring, to assess areas and rates of change, is most often conducted using satellite remote sensing. The combination of ground-based plot biomass measurements with satellite remote sensing analyses is a powerful combination for measuring and

monitoring REDD+ carbon stock and carbon stock changes. Measurement and monitoring of REDD+ carbon is not dependent on ground only or remote sensing only techniques, but a combination of the two.

Integrating community level plot measurement data with geo-spatial analysis (GIS and remote sensing) supports REDD+ measurement and monitoring requirements for an MRV system. The combination recognizes the opportunity to integrate community level abilities with professional, expert analysis.

11. *Develop advanced MRV systems that include other environmental and social data (beyond carbon) that scale*: REDD+ may focus primarily on greenhouse gas emission fluxes, but also must be cognizant of additional environmental and social co-benefits. Forest Carbon MRV systems supported by governments should include additional data management and reporting functions to include environmental and social data. Such systems should not be scale dependent and should provide support from local level to national and regional level REDD+ activities. Scaling up from project level to national or even regional level analyses will no doubt incur costs. Use of Internet-enabled GIS, databases, and remote technologies together with the use of remote sensing satellite data can reduce the costs associated with a 'boots-on-the-ground' approach.

    National REDD+ MRV systems can be supported through online technologies and include more than just carbon accounting. Such systems can be scalable and flexible to support a variety of national-needs objectives and implementation plans.

Conclusion

Community-participatory forest measurement and monitoring is not, by itself, a guarantee for a successful REDD+ intervention, either in terms of reporting emission reductions or in terms of equitable distribution of REDD+ monetized benefits. That communities are willing to participate and even learn about broader issues such as carbon cycle science or calculating carbon and CO_2 in biomass is, however, an indicator and perhaps a bell-weather for the potential for success. A resonating view among those skeptical about REDD+ implementation identify the risk that

REDD+ may marginalize local people, hinder their access and rights to forests and forest resources, and negatively impact local livelihoods. These are legitimate concerns, even in light of discussions about safeguards working to ensure "the full and effective participation of relevant stakeholders, in particular indigenous peoples and local communities" (UNFCCC CP 16 2010). For a review of these concerns see Ribot and Larson (2012), who argue that, "[t]o make forestry policy emancipatory, strong social protections or safeguards are still needed that require REDD and other interventions to support and work through local democratic institutions."

This SUMERNET project does not focus on analyzing the various forest policies and implementation of these policies in the three countries that can indeed impact how local people benefit or are marginalized in such activities as REDD+. Rather, the research team targeted the practical implementation of engaging communities in the tasks of measuring biomass in forested areas and the collection of field data that could be integrated with satellite remote and GIS analysis in a REDD+ MRV system. The lessons learned from a comparative point-of-view provide insights into the differences and similarities between (1) each country's set of stakeholders or participants, (2) the local communities themselves, and (3) national aspects of governance, culture, and history.

We recognize that there is no one-size-fits-all recipe to ensure community participation in measuring and monitoring forest biomass for REDD+. Each community is different. In the three case study areas in Laos, Thailand, and Vietnam, there are differing levels of community cohesion, tenure and forest access rights, and relationships with local agencies, infrastructural development, knowledge of climate change, all embedded in different governance structures and cultural traditions and histories. There are, however, a number of common, general key elements to remain cognizant of when trying to engage local communities in REDD+ (these are noted above in some detail in the discussion section).

Community-participation in any collaborative activity that includes multiple stakeholders is only half or one side of a project. Equally important are the roles of the other stakeholders (local government agency staff, university or NGO staff, who also might be involved; even national agency supervisors and directors who often set the agenda for local agency staff). The importance of the relationships between stakeholders or project participants cannot be over-emphasized. Social capital between community members and between local agency staff and communities is equally as important as tenure and access rights, transparent policies and enforcement, and an MRV system for determining success in a REDD+ intervention. Success for community participation in REDD+ involves long-term thinking about more than just climate mitigation.

Notes

We acknowledge colleagues at ICRAF-VN who graciously partnered with us. The research was also financially supported, in part, by the Asia-Pacific Network for Global Change Research (APN): EBLU2010-04NMY(C)-SKOLE and EBLU2011-02CMY(C)-SKOLE and the Global Environment Facility, Carbon Benefits Projects.

1 This SUMERNET project has compiled reference material regarding 'Community-based Measurement and Monitoring' and 'REDD/REDD+ MRV.' These documents provide summaries of reference material with links to the full documents on-line. These two documents can be accessed at: http://www.goes.msu.edu/sumernet/docs/ComCarbMRV.pdf and http://www.goes.msu.edu/sumernet/docs/REDDPlus.pdf (http://www.goes.msu.edu/sumernet/back_material.html).

2 See map in ibid.

3 See carbon maps in ibid.

16

Getting to the Source: Community Planning and Costing of Adaptation Interventions in Vietnam and Lao PDR

Caitlin Corner-Dolloff and Julian Moll-Rocek

The threat that climate change poses to livelihoods is increasing around the globe, disproportionately affecting poor rural farmers in developing nations (Morton 2007). Agricultural systems are especially susceptible to changes in rainfall and temperature that could result in productivity shifts and threaten food security. The majority of the populations of both Vietnam and Lao PDR (Laos) are highly reliant on natural resources and agriculture for their livelihoods, and hence more vulnerable to climate change, which presents an urgent need to identify adaptation actions (Eastham et al. 2008; Yu et al. 2010). This need has been recognized internationally and been met with a multitude of development initiatives and funding programs. The national governments of both Laos and Vietnam have created national climate change legislation to channel these newly available funds (OPM 2011; WREA 2009).

Despite such recognition of climate change and its effects on agricultural systems, there is no clear understanding of how to channel such adaptation funds into locally appropriate, efficient, and effective interventions (Adger et al. 2005; OECD 2009). Adaptation aims to achieve resilience, viewed here as "the capacity of a system to experience shocks while retaining essentially the same function, structure, feedbacks, and therefore identity" (Walker et al. 2006), and the ability of a system to adapt using the resources available within it (Adger et al. 2011).

Including individuals and communities affected by adaptation planning in some capacity in decision-making processes is critical to empowering communities to determine which adaptation actions fit within local perceptions of identity and what changes are acceptable. This chapter demonstrates that, while engaging communities in adaptation planning is resource intensive, community involvement is necessary to prioritize appropriate adaptation interventions and can at times improve costing assessments, with the ideal level of engagement dependent on the type and scope of the intervention.

Selecting appropriate climate change adaptations is a challenge due to the variety of possible paths to building resilience in a system. Uncertainty about future conditions further complicates the selection process. While rural livelihoods in Vietnam and Laos will be affected by climate change, estimates of impact vary, and other drivers of changes may be more profound in shaping the future and must also be incorporated in planning (Eastham et al. 2008; Lefroy et al. 2010). Action must be taken in the face of these uncertainties. Selection of 'no-regrets options'—those that yield benefits regardless of climate change impacts—based on capacities that address current climate variability, can be an effective way forward (Lefroy et al. 2010; Ranger and Garbett-Shiels 2011; Vermeulen et al. 2013).

A rationale for the prioritization of adaptation action is also needed. The Participatory Social Return on Investment (PSROI) framework used in this study provides a community level process for identifying, prioritizing, and costing adaptation interventions. To ensure that adaptation interventions are locally appropriate, the framework takes into account economic, social, and environmental contexts. Adaptation options are selected by communities based on what actions will be effective at achieving future desired states, which are clarified during the PSROI process, as well as for the capacity of such actions to build on existing strengths and assets (Sova et al. 2012). The impacts of prioritized adaptation interventions are then forecast. These assessments can assist in grounding higher-level adaptation planning and funding support.

Assessing impact requires a clear understanding of the expected costs and benefits required to plan efficient interventions. Most adaptation costings are top-down processes. Grounding assessments using locally informed cost–benefit estimates, such as in PSROI, may reveal more accurate expectations of project efficiency than national-level estimates. Governments are already aware that the participation of affected stakeholders is critical in policy decisions, but because full collaboration at the community level is not always possible due to limited resources, protocols for community engagement in adaptation planning and costing are needed.

In order to elucidate the importance of local participation across the adaptation planning process, this study applies the PSROI framework to four case studies in Vietnam and Laos to address the following research questions: (1) When and how much community involvement is required for adaptation prioritization? (2) When and how much community involvement is required to accurately cost adaptation interventions?

Methods

Study sites

The study was carried out in four villages in 2012 (Fig. 16.1), two in different districts of Yen Bai province, Vietnam, and two in different districts of Savannakhet province, Laos.

Fig. 16.1 PSROI field sites

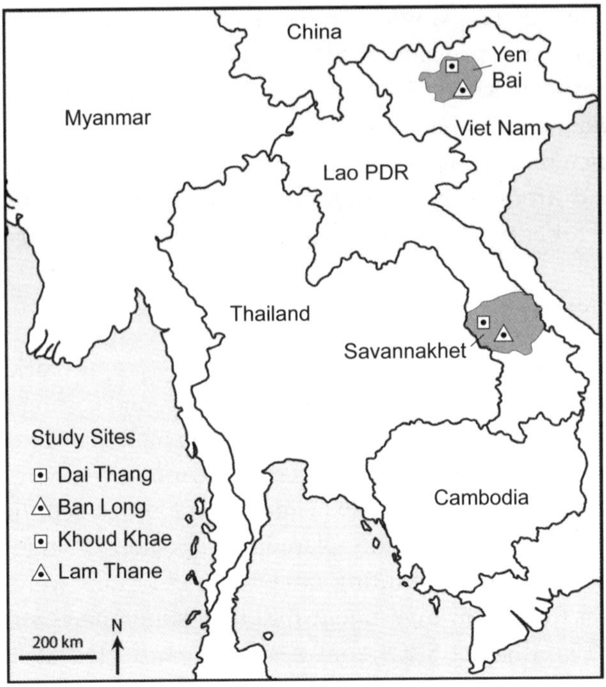

Yen Bai province is located in the center of the northern uplands of Vietnam. The two study sites in Yen Bai were Ban Long village (199 households), Son Thinh commune, Van Chan district and Dai Thang village (85 households), Dai Phac commune, Van Yen district. Yen Bai province was selected because of its high vulnerability to climate change due to environmental and economic factors. Yen Bai is a climate hazard hotspot for droughts and prolonged cold periods (McElwee 2010; Yusuf and Francisco 2009) and has high incidences of poverty and low education levels compared to the national average (FAO 2012). Over 65 percent of the terrain in the province is upland (FAO 2012), with numerous microclimates, highlighting the need for localized adaptation planning. Land use patterns have led to natural disasters such as flash floods and landslides, which are expected to be exacerbated by climate change (McElwee 2010).

Two-thirds of Savannakhet province in Laos comprises of lowlands while the remainder is mountainous (Barrios 2008). Recognized for its vulnerability to climate change, Savannakhet has become the focus of a number of agriculture and climate change projects (see e.g. ACIAR 2013; UNDP 2010). Current challenges to agriculture include drought, flooding, and pests and diseases, which have decreased rice yields and led to unpredictable farmer income (CSIRO et al. 2011). The two study sites in Savannakhet were Khoud Khae village (114 households), Outhomphone district, and Lam Thane village (93 households), Champhone district. Critical agricultural challenges include drought in Outhomphone and both flooding and drought in Champhone (PAFO 2012); emigration is also a major issue for agricultural initiatives in both sites (Manivong et al. 2012).

Study design

The study design was based on the PSROI framework, utilizing participatory processes for prioritizing, planning, and costing community level agricultural adaptation interventions. The methods are described in detail in Sova et al. (2012) and only briefly summarized here with some modifications noted. Local contexts and visions of resilient futures were conceptualized in a participatory planning workshop ('P' in PSROI, Track One, Fig. 16.2) and the resulting adaptation priorities were integrated into a Social Return on Investment (SROI) costing analysis (Track Two, Fig. 16.2). Two distinct SROI analyses were conducted at the national and local levels for each intervention (one for each study site) to identify how community involvement can shift assessments of the value of interventions.

Fig. 16.2 PSROI methodological framework (modified from Sova et al. 2012)

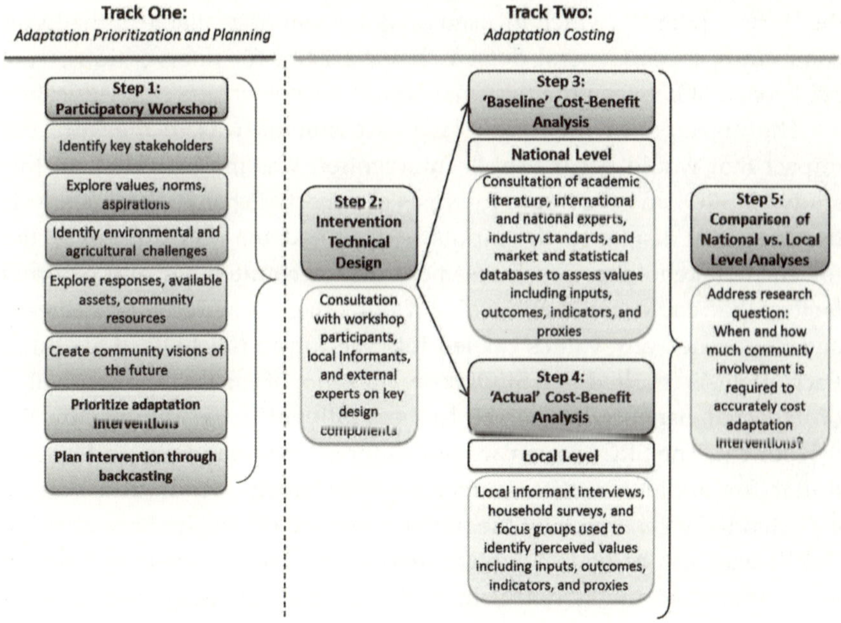

Track One encompassed the participatory community workshop (Step 1), which focused on a visioning approach, as opposed to a problem-based approach, to identifying adaptation options. The critical initial step was the selection of 40 participants, based on a stakeholder analysis of the community, to represent a broad spectrum of social differentiation. Participants were divided into three groups, each identifying and either discussing or voting on agricultural challenges, responses, and community assets. Ultimately one agricultural adaptation theme or specific action was prioritized in each group, based on the community's vision of a desired future. A plan was created to achieve priority interventions that utilized existing community strengths. Participants discussed interventions and where possible prioritized one from the workshop to feed into Track Two. Track Two was a process for fine-tuning intervention design and establishing SROI cost-benefit analyses for the prioritized interventions. Technical design (Step 2) of the interventions was established at the workshop and through consultations with local and external experts, integrating climate-proofing components where applicable. Two SROI analyses were conducted for each intervention to assess the relevance of community involvement in costing: the 'Baseline' (Step 3), which limited data collection to information sources available at the national level, and the 'Actual' (Step 4), which focused on data from or available at the local/ community level. General design characteristics for interventions were held constant between the two analyses.

The objective of the Track Two assessments was to measure the impact that would occur if each intervention was implemented. In this study all SROI analyses were forecasts of expected change. Each analysis transparently displayed the inputs needed and the outcomes expected for each stakeholder. The assessment also accounted for any external factors that could cause the claimed outcomes to increase or decrease and discounted any values carried into the future (Nicholls et al. 2012). Each analysis resulted in a SROI score, or ratio, of the total present value of expected benefits (outcomes less operational costs) divided by the investment. The SROI scores are useful for donors aiming to understand dollar for dollar (or other currency) how much social value can be expected to be generated by the initial investment. The Net Present Value (NPV) was also provided, which denotes the present value of benefits less investment and is useful in understanding the aggregate level of benefits derived from an intervention. Social value means that social, environmental, and economic outcomes were valued when possible using proxies when direct market values were not available. The investment was defined as all upfront costs incurred in year 0 and operational costs in year 1, because it was assumed that the project would have to be self-sustaining after initial project funding. Only real costs, as opposed to opportunity costs, were included as investment.

The study also collected community perceptions of changes in weather and environmental conditions through transect walks with village leaders prior to the workshop, semi-structured interviews, and focus groups when necessary.

Results and discussion

Perceptions of environmental change

Understanding perceived changes in weather and environmental conditions can provide important insights for adaptation planners. This can clarify why certain interventions are favored by stakeholders and how concerns about climate change fit into the broader set of community priorities. Technical climate change information and expert analyses of projected effects can also be useful to adaptation planning, but such information is not always available and often has high levels of uncertainty.

All study site communities had perceived changes to their environment and local weather patterns within the past 5–10 years (Table 16.1). Increased temperatures and unpredictability were the major changes in weather occurring across all study sites. In Vietnam, perceived changes in weather and environmental conditions were similar in both communities. Farmers noted they are already autonomously adapting to climate change through actions such as incorporating new farm management practices. Local perceptions of change were often supported by scientific studies. In Laos, perceptions of increased temperature and unpredictable precipitation link with historic regional trends and also with projections that future climatic conditions will be highly variable, especially in terms of the amount and distribution of rainfall between and within years (Lefroy et al. 2010). Both communities appear to be experiencing opposite precipitation phenomena, but both are experiencing late starts to the rainy season, matching meteorological records. Flooding is seen as a major environmental change in Lam Thane, which could be due to a number of climate and land use factors.

Table 16.1 Perceptions of change

Country	Village	Weather	Environmental
Vietnam	Dai Thang	Hotter (69%) Unpredictable weather (34%) Increased sun (34%) Increased storm intensity (17%)	Deforestation Soil degradation
	Ban Long	Less rain (46%) Hotter (35%) Unpredictable weather (31%) Higher frequency and duration of droughts (27%)	Deforestation Soil degradation Increased pollution
Lao PDR	Khoud Khae	Delayed rainy season (77%) Less rain (67%) Unpredictable weather (63%) Hotter in dry season Shorter periods of cool weather	Soil degradation Decreased NTFP
	Lam Thane	Heavier rain (37%) Delayed rainy season (27%) Hotter (23%) Unpredictable weather (17%)	Flood height increasing (67%) Longer periods of flooding (40%)

Source: Compiled by authors.

Table 16.1 displays the main changes in weather and environmental conditions perceived by study communities. The percentages represent the number of individuals who indicated changes out of the total number of survey respondents. All other results are drawn from focus group discussions.

Adaptation prioritization and planning

Track One focused on planning adaptation actions that emerged from a vision of a desired future in each workshop. This vision was reached through mapping current agricultural challenges to highlight systems components that are less resilient. Historic responses to challenges present stories of successful coping and initiate conception of actions that draw on local assets. This section presents a selection of results from the Track One workshops in each study site (Table 16.2) and discusses five key findings that support the need for community engagement in prioritizing and planning adaptation interventions. Links to perceptions of environmental change are assessed below. In Table 16.2, challenges were ranked by the percentage of votes given by all workshop participants. The top adaptation priority from the three groups in each workshop are listed, with the overall priority from each workshop in bold.

Table 16.2 Track one workshop outcomes

Country	Village	Agricultural Challenges	Adaptation Priorities
Vietnam	Dai Thang	Animal and crop pests and diseases (25%) Lack of financial capital (15%) Low market prices for agricultural products (14%)	**Cinnamon forest development** Increased rice productivity Reduced rice pests and disease
	Ban Long	Lack of financial capital (33%) Water storage and pollution (21%) Soil degradation (18%)	**Improved livestock raising: semi-intensive pig facility** Clean environment Improved rice seed varieties
Lao PDR	Khoud Khae	Lack of water (17%) Lack of financial capital (17%) Rice disease (13%)	**Semi-intensive aquaculture: fish ponds** Irrigation: water gate (two groups)
	Lam Thane	Natural disasters – flood and drought (28%) Poor road quality (16%) Lack of financial capital (13%)	**Irrigation: water gate** Irrigation: canal system Irrigation: cement dam

Source: Compiled by authors.

Dai Thang village

Developing the cinnamon forest was prioritized slightly over other adaptation options due to its advantage of building on local assets; it was viewed as a way to improve income, and address key challenges (Fig. 16.3). It also directly counters local environmental changes and was perceived to assist with hotter temperatures and unpredictable precipitation. Intercropping cassava with cinnamon was proposed as a way of addressing the lack of income from cinnamon planting in the initial years. This cultivation method is both prevalent and unique to this area of Yen Bai.

Fig. 16.3 Dai Thang adaptation priorities

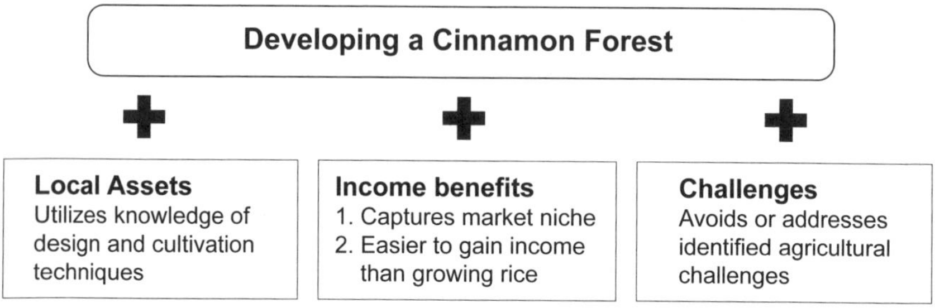

Ban Long village

In the overall workshop discussion three priorities were viewed as potential adaptation pathways. Given local strengths and constraints, a semi-intensive pig raising facility, to be co-managed by a number of households, was prioritized (Fig. 16.4). The pig raising intervention could begin immediately and buy time needed for additional planning and raise funds to facilitate changing to improved rice seed varieties. Both of these interventions were seen as necessary first steps to broader environmental adaptation.

Fig. 16.4 Ban Long adaptation priorities

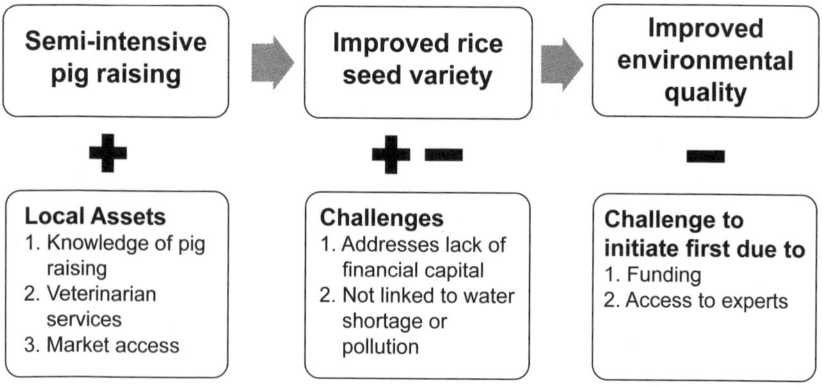

Khoud Khae village

The fish pond intervention was selected by workshop participants as it was perceived as more feasible than the water gate intervention (Fig. 16.5). This is mostly due to local topography and concerns about governing shared resources. Fish pond development was expected to address livelihood concerns as well as increase water access, reducing risks from unpredictable rainfall patterns that can affect rice and vegetable cultivation.

Fig. 16.5 Trade-offs between top adaptation priorities in Khoud Khae

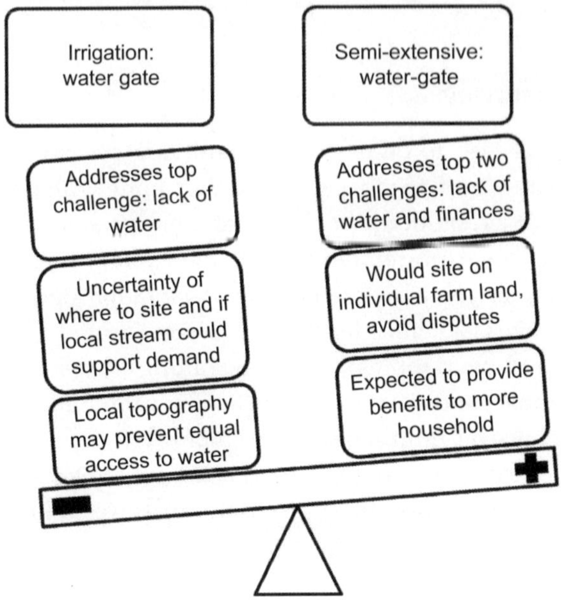

Lam Thane village

Despite challenges with flooding, irrigation aimed at improving dry season cultivation was the focus for adaptation prioritization (Fig. 16.6). All groups in the workshop unanimously prioritized irrigation as the most important adaptation intervention, and participants in one group even refused to vote on any other adaptation options to emphasize this priority. Different irrigation options were discussed in each group including a cement dam to replace an earth dam currently used to store water in a local reservoir; a water gate; and connecting the village to a nearby government canal system from a reservoir in the district.

Fig. 16.6 Misalignment of primary challenge and adaptation priority in Lam Thane village

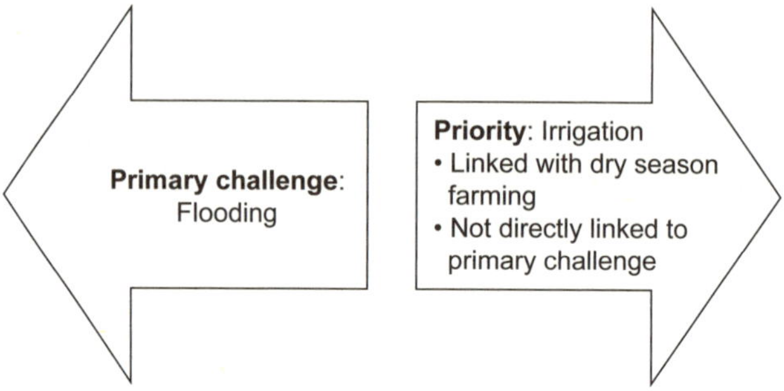

Adaptation pathways

Local input was seen to be useful in identifying system characteristics, causal pathways, and feedback loops. Ban Long village exemplifies this in the three adaptation priorities that were presented as pathways, demonstrating local understanding of the system and the need for adaptation options to change over time. Adaptation options were seen as stepping-stones, where interventions were selected for their ability to be nested and overlapping, with the shifting of system components creating favorable conditions for future action. Community participation can therefore uncover multiple stressors that result in various challenges and multiple consequences of prioritized actions (O'Brien et al. 2004). This enhances the identification of appropriate measures and enhances the likelihood that actions will be effective.

Supplementing problem identification

Development planning techniques often approach problems as the starting point for identifying solutions. These case studies indicate that this approach may not be suitable as perceptions of environmental change and challenges did not always correspond with prioritized adaptation plans. As noted in Table 16.3 sites in Vietnam were more closely related to each other in both terms of perceptions of climate change and challenges than any other site comparisons. Interestingly though, the Vietnamese sites had much more diverse adaptation priorities in comparison to each other rather than other across country sites. In contrast, the Lao sites, while having fewer similarities in perceptions of weather change and challenges, had similar adaptation priorities. Non-metric multidimensional scaling and chi-squared analyses across all sites demonstrated that communities with similar perceptions of changes in the weather also had similarities in the way they categorized and ranked challenges. Perceptions of environmental change were seen to assist in the identification of agricultural challenges. However, the use of only identified challenges and perceptions of environmental change could also mislead prioritization.

Table 16.3 Comparison of track one results within each study country

	Vietnam	Lao PDR
Perceptions of environmental change	Similar	Different
Agricultural challenges	Similar	Different
Adaptation priorities	Diverse	Similar

These results support the use of a visioning approach, which, in contrast to a reactive problem or needs-based approach, is proactive, encouraging communities to identify local strengths as well as generate visions of what is desired into the future (Evans et al. 2006). For example, the priorities selected in Laos that were not related to climate change were part of the future desired vision, which drew on local assets, successful past coping strategies, and community values. In Vietnam, similar challenges yet diverse priorities across sites also indicates that both communities utilized their created vision, not just identified challenges, to determine what would be important for building resilience. Anticipating local adaptation priorities based only on an understanding of local problems can provide a limited and unrealistic set of adaptation options. The community visioning approach also addresses ethical concerns, ensuring that those affected by interventions have the opportunity to incorporate their goals and values. External planners must ensure that communities remain empowered to determine when they want to maintain and strengthen their existing systems and when they want to strive for socioeconomic transformation.

Incorporating local strengths

Resilience, as defined in this chapter, emphasizes adapting through the use of local resources, strengths, and within local system realities. Many of the case studies demonstrate the selection of adaptation options that build on local strengths. In Ban Long village the semi-intensive pig raising facility was able to capitalize on local knowledge, with 96 percent of respondents (n=30) knowledgeable about pig raising, and 44 percent with experience of semi-intensive pig raising. Locally available resources, such as access to feed, veterinary services, and a nearby market, were also strengths that would assist in the operation. In Dai Thang village the community likewise had ample local knowledge of cinnamon cultivation and existing access to cinnamon markets that could be further developed. Cinnamon cultivation is also recognized for its suitability to the local steep terrain (Quang and Sato 2008). Similar local knowledge was embedded in the decision of participants in Khoud Khae village, Laos, who revised their priority from large-scale irrigation infrastructure to fish ponds to address issues of topography and equitable access. Thus, we see that working with local communities can assist in the selection of adaptation options that build resilience by capitalizing on local strengths and knowledge.

Barriers to adaptation

Identifying barriers to prioritized adaptation interventions is another important reason to involve communities in adaptation planning. A 'backcasting' planning technique was used in the workshop to assist with this, guiding participants to start planning from the goal, in this case the prioritized adaptation option, and plan backwards, identifying barriers to implementation. Understanding how challenges and barriers are perceived in relation to different interventions is crucial to understanding possibilities for adoption. This exercise when conducted in Ban Long village clarified the need to enact some adaptation options prior to others due to barriers to adoption. Soil quality, transportation, lack of labor availability, and lack of finances are examples of challenges described across cases that were constraints in adaptation plans. Including this detailed level of local understanding strengthens adaptation planning by identifying supplemental action needed and therefore increasing the likelihood of successful adaptation.

Adaptation costing

After adaptation priority setting, Track Two (Fig. 16.2) of the PSROI methodology involves two parallel costing approaches, one at the national ('Baseline') level and one at the local ('Actual') level to assess the importance of local involvement in adaptation costing exercises. An in-depth study of the Dai Thang village case provides insights into the comparative advantages and relative importance of these two approaches.

Mapping the impact of intercropping cinnamon and cassava

Dai Thang village prioritized a cinnamon (*Cinnamomum loureirii*) and cassava (*Manihot esculenta*) agroforestry system, replacing 50 ha of privately owned land currently used exclusively for cassava cultivation with a ten-year cycle of intercropped cinnamon and cassava. This system, a checkerboard of cinnamon seedlings planted among cassava plants, was costed with the following key assumptions:

- All cinnamon intercropping would occur on privately owned land currently under cassava cultivation.
- All cinnamon products are sold at a fixed market price.
- No transaction costs were included.
- Only actors within the first level of contact beyond the farm gate were included as potential stakeholders (i.e. this is a community-centric analysis as opposed to a value-chain analysis).

The 'Actual' SROI score, based on local information, was greater by a factor of 3 when compared to the 'Baseline' score, calculated using secondary literature and experts and information available at the national level, i.e. the 'Actual' estimates that the intervention will produce roughly US$2.30 more per dollar invested than the 'Baseline' estimate (Table 16.4). The NPV estimate for 'Actual' was more than 12 times larger than that for Baseline assessment. Two key factors lead to these differences in scores: differences in design; and, differences in valuation (Table 16.5).

Table 16.4 Summary of differences in 'baseline' and 'actual' scoring in Dai Thang village, Vietnam

	SROI score (Total present value/Investment)	NPV (Total present value/ investment) million VND
'Baseline'	1.24	254
'Actual'	3.51	3,133
Difference	2.27	2,879

Table 16.5 Key differences in the 'baseline' and 'actual' inputs and outcomes in Dai Thang village, Vietnam

Inputs	'Baseline'	'Actual'
Stakeholders identified	Community members, fertilizer sellers and bark buyers	Community members, bark buyers, fertilizer sellers, harvesting company, local cinnamon oil producer
Density of cinnamon seedlings	2650/ha	5907/ha
Amount of labor for cassava	225 man days/ha	130 man days/ha
Outcomes	**'Baseline'**	**'Actual'**
Value of jobs	Not included, no labor would be paid	Harvesting jobs are paid jobs
Value of Cinnamon Oil	Costing data not available	High value product, increases value of intervention during years 5–10

The key stakeholders in both the 'Baseline' and 'Actual' interventions were members of the community. Other stakeholders that did not have inputs associated with the intervention, but would experience outcomes, such as fertilizer sellers and cinnamon bark buyers, were included in the analyses, but the community members were the focus of the results displayed. Overall, community members' inputs and outcomes followed similar patterns in both levels of analyses, although the 'Actual' costing revealed a higher level of detail and consistently higher outcomes (Fig. 16.7a). Further, sensitivity analyses revealed that the 'Baseline' was much more sensitive to variations in labor costs because of a higher estimation of the labor required for cassava harvesting and in general a lower level of detail, meaning any single valuation change had a greater effect (Fig. 16.7b).

Fig. 16.7a Projected inputs and outcomes from the cinnamon intercropping intervention in Dai Thang, comparing 'actual' and 'baseline' results per year

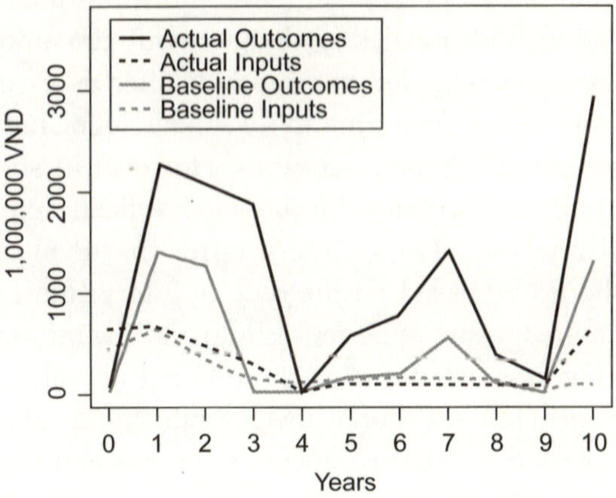

Fig. 16.7b Sensitivity analysis varying the value of wages, identified as the most important value, from 60,000 to 130,000 VND, in both 'actual' and 'baseline' costing projections

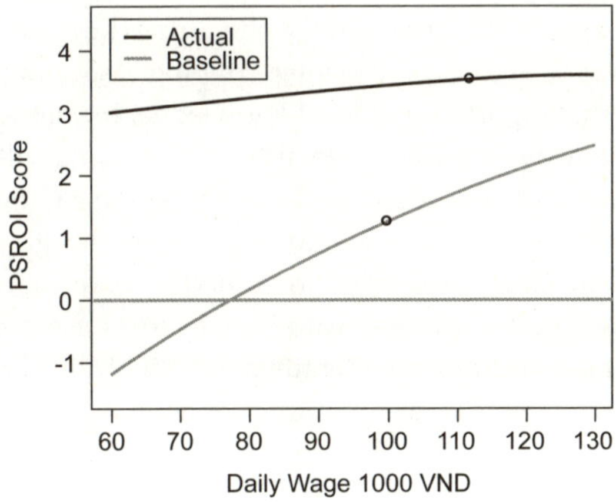

Lessons learned from costing adaptation

Local knowledge

The cinnamon and cassava intercropping case demonstrates that when the community has knowledge or experience relevant to the intervention that experts at the national level are unaware of, a comprehensive participatory approach can lead to much greater levels of detail and starkly differing costing results. There are two primary reasons for the differences we see in the scores of the 'Baseline' and 'Actual': local traditional knowledge, and expert specialization leading to a myopic understanding of the intervention. In the case of Dai Thang, the prioritized intervention relied heavily on existing traditional knowledge about cinnamon cultivation. Local traditional knowledge led to a more detailed assessment of inputs and outcomes, or impact, including more stakeholders, higher resolution and distinct valuations of their activities. However, it should be noted that a full participatory costing implies a significant expense for data collection and analysis. Thus, determining the level of community involvement should be based on the type of intervention (how relevant community knowledge and experiences are) and the information already available. Costing assessments from the other study sites not presented in this chapter provide additional insight into the level of community involvement needed to ensure accurate assessments of adaptation interventions.

Intervention design

Technical design is an area where the importance of community involvement was seen to vary. The intercropping case study above clearly demonstrated the importance of local knowledge. In contrast, the costing of the semi-intensive pig raising facility in Ban Long village included some inputs that were not identified directly by the community, such as a two-tiered water system that would improve the health of pigs. There were also certain inputs that were identified in both the 'Baseline' and 'Actual' but assigned divergent values. The costs associated with the construction of the facility were identified as roughly 40 percent higher in the national design, which was optimized for animal health and durability rather than minimizing costs. In this case, national experts presented alternative design options, which may be desired locally. Therefore, in cases where communities are attempting to engage in a new activity, or even a substantially new scale of an existing activity, engaging external specialists in a collaborative process is imperative to ensure it will be cost-effective and appropriate.

The diversity in designs offered by external specialists and communities often create opportunities to optimize the intervention during implementation. In both the semi-intensive pig raising facility and the fish pond intervention, feed systems were specific components where integrating external and community dialogues would have assisted in maximizing efficiency of the systems. In the case of pig raising, it was seen that the community projected lower costs by using local inputs for feed, whereas commercial feed, with higher prices, was recommended by external experts. For the fish pond intervention the community selected the use of commercial feeds, whereas national experts were designing and advocating for lower cost community-generated feed options. With feed systems in these cases ranging from 40–90 percent of total input costs (before discounting) this demonstrates a crucial area where integration is needed between community and external information to effectively design and accurately cost. When major design changes are necessary, which can also be applied to climate proofing interventions, the desirability of the intervention to the community may shift, requiring a reprioritization of actions. Early incorporation of external feedback on design is crucial to ensure communities fully understand the implication of prioritized actions.

Identifying outcomes

The need for community involvement in identifying outcomes of interventions varies depending on community exposure to selected adaptation interventions. Workshop results demonstrated that communities can identify feedback loops between challenges and multiple effects from one stressor. These links may not be anticipated without local knowledge. In the case of the fish ponds some outcomes were only noted by the community, such as the development of a local market (97 percent of respondents), damage to land from construction (50 percent), lower local fish prices (90 percent), and increased livestock grazing in the forest (97 percent). These are important social, economic, and environmental outcomes that would have been missed without community involvement. One challenge experienced in all communities though was consistently identifying appropriate indicators for outcomes without direct market values. For example, the fish-ponds were seen to be associated with improved human and livestock health, but identifying the most appropriate indicator for rigorous evaluation of this change was a challenge. Given that adaptation strategies often have effects beyond those intended, capturing broad outcomes with communities can assist in costing what matters and incorporating complexity and feedback of socioeconomic systems. Technical indicator selection can be sufficiently conducted at a higher levels of analysis with local verification of applicability. That said, even when a higher level costing of inputs and outcomes is considered relatively accurate, it is important to remember that SROI costing requires an accurate measurement not only of the activity, but also of the change that the adaptation action facilitates. Therefore, community level input is necessary to accurately establish the change between the future and current systems.

Cross-level and cross-sector assessment

A major lesson that emerged from the case studies is that interventions with landscape level implications require cross-level planning and costing regardless of the extent of local knowledge and the ability of communities to cost accurately. Irrigation interventions, such as the construction of a water gate or connection to a district canal system as proposed in Lam Thane village, will have inputs and outcomes associated with stakeholders at multiple levels outside the community where the adaptation prioritization originated. In this case, information required to cost the inputs was not fully available at the community level, further justifying cross-level interactions for multiple components of costing. Similar to the finding that top-down planning without local input is not appropriate, it was found to be equally inappropriate to plan and cost some adaptation interventions in isolation at the community level, especially as adaptation in one community could cause maladaptation in another.

Cross-sector analyses can also be critical to establishing the impact of adaptation interventions. The fish pond in Khoud Khae is a good example of an adaptation intervention that interacts with a number of areas that are considered distinct sectors or domains: aquaculture, vegetable cultivation using water from the pond, rice seedling harvesting, livestock benefits, and forestry effects from land use displacement, not to mention the health, livelihood, and social network outcomes identified. Costing these interventions required cross-sector analyses at the national and local levels to access the required information in enough detail to obtain cost estimates. This is a useful finding for adaptation planners and policymakers, because it emphasizes the need for planning and costing processes that bring together different ministries, divisions, and academic specialists, and different sectors of communities and local key information to channel adaptation funds efficiently, effectively, and appropriately to build resilience at the community level.

Limitations

One important limitation is that clearly identifying whether 'Actual' or 'Baseline' analyses are more accurate cannot be fully evaluated in a forecasting scenario. Uncertainty of visioning techniques and stakeholder-based selection of interventions also means that selected adaptation could be ineffective and create undesired or unintended consequences (Shackley & Deanwood 2003). Lessons drawn from assessing local participation in prioritization and effects of local information in costing could be further assessed by conducting an evaluation assessment after implementing the interventions. Bias in forecasting studies may occur in order to inflate benefits to make an intervention appear more appealing. To address these inherent distortionary incentives, bias reducing surveying techniques, such as triangulation through strategic questions and different interviewers, are needed throughout the data collection process.

Another limitation is that differences in parameters used to define 'investment' versus 'operational' costs can greatly vary the score. The NPV avoids this issue by simply subtracting all costs from all benefits. Thus, the inclusion of the NPV helps to provide a more complete understanding of benefits generated. These discrepancies are clearly portrayed in the cinnamon example, where small differences in the SROI score obscure enormous differences in scale, as revealed by the NPV. Furthermore, as noted by Arvidson et al. (2010), the SROI score cannot be compared across sites because of the subjective nature of what is selected for valuation and of valuing non-market value goods, which varies across cultures and contexts.

Refined framework and research needs

Within the context of a changing climate and the need to distribute available funds for climate change adaptation, processes are required that can appropriately, efficiently, and effectively prioritize, plan, and cost interventions. Fig. 16.8 summarizes a framework for such a process, with this study exploring when and how much community level involvement is needed at different stages. Findings from piloting PSROI demonstrate participatory community level processes are needed as an initial filter for prioritizing locally appropriate adaptation interventions. Costing of interventions is needed to ensure efficiency, with varying levels of local participation (represented by gradient) required to improve accuracy of assessments, the amount determined by the information available to national planners, intervention design characteristics, and local knowledge related to interventions. The costing forecasts can be used for planning decisions and to establish monitoring and evaluation systems to measure the establishments of desired outcomes. Adaptation actions or pathways are established to build community resilience. Feedback loops of monitoring and evaluation and adaptive management involving decision-makers, communities, and other stakeholders is needed to ensure effective outcomes leading to community resilience. This entire process is ideally completed with some level of integration between actors across levels and sectors, aiming to bridge the gap between national and local level adaptation planning processes.

Fig. 16.8 Conceptual map of adaptation planning incorporating study results

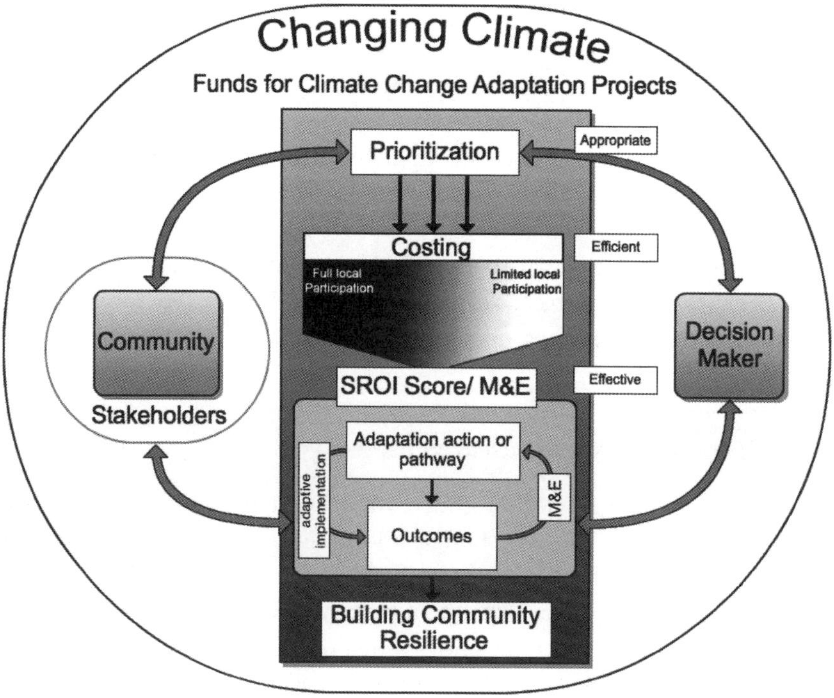

Further research is needed on tools and procedures to support the allocation of adaptation funds. Simple tools are needed to assess the impact of multiple interventions that interact in time and space, and provide efficient portfolios of adaptation options to invest in. As seen in this study, communities were often lacking information regarding new adaptation options, technical design, or projected climate change impacts. Further development of methodologies for integrating these types of specialized information into participatory planning methods is needed. This proposed framework could help guide such research as well as the design of adaptation assessment processes on the ground.

Conclusions

Local contexts were seen to play a crucial role in adaptation prioritization, justifying the need for linked national and local adaptation planning processes. While it is costly and time consuming to engage communities on the ground, this can provide critical information. The findings from the cases here demonstrate that engaging communities in processes to prioritize adaptation options is necessary for identifying appropriate interventions that will build resilience. The level of community involvement needed to accurately cost adaptation options was seen to vary depending on the prevalence of research, the type of intervention, and local knowledge. Local input in costing is useful for integrating local knowledge into design, identifying stakeholders, verifying costs of variables that are highly influential to the analysis, identifying expected outcomes and appropriate indicators, and establishing a baseline to measure the change caused by an intervention. However, 'full' community involvement in costing may not always be justifiable given limited resources for planning activities. The findings of this study can be considered initial contributions to establishing a full set of guidelines for adaptation decision-makers about when and how much local level information should be incorporated in planning and costing activities.

Note

The authors would like to acknowledge Dr. Rod Lefroy, Dr. Andy Jarvis, the mentors at SUMERNET, reviewers, and editors for their guidance. This project was conducted through partnerships lead by Dr. Nguyen Van Bo, President of the Vietnamese Academy of Agricultural Science, and Dr. Thavone Inthavone, Acting Director of the Agriculture and Forestry Policy Research Centre within the Laos National Agriculture and Forestry Research Institute. Chase Sova, Cu Thi Le Thuy, Abrar Chaudhury, Dr.

Ariella Helfgott, Bui Huy Hop, Somsameu Douangdavong, Phuthasone Ouanesamone, and the research teams are acknowledged for their contributions to developing and implementing the project. Our sincere appreciation to colleagues at CIAT for their guidance. Finally we thank the local authorities, village leaders, individual farmers, and others who participated in this research for their valuable contributions.

III

Synthesis

17

Place-based Lessons for Regional Economic Development and Sustainability

Louis Lebel, Chu Thai Hoanh, and Chayanis Krittasudthacheewa

The pursuit of sustainability takes place at multiple levels. At the global level key concerns are environmental changes such as the loss of biodiversity, disruption of the nitrogen cycle, and greenhouse gas emissions driving climate change; in addition, there are socioeconomic issues around the sharing of benefits and risks created by the international economy. At the local level concerns are often related to how land, water, and other natural resources are used, and whether local ecosystems will be able to continue to support local livelihoods and protect residents. At intermediate levels—from large sub-national to supra-national regions—concerns include transboundary environmental impacts and the economic risks and opportunities created by the movement of natural resources, people, money, and goods across shared borders.

A first key message of this book is that to understand the drivers, opportunities, and constraints on the sustainability of the Mekong region it is important to take into account their linkages at various levels. Global economic and environmental changes drive and greatly constrain what nations and regions can do; likewise, local environments, capacities, and development are affected by, but also profoundly influence, the outcomes of, national policies and regional cooperation. Many promising policies and planning interventions identified in this book are at the sub-national levels. This suggests that one hitherto under-recognized source of policy ideas for pursuing regional sustainability are bottom-up and decentralized initiatives.

A second key message is that insufficient progress on regional sustainability could undermine the climate resilience of societies in the Mekong region. Unsustainable development creates or deepens vulnerabilities, making it more difficult for households and societies to effectively respond to multiple challenges, including those arising from climate change and other global and regional integration processes. Unsustainable development also makes it harder to pursue low carbon and green growth in one place without impacting or trading-off benefits and costs in other sectors and places. Reducing climate risks requires attention to both adaptation and mitigation.

A third and final message is that the impacts of economic integration, urbanization, and climate change vary greatly among places and communities. This implies the need to pay attention to environmental and social contexts when considering development policies. The longer-term burdens and risks, and not just immediate benefits, of interventions should be factored in. Insights on these issues are only possible through regional comparative studies such as those undertaken by the teams contributing to this book. Multi-site transborder studies provide a much stronger test of generalized policies than individual studies in a single country.

In recent years there has been a lot of enthusiasm expressed for regional economic integration—through the Greater Mekong Sub-Region (GMS) initiatives and more broadly as part of the 2015 ASEAN Economic Community (Das 2012). The analyses in this book underline the urgent need for policies and plans that put sustainability as a core objective and measurable target, and not just as vague, feel-good principles in the preamble. The evidence and arguments that lead to these conclusions are laid out in the rest of this chapter.

Regional economic integration, trade, and livelihoods

In most public discussions globalization and economic development are conflated with the impacts of regionalization. Transboundary flows of goods and investment within the Mekong region are growing but not necessarily large relative to domestic and extra-regional flows (Lebel et al. this volume). Thailand, with an open and industrialized economy, is *relatively* less dependent on transboundary flows with neighbors than Laos or Myanmar, but even so draws significant gas, electricity, and labor resources from those countries. Economic conditions have large impacts on transboundary flows. Flows create needs for, and are stimulated by, road and communication infrastructure. Studies of the fish trade, contract farming, and foreign trade and investment illustrate some of the key sustainability issues.

The fish trade is an important but underestimated source of employment and income for the rural poor in Lao PDR (Laos) and Cambodia. Studies have estimated the value of the fish trade for the two countries at 13 percent and 9 percent of GDP respectively (Glemet et al. this volume). Policymakers tend to underestimate the importance of the cross-border fish trade in the Mekong region because fishing is often a secondary activity done in parallel with farming; the informal nature of the trade also makes it difficult to monitor and control. Although a fairly small number of people are directly involved in fish trading, the jobs are of relatively high value and important to households and local communities.

Some trade routes for fish have changed recently as new consumer markets open up along with new infrastructure. Up until 2004, for instance, much of the wild fish captured in northern Cambodia or southern Laos was transported north to a market in Pakse, Laos, then sold directly across the border into Ubon Ratchathani in Thailand. But with the growth of Vientiane and improved access roads, now much of this catch is transported for sale in the capital intead (ibid.). This example illustrates how economic development can redirect transboundary flows.

The fish trade has been neglected in most discussions of transboundary trade as well as fish conservation in the Lower Mekong Basin (ibid.). Moreover, as fish tend to migrate across borders, the trade is not adequately monitored or regulated within or among countries, making it difficult to respond to threats to the sustainability of fisheries across borders. The importance of the transboundary fish trade to livelihoods needs to be brought to the attention of regional bodies and interests represented in deliberations and negotiations around river uses, especially given the intense debates over the magnitude of the livelihood impacts of hydropower dams on the Mekong.

Contract farming has been mooted as a way to reduce poverty and enhance trade among neighboring countries. It is important to distinguish local smallholder farming under contract from those accompanied by large-scale land concessions. As many as 1.1 million ha or 5 percent of the land area of Laos is now under concessions or leases (Schönweger et al. 2012). Foreign direct investment—largely from China, Thailand, and Vietnam—accounts for as much as 72 percent of the land granted. The primary use of the land invested in by China and Vietnam is for mining and tree plantations, whereas in the case of Thailand it is largely for agriculture. The agricultural crops being grown tend to be non-food (or multi-use) and export-oriented, like sugarcane and jatropha. Domestic investors hold 65 percent of projects, but the land areas involved are on average much smaller than those under foreign investment. Most of the land under foreign investment is in wealthier, lowland areas with good access to infrastructure—these are not marginal areas in which land acquisition might be argued to directly help with poverty alleviation (Heinimann and Messerli 2013).

Across a range of models in four countries, studies of rice and sugar farming in this book demonstrate that contract farming can bring significant benefits to rural households. Research on rice and sugar farmers found that current contract farmers were 1.4 times as likely as non-contract farmers to have experienced raised profits per hectare in the past five years (Voladet et al., this volume). Contract farming was perceived to increase workloads, but not to influence social cohesiveness, or to produce more adverse impacts on the environment than non-contract farming. In other dimensions, there were substantial differences among locations, implying that contract farming mechanisms need to be adjusted to local physical and social conditions (Prachvuthy et al. 2013). First, rural development policies should help expand access to contract opportunities for poorer, less-skilled households. Second, monitoring and enforcement of regulations should be standardized. Third, the bargaining power of farmers in their relations with companies can be improved by strengthening farmers' associations or by offering company shares to farmers (Voladet et al. this volume).

The social and environmental impacts of China's trade and investment in the region are a popular topic of discussion. Over the last decade China and the ASEAN countries have signed a series of multilateral and bilateral agreements to liberalize trade. In most years all Mekong countries are in a trade deficit with China—Laos being an exception in 2010–11. China's outward FDI flows to the Mekong countries increased 1.7 times between 2003 and 2011, yet it represented just 9 percent of total inward flows. As will be discussed further below, this trade and investment has had the effect of reducing emissions within the Mekong region (Hu and Cao, this volume).

Another significant flow that integrates the economies of the region is labor migration. The largest flows are to Thailand (Lebel et al., this volume). These flows reflect the unequal levels of economic development in the region (Caouette et al. 2006). Migrants contribute to development in both destination and source locations—the latter through remittances. Many of the flows between source and destination have been irregular. International cooperation on migration issues other than trafficking has been limited and mostly bilateral (MMN and AMC 2013).

The findings from the SUMERNET studies in this book are in agreement with previous research on regional economic integration that suggest real opportunities to address uneven development through better cooperation (Menon 2012; Verbiest 2013). At the same time, these studies also raise questions about the actual level of integration and cooperation taking place and who stands to benefit. Large-scale water infrastructure, in particular, poses profound threats to ecological sustainability, livelihoods, and human health, raising governance challenges at multiple levels (Molle et al. 2009; Grumbine et al. 2012; Ziegler et al. 2013). This book extends these concerns beyond water/hydropower issues and reaffirms the importance of a multi-level perspective on development planning and decision-making.

Urbanization

The increasing mobility of money and people is reflected most strongly in the rapidly growing urbanizing areas of the region, especially, but not only, around the national capitals and major cities. In 2010 around 31 percent of the region's population lived in urban areas; in the next two decades this is expected to increase to 44 percent. Major urban areas dominate the economic output of countries: the Bangkok Metropolitan Administration, for example, accounts for 50 percent of Thailand's GDP. Key sustainability challenges include ensuring safe and healthy living environments for all while taking full advantage of the opportunities that urbanization creates for gains in efficiency in providing services. Moving towards more sustainable cities will be a highly political process as it must deal with many vested and privileged interests (Middleton and Krawanchid, this volume).

The connections between cities and their hinterlands imply that sustainable urbanization will also be an important dimension of regional sustainability (ibid). The urban hinterland is a transitional landscape in which urban and rural land-uses and socioeconomic processes strongly interact. From the perspective of hinterland populations, urbanization has both positive and negative impacts. Studies around the very different cities of Khon Kaen in Thailand and Vang Vieng in Laos identified many shared perceptions about the benefits of urbanization, for example, increased opportunities for women's education, employment, and participation in local administration (Thongyou et al. this volume). Many of the adverse impacts identified were environmental, but the specifics varied a lot among subjects and places. For example, in Vang Vieng, smoke, dust, and smells caused by cement factories, stone grinding, and blasting were issues of concern.

New approaches to governance are needed to improve coordination between the administrations of major municipalities with their hinterland; this is demonstrated in the case of Khon Kaen where there is significant fragmentation in planning and budgeting as a result of Thailand's decentralization policy. In Laos over-centralized governance structures create the opposite problem of insufficient local, context-specific inputs into planning, suggesting the need for an intermediate planning level. Increasing the representation of residents and civil society organizations in planning committees and processes is also recommended (ibid.).

Other studies have come to similar conclusions about the need for regional planning that properly links agriculture and urban planning. In Nonthaburi on the outskirts of the Bangkok Metropolitan Area a key issue is how the density of development affects the transport of wastes and landfill material flows in a landscape criss-crossed by not only roads but also irrigation canals (Hara et al. 2010). Failure to consider irrigation canals in a rapidly urbanizing hinterland can disrupt remaining agricultural activities and create conflicts between farmers and new residents (Davivongs et al. 2012).

Tourism provides opportunities for cities to become more sustainable. In Hue, Vietnam, Garden Houses were promoted for tourism; in Chiang Mai, Thailand, non-motorized transport in a cultural precinct with temples and tourist attractions was promoted (Kumar et al., this volume). These case studies show that local cultural heritage can be the basis for more climate-compatible development: reducing emissions on one hand, and vulnerability on the other, while creating employment for low-income earners (Kusakabe et al. in prep.).

Attention also needs to be given to the roles that cities play in key commodity chains and inter-city networks at supra-national and global levels. Infrastructure—road, rail, electricity, and communication—plays a key role. Ho Chi Minh City, for instance, is expanding its industrial infrastructure as it urbanizes, providing a complex set of opportunities and risks for migrants from rural areas (Vind and Fold 2010). In both Phnom Penh and Vientiane improved infrastructure is promoting economic activity and linking it to cross-border markets (Walsh and Amponstira 2013).

Taken together the studies in this book highlight how urbanization influences sustainability at local, regional, and global levels. Urbanization creates incentives and opportunities for efficiencies, but it also drives lifestyles based on increased consumption and hence more environmental challenges arising from the sheer density of human activities. Resource use in cities, through ecological footprints (e.g. water, carbon, and labor), also has implications for environmental quality in hinterlands, transboundary flows, and global emissions. Finally, urbanization also shifts power geographically, creating incentives and interests in both new urban centers and their hinterlands to which politicians and policies may respond.

Conservation of ecosystem services

One reason for the continued degradation of natural resources across the Mekong region is that the value of environmental services is not well-recognized in markets, and cost–benefit analyses of policies neglect their impact on ecosystems. Ecosystem services need to be taken into account in planning and policy for sustainable development. Several strategies are suggested (Janekarnkij and Polpanich, this volume). First, reverse ecosystem degradation and restore and enhance ecosystem services for pro-poor economic growth. Second, carefully evaluate trade-offs among ecosystem services as well as against other objectives. Third, include ecosystem services into assessment procedures. Fourth, enhance research, education, and communication of ecosystem knowledge. Finally, economic incentives can be used to help protect, restore, and enhance both ecosystems and livelihoods.

After more than a decade of projects and several studies conducted in the Mekong region, it is still not clear under what conditions and to what extent payment for environmental services (PES) can contribute to both conservation and poverty alleviation (Neef and Thomas 2009). Two chapters in this book provide additional evidence and insights. To address widespread deforestation and forest degradation in Vietnam the government piloted a payment scheme for forest environmental services (PFES) in Lam Dong and Son La provinces. Research shows that the pilot project made a significant contribution to forest protection and household incomes (Bui et al. this volume). Participating in the scheme yielded on average 32 percent of a household's total income and reduced the proportion of households below the poverty line. Distribution of benefits of the scheme, however, was uneven, with Kinh-headed households benefiting more than ethnic minority ones. On the positive side, participation in the PFES scheme has raised awareness about the importance to businesses and livelihoods of forest-related ecosystem services among hydropower, water supply, and tourism companies.

A review of previous projects with PES-like features in Cambodia identified factors likely to contribute to success as low administrative costs, transparency, and active local participation (Nyda and Sopheak, in prep.). Barriers to successful PES were in the unequal sharing of benefits, lack of capacity to monitor and sanction, and low communication skills. In appropriately chosen sites it was felt that PFES schemes could reproduce beneficial results like those documented in Vietnam (Bui et al. this volume).

Apart from hydrological services forests also play a role in mitigating climate change though sequestration and storage of carbon in biomass and soils (Samek et al., this volume). Reducing Emissions from Deforestation and Forest Degradation–plus (REDD+) projects aim to reduce net emissions from land-use activities. The involvement of local communities is important to the success of such projects to make them sustainable and beneficial. The involvement of residents in measurement and monitoring forest carbon is one way for them to become co-managers of these ecosystems. Studies conducted in Laos, Thailand, and Vietnam presented in this book demonstrate that it is possible to integrate such community-based measurement activities with remote sensing analysis in a flexible but still reliable way (ibid.).

An independent study in Bac Kan province in Vietnam suggests that combining service payments from REDD+ projects with income from forestry and agroforestry goods increases sustainability and long-term environmental benefits (Hoang et al. 2013). Existing government policies that aim to achieve both conservation and rural development objectives as well as pay directly for environmental services make a suitable foundation for such an approach. The authors also argued that a bottom-up, participatory approach was crucial to effective REDD+ projects consistent with the findings above.

Studies in this book and elsewhere reaffirm the importance of engaging local resource users and residents on a level-playing field in the management of ecosystem services. When forest policies favor a narrow set of state and private interests, there is a substantial risk that the marginalized poor subsisting on forest resources will not benefit from REDD+ projects (Ribot and Larson 2012). Sustainability is easier to pursue when the benefits of protecting ecosystems are understood and fairly shared. Co-management approaches to watershed management, however, are still far from common where top-down zonation and regulatory approaches are favored by governments even though they are often ineffective in meeting stated objectives such as conservation and poverty alleviation (Lebel and Daniel 2009).

Energy security and emissions

Energy use in the Mekong region has grown tremendously in the last few decades. Electricity consumption in Thailand quadrupled between 1990 and 2010. In 2010, Thailand consumed 43 percent of the electricity used in the GMS region and produced 69 percent of the energy-related CO_2 emissions (ADB 2013). At the same time more than a fifth of the population in the Mekong region has no access to electricity (Li and Vijitpan, this volume). Securing safe energy for all is a major sustainable development challenge.

Energy security is used to justify many activities in the Mekong region, in particular, hydropower. The mainstream discourse focuses on large-scale hydropower dams—increasingly justified as clean energy (Kaisti and Käkönen 2012). Alternative ways to improve energy security such as managing and reducing demand and meeting supply with smaller-scale and other forms of renewable energy deserve greater attention.

Biomass fuels are still very important energy sources for many households, representing 40 percent of the energy mix across the region, with adverse effects on health from indoor air pollution. One possible solution is to pursue community-based low carbon economies. Such communities would depend on sustainable supplies of low-carbon energy, enlarge carbon sinks, minimize disposal of wastes, and create green businesses and jobs (Li and Vijitpan, this volume).

An example towards such an approach comes from the tourism sector of Chiang Mai municipality, Thailand, and Hue city in Vietnam (Kumar et al., this volume). Tourism contributes substantially to the local economies of these cities, both of which have invested in infrastructure to support the sector. A study of the GHG emissions associated with the activities of tourism service providers in each city provided a basis for consultation with government and private sector stakeholders to develop emission reduction strategies compatible with existing tourism policies. Chiang Mai municipality focused on non-motorized transport and Hue on its Garden Houses—both activities which reduce GHG emissions and create green jobs for lower-income residents. At the same time it is recognized that these interventions only address a small fraction of current tourism-related emissions, most of which are related to transport, especially air travel. Further research is also needed to improve understanding of why such alternatives are downplayed in many policy debates even though they are key to pursuing climate-compatible development in less developed countries such as Laos (Kaisti and Kakonen 2012).

Transboundary flows redistribute emission burdens. Direct investment and trade by China in the Mekong region is having different impacts on CO_2 emissions: investments increase, whereas trade decreases emissions (Hu and Cao, this volume). As the volume of trade is much larger than investments, the net impact from China is to reduce emissions in the Lower Mekong countries—because the emissions associated with imported goods 'belong' to China. This net impact on emissions, however, is modest given that Chinese trade and investment only gives rise to 4 percent of total domestic emissions in the other countries.

Green investment policies could help further reduce pollution emissions in Mekong countries. This requires implemented standards in these countries along with policies in investing countries that guide banks and companies. Likewise green trade policies may also be encouraged through regional cooperation on sustainable development (ibid.). ASEAN, for instance, has developed a series of action plans on energy cooperation. GMS parties have signed agreements on power trade and developed a regional energy strategy (Li and Vijitpan this volume). Modeling studies suggest that regional trade even with emission constraints can improve energy security (Watcharejyothin and Shrestha 2009). Conversely, improvements in energy efficiency alone in the cases of Thailand and Vietnam are not sufficient to improve energy security for more than a decade or so in the future (Selvakkumaran and Limmeechokchai 2013). Many infrastructural and institutional barriers to sustainable regional energy trading remain while the drivers of increasing energy demand show few signs of slowing. Energy security and energy poverty—a lack of secure access to reliable and safe energy—remain critical issues for sustainable development in the Mekong region.

Changing climate risks

The climate of the Mekong region is strongly influenced by the Asian Monsoon. In many locations away from the core tropical zone there are challenges of both too much and not enough water in the same year. With still a large part of the population living in rural areas and depending on agriculture and fishing livelihoods, effectively managing climate risks is important to human well-being and sustainable economic development. While temperatures and the sea-level are expected to rise, there are still significant uncertainties as to how the climate will change and with what impacts across the region (Lacombe et al. 2012).

Baseline studies in the delta areas of Vietnam, Cambodia, and Thailand have found that people whose livelihoods are more vulnerable to climate-related hazards are more concerned about climate change (Chinh et al. in prep.). Differences among sites in terms of knowledge and perceptions of risk were large. Men were more likely than women to have heard of and be concerned with climate change. As might be expected, those who had been greatly affected by recent floods or had directly observed climate variability were more concerned. Individuals with more years of schooling were more likely to think that climate change will adversely impact future generations. Understanding existing perceptions and knowledge about climate and climate change is helpful to designing interventions to further raise awareness, especially among vulnerable communities.

Effective risk communication is another key step in successful adaptation to climate change (Lebel et al. 2013). A study comparing the use of SMS, video, and 'Talking Farmers' suggests that the latter was the most effective in the rural populations studied, but it is also the most costly (Chinh et al. this volume). Two-way communication, however, is not always a realistic option, so effective messaging using one-way media also needs to be developed. All methods of communicating can be improved by tailoring content to specific audiences—which, in turn, implies the need to actively involve representatives from the target audiences themselves in the design of messages and selection of media formats.

A comparative study of rice growing in Laos and the Philippines found that poor farmers were highly vulnerable to extreme climate events. Projected impact of climate change based on downscaled scenarios in Laos and the Philippines showed that impacts were likely to vary by ecological zone, location, and cropping season (Peñalba et al. this volume). In three of the four study sites, a combination of increased temperature and rainfall is expected to increase yields, but in the fourth site the impact would be a decline in rice yields. Food availability and access patterns were predicted to be similarly affected. Impacts on livelihoods were anticipated to be highest on those who depend on farming for income whereas those with more diversified income sources were more resilient to climate change. In rainfed areas, there were fewer options to adapt with alternative livelihoods or technologies than in irrigated lowlands, underlying the importance of ecological settings.

These findings are similar to previous studies done in Thailand and Laos. Chinvanno et al. (2008a), for example, found mild positive impacts on average due to elevated CO_2, increased rainfall, and modest projected temperature increases. Farmers in that study regarded the two most important climate risks as prolonged midseason dry spells after sowing or transplanting seedlings and flooding near harvest time (Chinvanno et al. 2008b). Taking into account these risks implies much more mixed impacts from climate change. Other studies of rice farming suggest there may be a trade-off between reducing the risks of crop failure and maintaining high yields (Felkner et al. 2009).

Unusual floods in the Mekong Delta, as in 2000, have an enormous impact, especially on households along the Cambodian–Vietnamese border. On both sides of the border the vulnerabilities of poor households, as measured by two different indices, were much higher than for wealthier households (Can et al. this volume). Farmers in An Giang, Vietnam, and Kandal, Cambodia, reduce their vulnerabilities to floods by measures such as strengthening their houses and moving their children to safe places. In Vietnam policies for living with floods have been implemented and moderately successful. This area of the Mekong Delta is very productive and important to food security but also at high risk from the interaction of rising sea levels caused by climate change with large seasonal floods (Dinh et al. 2012).

Although farmers have always had to deal with a degree of climate variability, extreme weather, and uncertainties in water supply and markets, climate change represents a significant challenge. Risk management can and often needs to be improved. Funds for development or adaptation, however, are not unlimited. Understanding expected costs and benefits from a proposed climate adaptation plan can help with allocation decisions (Corner-Dolloff and Moll-Rocek, this volume). Community involvement in planning is thus important to appropriately prioritize and value adaptation actions, for instance, the level of local versus external expert knowledge relevant to the particular adaptation technology. Case studies in Laos and Vietnam found local engagement to be useful in prioritizing adaptation options because it helped identify the most appropriate options for building climate resilience. The results for costing were also much more context-specific, depending, for instance, on information on the levels of previous research and of local experience and/or knowledge of particular technologies.

The ease with which places, countries, and the region will be able to adapt to a changing climate depends greatly on the sustainability of development approaches. The key feature of climate change as a global phenomenon is that it is accelerating. Development strategies need to be as robust as possible; they should seek to build resilience to a wide range of possible future climates. However, many current approaches to economic development in the Mekong region are not reducing climate risks.

Outstanding and emerging policy issues for further research

Regional economic integration has the potential to reduce environmental degradation and inequalities through its impacts on trade, investment, and labor mobility. Studies reported in this book, however, often found that these impacts varied significantly among study sites, countries, and climatic conditions. Local ecosystems, livelihood portfolios, gender relations, and differences in perception and culture all can have bearings on outcomes. This cautions against 'one-size-fits-all' approaches to development policy and planning. It also suggests that sustainability at the regional level is unlikely through national and international policies and processes alone—diverse local and place-specific experiments and responses are also needed.

Many of the case studies presented in this book were innovative: participatory costing of adaptation projects; contract farming for high value organic products as part of a social enterprise; and, pilot schemes that pay for forest environmental services. Such practices and approaches are still not the norm. To address uneven and unsustainable development mechanisms are needed whereby successful local pilot experiments more readily inform national and regional agendas.

One promising innovation of this project was the requirement for research groups to engage boundary partners. The idea was that doing so from the start would produce more useable findings. Boundary partners often included individuals in local government or local branches of line ministries. In some projects, partners became closely involved in the research process and followed up with their own activities. Our studies show that research which engages with local policy and planning processes is both feasible and can be influential. They also show that three core areas of emerging or outstanding policy problems require more comparative and critical research.

- *First, governments should focus more attention on the costs and burdens to different social groups of major policies and large infrastructure projects and not just the benefits with which proponents justify them.* The studies in this book show that policies that support integration, such as harmonization of standards, reductions in tariffs, freer movement of labor, and the liberalization of foreign investment can have unanticipated consequences on the environment and disadvantaged groups.
- *Second, ensuring that the importance and values of ecosystems to local development are fully acknowledged in policy and planning at all levels.* Regulating and supporting services are often not as well as understood as more immediate benefits like crop yields. Policy-relevant research and community-based engagement are needed to help develop knowledge and tools to support local development that is sustainable. More work is needed to assess and design incentives to protect and manage ecosystem services in diverse landscapes.

- *Third, climate compatible development that promotes human development and ecological sustainability, while taking into account needs to both mitigate and adapt to climate change.* Science can help improve understanding of development options and their consequences, for example, with respect to energy sources and urbanization patterns. Climate compatible development should reduce climate risks. It should also be concerned with issues of fairness, for example, around who pays and who benefits. More research is needed on how benefits and risks in adaptation and mitigation programs are governed and dealt with in the Mekong region.

This book underlines the importance of comparative, place-based, and engaged research at multiple levels to understanding how to ensure sustainability in the Mekong. Places and people differ from each other in their vulnerabilities, capabilities, and aspirations. Global, supra-national and local drivers of change interact in complex ways, creating substantial governance challenges, but also stimulating innovation and creating opportunities for sustainability.

List of Figures and Tables

Boxes

Figures

Tables

Bibliography

Adebambo, S.O., and S.F. Yetunde. 2010. "Evaluation of Non-Motorised Transport (NMT) System in the Movements of Market Wares in a Medium Sized City of a Developing Economy." *American Journal of Scientific Research* 11: 72–85.

Adger, W.N. et al. 2011. "Resilience Implications of Policy Responses to Climate Change." *Wiley Interdisciplinary Reviews: Climate Change* 2, 5: 757–66.

Adger, W.N., N.W. Arnell, and E.L. Tompkins. 2005. "Successful Adaptation to Climate Change across Scales." *Global Environmental Change* 15, 2: 77–86.

Adger, W.N., and P.M. Kelly. 2001. "Social Vulnerability and Resilience." In *Living with Environmental Change: Social Vulnerability, Adaptation and Resilience in Vietnam*, ed. W.N. Adger, P.M. Kelly and Nguyen Huu Ninh. London: Routledge.

Agence de l'Environnement et de la Maîtrise de l'Energie (ADEME). 2009. *Bilan Carbone®: Companies-Local Authorities-Regions, Methodology Guide, Version 6, objectives and accounting principles*. Paris: ADEME.

Agricultural and Food Marketing Association for Asia and the Pacific (AFMA). 2011. *Proceedings of the Workshop on Modernizing the Asian Rice Industry, Bangkok*. http://www.afmaasia.org/publication/Asian_Rice_Proceedings_low_resolution.pdf.

Aguilar, A.G. 2008. "Peri-urbanization, Illegal Settlements and Environmental Impact in Mexico City." *Cities* 25: 133–45.

Agyeman, J., R.D. Bullard, and B. Evans. 2003. "Introduction. Joined-up Thinking: Bringing Together Sustainability, Environmental Justice, and Equity." In *Just Sustainabilities: Development in an Unequal World*, ed. J. Agyeman, R.D. Bullard and B. Evans, 1–18. London: Earthscan.

Ahmad, N., and A. Wyckoff. 2003. *Carbon Dioxide Emissions Embodied in International Trade of Goods*. OECD Science, Technology and Industry Working Papers. Paris: OECD.

Anastasiadou, C., and N. de Sausmarez. 2006. "The Role of Regional Trading Blocs in the Development and Management of Tourism: An Analysis of the European Union and the Association of Southeast Asian Nations." *International Journal of Tourism Research* 8: 317–32.

Angelsen, A., S. Brown, C. Loisel, L. Peskett, C. Streck, and D. Zarin. 2009. *Reducing Emissions from Deforestation and Forest Degradation (REDD): An Options Assessment Report, Prepared for the Government of Norway*. Washington, D.C.: Meridian Institute. http://www.REDD-OAR.org.

Asian Centre for Biodiversity (ACB). 2011. *ASEAN's Rich Biodiversity*. http: //www.aseanbiodiversity.org.

Asian Development Bank (ADB). 2000. "Environments in Transition Cambodia, Lao PDR, Thailand, Viet Nam." ADB, Program Department West, Apr. 2000.

———. 2003. "Technical Assistance (Financed by the Poverty Reduction Cooperation Fund) to The Kingdom of Cambodia for the Participatory Poverty Assessment of the Tonle Sap." Phnom Penh: ADB.

———. 2005. *The Greater Mekong Subregion Tourism Sector Strategy*. Manila: ADB.

———. 2007. *Biofuel and Renewable Rural Energy Initiative in the Greater Mekong Subregion*. Manila: ADB.

———. 2008a. *Managing Asian Cities: Sustainable and Inclusive Urban Solutions*. Manila: ADB.

———. 2008b. *Greater Mekong Subregion: Maturing and Moving Forward*. Evaluation Study. Manila: ADB.

———. 2008c. "Preparing National Guidelines for Eco-Compensation in River Basins and a Framework for Soil Pollution Management: China, People's Republic of." Technical Assistance Report Project No. 42024, December 2008. Manila: ADB.

———. 2009a. *Building a Sustainable Energy Future: The Greater Mekong Subregion*. Manila: ADB.

———. 2009b. *The Economics of Climate Change in Southeast Asia: A Regional Review*. Manila: ADB.

———. 2009c. "Migration in the Greater Mekong Subregion." Background Paper for the Fourth Greater Mekong Subregion Development Dialogue, 5 May 2009, Beijing. http://www.adb.org/Documents/Reports/Migration-GMS/Migration-GMS.pdf.

———. 2009d. "Greater Mekong Subregion Economic Cooperation Program. Report of the 15th Ministerial Conference, Cha-Am, Petchburi Province, Thailand, 17–19 June 2009." http://www.adb.org/Documents/Conference/Mekong/gms-mc-15.pdf.

———. 2009e. "GMS Cross Border Trade and Contract Farming in the Greater Mekong Subregion Economic Cooperation Program Technical Meeting." http://www.adb.org/publications/proceedings-technical-meeting-gms-cross-border-agriculture-trade-and-contract-farming.

———. 2010a. "Joint Ministerial Statement. GMS in the Next Decade: New Frontiers of Cooperation. 16th Ministerial Meeting, 20 Aug. 2010, Hanoi." Manila: ADB.

———. 2010b. "Towards Sustainable and Balanced Development: Strategy and Action Plan for the Greater Mekong Subregion North–South Economic Corridor." Manila: ADB.

———. n.d. "Valuation of Ecosystem Services of Biodiversity Conservation Corridors in Cambodia. Greater Mekong Subregion Biodiversity Conservation Corridor Project." http://www.asiandevbank.org/Documents/RRPs/REG/40253/40253-02-reg-oth-05.pdf.

———. 2012. *The Greater Mekong Subregion at 20: Progress and Prospects*. Manila: ADB.

———. 2013. "Greater Mekong Subregion Environment Operations Centre." http://www.gms-eoc.org/gms-statistics/gms.

ADB and Making Markets for the Poor (ADB/M4P). 2005a. "Linking Farmers to Markets through Contract Farming: Proceedings of an M4P/AnGiang University Workshop, 27–34." http://www.value-chains.org/dyn/bds/docs/486/M4P%20Contract%20 Farming%20in%20Vietnam%202005.pdf.

———. 2005b. "30 Cases of Contract Farming: An Analytical Overview. Asian Development Bank Viet Nam Resident Mission: Making Markets Work Better for the Poor." 1115/XB-QLXB. http://www.markets4poor.org/m4p2/filedownload/30%20 cases_CF_eng.pdf.

ADB, and International Food Policy Research Institute (IFPRI). 2009. *Building Climate Resilience in the Agriculture Sector in Asia and the Pacific*. Manila: ADB.

———. 2012. *Overview: Greater Mekong Subregion Economic Cooperation Program*. Manila: ADB.

Arthur, R.I., and R.M. Friend. 2011. "Inland Capture Fisheries in the Mekong and Their Place and Potential within Food-Led Regional Development." *Global Environmental Change* 21: 219–26.

Arvidson, M., F. Lyon, S. Mckay, and D. Moro. 2010. "The Ambitions and Challenges of SROI." TRSC Briefing Paper 49, Third Sector Research Centre, University of Birmingham.

ASEAN Centre for Energy (ACE). 2009. *ASEAN Plan of Action for Energy Cooperation (APAEC) 2010–2015*. Jakarta: ACE.

Association of Southeast Asian Nations (ASEAN). 2008. *ASEAN Economic Community Blueprint*. Jakarta: ASEAN Secretariat.

———. 2012. *Tourism Statistics*, 14 Aug. 2012. http://www.asean.org/news/item/tourism-statistics.

Atici, C. 2012. "Carbon Emissions, Trade Liberalization, and the Japan–ASEAN Interaction: A Group-wise Examination." *Journal of the Japanese and International Economies* 26, 1: 167–78.

Australian Agency for International Development (AusAID). 2010. *Economic Benefits of Trade Facilitation in the Greater Mekong Subregion*. Report prepared by the Centre for International Economics, Aug. 2010. Canberra: AusAID.

Australian Centre for International Agricultural Research (ACIAR). 2013. *Vietnam: Medium-term Strategy*. http://aciar.gov.au/country/Vietnam.

Bader, N., and R. Bleischwitz. 2009. *Study Report Comparative Analysis of Local GHG Inventory Tools College of Europe and Institut Veolia Environnement*. http://www.institut.veolia.org/ive/ressources/documents/2/491,Final-report-Comparative-Analysis-of.pdf.

Bai, X., and H. Imura. 2000. "A Comparative Study of Urban Environment in East Asia: Stage Model of Urban Environmental Evolution." *International Review for Environmental Strategies* 1, 1: 135–58.

Bai, X. et al. 2010. "Urban Policy and Governance in a Global Environment: Complex Systems, Scale Mismatches and Public Participation." *Environmental Sustianability* 2: 129–35.

Baird, I.G. 2010. "Land, Rubber and People: Rapid Agrarian Changes and Responses in Southern Laos." *Journal of Lao Studies* 2: 1–47.

Bakker, K. 1999. "The Politics of Hydropower: Developing the Mekong. *Political Geography* 18: 209–32.

Balisacan, A.M. et al. 2005. "Rural Poverty in Southeast Asia: Issues, Policies, and Challenges." *Asian Journal of Agriculture and Development* 2, 1&2: 25–38.

Bangkok Post. 2010a. "Cane Farmers Call for End to Price Controls: Letting Market Work would End Smuggling." *Bangkok Post*, 27 Nov. 2010.

———. 2010b. "Khon Kaen Sugar Sets to Double Sugarcane Output." *Bangkok Post*, 27 Jan. 2010.

Baran E., and C. Myschowoda. 2008. "Have Fish Catches been Declining in the Mekong River Basin." In *Modern Myths of the Mekong: A Critical Review of Water and Development Concepts, Principles and Policies*, ed. M. Kummu, M. Keskinen, and O. Varis, 55–64. Helsinki: Helsinki University of Technology.

———. 2009. "Dams and Fisheries in the Mekong Basin." *Aquatic Ecosystem Health & Management* 12: 227–34.

Baran, E., T. Jantunen, and Chong C.K. 2007. *Values of Inland Fisheries in the Mekong River Basin*. Penang: WorldFish Center.

Barkey, D.L., M. Henri, and S. Bao 1996. "Identifying 'Spread' versus 'Backwash' Effects in Regional Economic Areas: A Density Functions Approach." *Land Economics* 72, 3; 336–57.

Barrios, E.B. 2008. "Small Area Estimation (District Level) of Rice Production and Demand Situation in Lao PDR in 2008." Report for TCP/LAO/3202. Vientiane: GoL and FAO.

Bates, B.C., Z.W. Kundzewicz, S. Wu, and J.P. Palutikof, eds. 2008. "Climate Change and Water." Intergovernmental Panel on Climate Change Technical Paper, IPCC Secretariat, Geneva.

Baumert, A. K., T. Herzog, and J. Pershing. 2005. *Navigating the Numbers: Greenhouse Gas Data and International Climate Policy*. Washington, D.C.: WRI. http://pdf.wri.org/navigating_numbers.pdf.

Be, T.T., B.T. Sinh, and F. Miller, eds. 2008. "Challenges to Sustainable Development in the Mekong Delta: Regional and National Policy Issues and Research Needs." Bangkok: SUMERNET, SEI – Asia.

Becken, S., and M. Patterson. 2006. "Measuring National Carbon Dioxide Emissions from Tourism as a Key Step towards Achieving Sustainable Tourism." *Journal of Sustainable Tourism* 14, 4: 323–38.

Berkes, F. 2007. "Understanding Uncertainty and Reducing Vulnerability: Lessons from Resilience Thinking." *Natural Hazards* 41: 283–95.

———. 2009. "Evolution of Co-management: Role of Knowledge Generation, Bridging Organizations and Social Learning." *Journal of Environmental Management* 90: 1692–702.

Birkmann, J. 2006. "Measuring Vulnerability to Remote Disaster-resilient Societies: Conceptual Frameworks and Definitions." In *Measuring Vulnerability to Natural Hazards: Towards Disaster Resilient Societies*, ed. Jörn Birkmann, 9–54. Tokyo: United Nations University Press.

Birkmann, J., M. Garschagen, F. Kraas, and N. Quang. 2010. "Adaptive Urban Governance: New Challenges for the Second Generation of Urban Adaptation Strategies to Climate Change." *Sustainability Science* 5: 185–206.

Bockstael, N.E., and K.E. McConnell. 2007. *Environmental and Resource Valuation with Revealed Preferences: A Theoretical Guide to Empirical Models*. Dordrecht: Springer.

Boholm, Å. 2008. "New Perspectives on Risk Communication: Uncertainty in a Complex Society." *Journal of Risk Research* 11: 1–3.

Boossabong, P., and M. Taylor. 2009. "Impact of Trade Liberalization on the Agricultural Sector and Adjustment Policy: The Case of Shallot Plantations in Northeastern Thailand." *J. Soc. Sci.* 30: 323–37.

Boulay A., L. Tacconi, and P. Kanowski. 2012. "Financial Performance of Contract Tree Farming for Smallholders: The Case of Contract Eucalypt Tree Farming in Thailand." *Small-Scale Forestry*. DOI 10.1007/s11842-012-9201-7.

Boyle, T.B. 2004. "Urbanization: An Environmental Force to be Reckoned With." *Population Reference Bureau*, www.prb.org/articles/2004.

Bradshaw, W.E., and C.M. Holzapfel. 2006. "Evolutionary Response to Rapid Climate Change." *Science* 312: 1477–8.

Braun, B. 2005. "Environmental Issues: Writing a More-than-Human Urban Geography." *Progress in Human Geography* 29, 5: 635–50.

Brees, I. 2008. Refugee Business: Strategies of Work on the Thai–Burma Border." *Journal of Refugee Studies* 21: 380–97.

Brockherhoff, M.P. 2000. "An Urbanizing World." *Population Bulletin* 55, 3: 3–44.

Broto, V.C., A. Allen, and E. Rapoport. 2012. "Interdisciplinary Perspectives on Urban Metabolism." *Journal of Industrial Ecology* 16, 6: 851–61.

Brown, S. 1997. *Estimating Biomass and Biomass Change of Tropical Forests: A Primer*. FAO Forestry Paper 134, Food and Agriculture Organization, Rome.

Bush, S.R. 2004. "Scales and Sales: Changing Social and Spatial Fish Trading Networks in the Siiphandone Fishery, Lao PDR." *Singapore Journal of Tropical Geography* 25: 32–50.

Buys, L., E. Miller, and K. van Megen. 2012. "Conceptualising Climate Change in Rural Australia: Community Perceptions, Attitudes and In(actions)." *Regional Environmental Change* 12, 1: 237–48.

Cacaud, P., and P. Latdavong 2008. "Fisheries and Aquaculture in the Lao PDR: A Legislative Review." Bangkok: FAO Asia Regional Office.

Cai J.N., L.N. Ung, S. Setboonsarng, and P.S. Leung. 2008. "Rice Contract Farming in Cambodia: Empowering Farmers to Move beyond the Contract toward Independence." Discussion Paper No. 109, ADB Institute (ADBI), Tokyo. http://www.fao.org/uploads/media/ADBI%20contract%20farming%20rice%20cambodia.pdf.

Cao, C., Hu Tao, and Jun Pang. 2011. "The Contribution of Export Virtual Pollution in Trade and the Potential in Energy-saving and Emission-reducing in 12th Five Year Plan." *Resources Science* 33, 9.

Caouette, T., R. Sciortino, P. Guest, and A. Feinstein. 2006. *Labor Migration in the Greater Mekong Sub-region*. Bangkok: Rockefeller Foundation.

Carney, D. ed. 1998. *Sustainable Rural Livelihoods: What Contribution Can We Make?* London: DFID.

Cash, D.W., J.C. Borck, and A.G. Patt. 2006. "Countering the Loading-dock Approach to Linking Science and Decision Making." *Science, Technology and Human Values* 31, 4: 465–94.

Castells, M. 1989. *The Informational City: Information Technology, Economic Restructuring, and the Urban Regional Process*. Oxford: Blackwell.

Castells, M. 1996. *The Rise of the Network Society*. Oxford: Blackwell.

Central Committee for Flood and Storm Control (CCFSC) 2011. Statistics for 1999–2008. Hanoi: CCFSC.

Chalamwong, Y., and R. Prugsamatz. 2009. "The Economic Role of Migration: Labor Migration in Thailand: Recent Trends and Implications for Development." *TDRI Quarterly Review* 24: 3–9.

Chambers, R., and G. Conway. 1992. "Sustainable Rural Livelihoods: Practical Concepts for the 21st Century." IDS Discussion paper 296, Institute of Development Studies, Brighton. http://www.ids.ac.uk/publication/sustainable-rural-livelihoods-practical-concepts-for-the-21st-century.

Chan, F.K.S., G. Mitchell, O. Adekola, and A. Macdonald. 2012. "Flood Risk in Asia's Urban Mega-deltas: Drivers, Impacts and Response." *Environment and Urbanization Asia* 3, 1: 41–61.

Chapagain, A., and A. Hoekstra. 2010. "The Green, Blue and Grey Water Footprint of Rice from Both a Production and Consumption Perspective." Value of Water Research Report Series No. 40, UNESCO-IHE Institute for Water Education, Delft.

Chaudhury, M. 2009. "Assessing the Protection of Forest-based Environmental Services in the Greater Mekong Sub-region." Working Paper No. APFSOS II/WP/2009/14, FAO, Regional Office for Asia and the Pacific, Bangkok.

Chen Zhi. 2011. "Backgrounder: ASEAN–China Free Trade Area." http://news.xinhuanet.com/english2010/china/2011-11/18/c_131254269.htm.

Chetthamrongchai, P., A. Auansakul, and D. Supawan. 2001. "Assessing the Transportation Problems of the Sugarcane Industry in Thailand." *Transport and Communications Bulletin for Asia and the Pacific*.

Chheang, V. 2010. "Environmental and Economic Cooperation in the Mekong Region." *Asia Europe Journal*. doi: 10.1007/s10308-010-0272-9.

Chiang Mai Municipality Office. 2011. *Data for the Year 2011*. Chiang Mai: Chiang Mai Municipality Air Quality and Noise Management Subdivision.

Chiang Mai Provincial Office. 2011. *Chiang Mai Summary Report*. http://www.chiangmai.go.th/meet_file/sarupCM2556.pdf.

Chinh, N.C., Y. Clarke, M.H. Nguyen, L. Lebel, B. T. Sinh, S. Sophat, and S. Boontaveeyuwat. n.d. "Factors that Influence Perception of Water-related Climate Change Risks and Climate Variability in the Mekong Region."

Chinvanno, S., S. Souvannalath, B. Lersupavithnapa, V. Kerdsuk, and N.T.H. Thuan. 2006. "Climate Risks and Rice Farming in the Lower Mekong River Countries." AIACC Working Paper No. 40, AIACC Project Office, International START Secretariat, Washington, D.C.

———. 2008a. "Climate Risks and Rice Farming in the Lower Mekong River Countries." In *Climate Change and Vulnerability*, ed. N. Leary, J. Adejuwon, V. Barros, I. Burton, J. Kulkarni, and R. Lasco, 333–50. London: Earthscan.

———. 2008b. "Strategies for Managing Climate Risks in the Lower Mekong River Basin: A Place-based Approach." In *Climate Change and Vulnerability*, ed. Leary et al., 228–46. London: Earthscan.

Clancy, S. 2010. "Pipeline Projects Fuel Expanding Economies in South East Asia." *Pipelines International*, June.

Coates, D. et al. 2003. Biodiversity and Fisheries in the Lower Mekong Basin. Phnom Penh: Mekong River Commission.

Coates, D., O. Poeu, U. Suntornratana, N. Thanh Tung, and S. Viravong, 2003. "Biodiversity and Fisheries in the Lower Mekong Basin." Mekong Development Series No. 2. Mekong River Commission, Phnom Penh. http://www.mrcmekong.org/assets/Publications/report-management-develop/Mek-Dev-No2-Mek-River-Biodiversityfiisheries-in.pdf.

Cohen, B. 2006. "Urbanization in Developing Countries: Current Trends, Future Projections, and Key Challenges for Sustainability." *Technology in Society* 28: 63–80.

Cohen, J. 1988. *Statistical Power Analysis for the Behavioural Science*, 2nd ed. New Jersey: Lawrence Erlbaum Associates.

Cohen, P.T. 2009. "The Post-Opium Scenario and Rubber in Northern Laos: Alternative Western and Chinese Models of Development." *International Journal of Drug Policy* 20: 424–30.

Collet, L., C. Corner-Dolloff, and R. Lefroy. Forthcoming. Crop Suitability and Climate Change Vulnerability Assessment of Vietnam. CIAT: Hanoi.

Combalicer, E.A., R.V.O. Cruz, S. Lee, and S. Im. 2010. "Assessing Climate Change Impacts on Water Balance in the Mount Makiling Forest, Philippines." *J. Earth Syst. Sci* 119, 3: 265–83.

Committee for Disaster Management (CDM) of Kandal. 2001. "Annual Disaster Management Report of Kandal 2000." (in Khmer).

Committee for Flood and Storm Control (CFSC) of An Giang. 2000. "Annual Flood and Storm Control Report of An Giang 2000." People's Committee of An Giang (in Vietnamese).

———. 2001; 2011. "Annual Flood and Storm Control Report of An Giang." People's Committee of An Giang. (in Vietnamese).

Commonwealth Scientific and Research Organization (CSIRO), National University of Laos (NUOL), and National Agriculture and Forestry Institute (AFRI). 2011. *Household Survey in Outhoumphone and Champhone District Savannakhet Province of Lao PDR. Developing Multi-scale Climate Change Adaptation Strategies for Farming Communities in Cambodia, Laos, Bangladesh and India*. Vientiane: GoL and CSIRO.

Cornford, J., and N. Matthews 2007. *Hidden Costs: The Underside of Economic Transformation in the Greater Mekong Subregion*. Melbourne: Oxfam Australia.

Costanza, R. et al. 1997. "The Value of the World's Ecosystem Services and Natural Capital." *Nature* 387 (May): 253–60.

Coxhead, I., and S. Jayasuriya. 2010. "China, India and the Commodity Boom: Economic and Environmental Implications for Low-income Countries." *World Economy* 33, 4: 525–51.

Dalal-Clayton, B., and S. Bass. 2009. *The Challenges of Environmental Mainstreaming: Experience of Integrating Environment into Development Institutions and Decisions*. London: IIED.

Dang, A.N. 2003. "Internal Migration Policies in ESCAP Region." *Asia–Pacific Population Journal*: 27–40.

Dang, N.M., M.S. Babel, and H.T. Luong. 2010. "Evaluation of Flood Risk Parameters in the Day River Flood Diversion Area, Red River Delta, Vietnam." *Natural Hazards*. doi:10.1007/s11069-010-9558-x.

Das, S. 2012. "A Critical Look at the ASEAN Economic Community Scorecard." *East Asia Forum*, 1 June. http://www.eastasiaforum.org/2012/06/01/a-critical-look-at-the-asean-economic-community-scorecard.

Dasgupta, S., B. Laplante, S. Murray, and D. Wheeler. 2009, "Climate Change and the Future Impacts of Storm-surge Disasters in Developing Countries." Washington, D.C.: Center for Global Development.

Davivongs, J. et al. 2012. "Neglected Canals: Deterioration of Indigenous Irrigation Systems by Urbanization in the West Peri-Urban Area of Bangkok Metropolitan Region." *Water* 4, 1: 12–27.

De Koninck, R., J. Rigg and P. Vandergeest. 2012. "A Half Century of Agrarian Transformations in Southeast Asia, 1960–2010." In *Revisiting Rural Places: Pathways to Poverty and Prosperity in Southeast Asia*, ed. J. Rigg and P. Vandergeest, 25–37. Singapore: NUS Press.

Deelan, L., and P. Vasuprasat. 2010. "Migrant Workers' Remittances from Thailand to Cambodia, Lao PDR and Myanmar." Bangkok: ILO.

Delfino, R.J., R.D. Lasco, and F.P. Lansigan. 2013. "Evaluation of Statistical Downscaling Model for Simulation and Downscaling Temperature and Precipitation in a Critical Watershed in the Philippines." Los Baños: University of the Philippines Los Baños.

Delforge, I. 2007. "Contract Farming in Thailand: A View from the Farm." A Report for Focus on the Global South, CUSRI, Chulalongkorn University. http://ideas.repec.org/p/ess/wpaper/id2340.html.

Department of Drainage and Sewerage (Thailand). n.d. "Water Quality Management and Remediation." http://dds.bangkok.go.th/wqm/English/inplementation.html.

Department for International Development (DFID). 1999. "Key Sheets for Sustainable Livelihood, Figure 1: Sustainable Livelihoods Framework." London: DFID. http://www.oneworld.org/odi/keysheets/.

Department of Environment, Food and Rural Affairs (DEFRA). 2007. *An Introductory Guide to Valuing Ecosystem Services*. London: DEFRA. http://ec.europa.eu/environment/nature/biodiversity/economics/pdf/valuing_ecosystems.pdf.

Deshingkar, P. 2006. "Internal Migration, Poverty and Development in Asia." Paper presented at Asia 2015 Conference, Chatham House, London.

Dietzenbacher, E., J. Pei, C. Yang. 2012. "Trade, Production Fragmentation and China's Carbon Dioxide Emissions." *Journal of Environmental Economics and Management* 64, 1: 88–101.

Dinh, N.Q., S. Balica, I. Popescu, and A. Jonoski. 2012. "Climate Change Impact on Flood Hazard, Vulnerability and Risk of the Long Xuyen Quadrangle in the Mekong Delta." *International Journal of River Basin Management* 10: 103–20.

Dodman, D. 2011. "Forces Driving Urban Greenhouse Gas Emissions." Current Opinion in Environmental Sustainability 3:121–5.

Doorne, S., I. Ateljevic, and Z. Bai. 2003. "Representing Identities through Tourism: Encounters of Ethnic Minorities in Dali, Yunnan province, People's Republic of China." *International Journal of Tourism Research* 5: 1–11.

Dugan, P. 2008. "Mainstream Dams as Barriers to Fish Migration: International Learning and Implications for the Mekong." *Catch and Culture* 14: 9–15.

Dugan, J.P. et al. 2010. "Fish Migration, Dams, and Loss of Ecosystem Services in the Mekong Region." *Ambio* 39, 4: 334–48.

Duong, Q.N. 2008. "Urbanization without Sprawl: Vietnam Experiences and Perspectives." Paper presented at "Urban Growth without Sprawl: A Way Towards Sustainable Urbanization", 44th ISOCARP Congress, Dalian.

Eastham, J., F. Mpelasoka, M. Mainuddin, C.Ticehurst, P. Dyce, G. Hodgson, R. Ali, and M. Kirby. 2008. *Mekong River Basin Water Resources Assessment: Impacts of Climate Change*. Canberra: CSIRO.

Eaton, C.S., and Shepherd, A.W. 2001. "Contract Farming: Partnerships for Growth." *AGS Bulletin* No. 145. Rome: FAO.

Economics for the Environment Consultancy (EFTEC). 2005. "The Economic, Social and Ecological Value of Ecosystem Services: A Literature Review." Final report for DEFRA. London: EFTEC.

Edame, G.E., B.E. Anam, W.M. Fonta, and E.J.C. Duru. 2011. "Climate Change, Food Security and Agricultural Productivity in Africa: Issues and Policy Directions." *International Journal of Humanities and Social Science* 1, 1: 205–23.

Energy Policy and Planning Office. 2011. *Energy Statistics of Thailand 2011*. Bangkok: EPPO.

Ekasingh, B., C. Sungkapitux, J. Kitchaicharoen, and P. Suebpongsang. 2007. "Competitive Commercial Agriculture in the Northeast of Thailand." Competitiveness Country Case Studies. Washington, D.C.: World Bank. http://siteresources.worldbank.org/INTAFRICA/Resources/257994-1215457178567/CCAA_Thailand.pdf.

Electricity Generating Authority of Thailand (EGAT). 2010. *Thailand's Power Development Plan 2010–2030*. Bangkok: EGAT.

Elliott, J.A. 2006. *Sustainable Urban Livelihoods: An Introduction to Sustainable Development*. London: Routledge.

Emerton, L. 2005. *Making the Economic Links between Biodiversity and Poverty Reduction: The Case of Lao PDR*. Colombo: World Conservation Union, Ecosystems and Livelihoods Group Asia.

Emrouznejad, A., E. Cabanda, and R. Gholami. 2010. "An Alternative Measure of the ICT-Opportunity Index." *Information and Management* 47: 246–54.

Etwire, P.M., R.M. Al-Hassan, J.K.M. Kuwornu, Y. Osei-Owusu. 2013. "Application of Livelihood Vulnerability Index in Assessing Vulnerability to Climate Change and Variability in Northern Ghana." *Research Journal of Environment and Earth Science* 3, 2.

Evans, P.B. 2002. "Introduction: Looking for Agents of Urban Livability in a Globalized Political Economy." In *Liveable Cities? Urban Struggles for Livelihood and Sustainability*, ed. P. Evans. Berkeley: University of California Press.

Evans, K., S.J. Velarde, R. Prieto, S.N. Rao, S. Sertzen, K. Davila, P. Cronkleton and W. de Jong. 2006. In *Field Guide to the Future: Four Ways for Communities to Think Ahead*, ed. Bennet E. and Zurek M. Nairobi: CIFOR, ASB, ICRAF.

Felkner, J., K. Tahzibaveya, and R. Townsend. 2009. "Impact of Climate Change on Rice Production in Thailand." *American Economic Review* 99, 2: 205–10.

Few, R. 2003. "Flooding, Vulnerability and Coping Strategies: Local Responses to a Global Threat." *Progress in Development Studies* 3, 1: 43–58.

Filimonau, V., J. Dickinson, and D. Robbins. 2013. "The Role of 'Indirect' Greenhouse as Emissions in Tourism: Assessing the Hidden Carbon Impacts from a Holiday Package Tour." *Transportation Research Part* A 54: 78–91.

Food and Agriculture Organization of the United Nations (FAO). 2007. *State of the World's Forests 2007*. Rome: FAO.

———. 2008a. *Climate Change and Food Security: A Framework Document*. Rome: FAO.

———. 2008b. "Emergency and Rehabilitation Programme Needs Assessment for the Cyclone Nargis Affected Areas: Agriculture (crops, livestock, fisheries, forestry)." Bangkok: FAO.

———. 2010. *Global Forest Resources Assessment 2010, Country Report Myanmar*. Rome: FAO.

———. 2011. "Fishery Value Chain Analysis in Cambodia." Rome: FAO.

———. 2011. *Forest and Forestry in the Greater Mekong Subregion to 2020: Subregion Report of the Second Asia-Pacific Forestry Sector Outlook Study*. FAO, Bangkok, http://www.fao.org/docrep/014/i2093e/i2093e00.pdf (accessed 11 Apr. 2013)

———. 2012. "Strengthening Capacities to Enhance Coordinated and Integrated Disaster Risk Reduction Actions and Adaptation to Climate Change in Agriculture in the Northern Mountain Regions of Viet Nam: Guidelines for Integrating Disaster Risk Reduction and Climate Change Adaptation into Agriculture Development Planning Plans in Phu Tho, Yen Bai, and Lao Cai Provinces." Project UNJP/VIE/037/UNJ.

Forsyth, P., S. Hoque, L. Dwyer, R. Spurr, T.V. Ho, and D. Pambudi. 2008. "Carbon Footprint of Australian Tourism." Technical Reports, Sustainable Tourism Cooperative Research Center. Gold Coast: STCRC.

Franks, P. n.d. "Promoting Equity in the Management of Protected Areas: New Evidence of the Need for Action. CARE International. http://www.povertyandconservation.info/docs/20080524-Phil_Franks_CARE_International2.pdf.

Friend, R.M. 2009. "Fishing for Influence: Fisheries Science and Evidence in Water Resources Development in the Mekong Basin." *Water Alternatives* 2: 167–82.

Friend, R.M. et al. 2009. "Songs of the Doomed: The Continuing Neglect of Capture Fisheries in Hydropower Development in the Mekong." In *Contested Waterscapes in the Mekong Region: Hydropower, Livelihoods and Governance*, ed. by F. Molle, T. Foran and M. Käkönen, 307–31. London: Earthscan.

Garden, P., and S.L. Nance. 2007. "Forums and Flows: Emerging Media Trends." In *Democratizing Water Governance in the Mekong Region*, ed. L. Lebel, J. Dore, R. Daniel, and Y.S. Koma, 157–76. Chiang Mai: Mekong Press.

General Statistics Office (GSO), Vietnam. 1992–2006. Statistical Data. Hanoi: GSO. http://www.gso.gov.vn.

Giersch, C. 2010. "Across Zomia with Merchants, Monks and Musk: Process Geographies, Trade Networks, and the Inner-East-Southeast Asian Borderlands." *Journal of Global History* 5: 215–39.

Gillenwater, M. 2008. "Forgotten Carbon: Indirect CO2 in Greenhouse Gas Emission Inventories." *Environmental Science & Policy* 11: 195–203.

Giok-Ling, O. 2009. "Industrialization and Environmental Management." In *Critical States: Environmental Challenges to Development in Monsoon Southeast Asia*, ed. L. Lebel, et al. Petaling Jaya: Strategic Information and Research Development.

Global Canopy Programme (GCP). 2010. *Forest Ecosystem Services*. http://www.globalcanopy.org.

———. 2011. *Global Canopy Programme*. Vietnam's News: Cultivating the nation's forests: http://sportsmoneyline.net/testblog/cultivating-the-nations-forests/.

GOFC-GOLD. 2010. A sourcebook of methods and procedures for monitoring and reporting anthropogenic greenhouse gas emissions and removals caused by deforestation, gains and losses of carbon stocks in forests remaining forests, and forestation. GOFC-GOLD Report version COP16-1, GOFC-GOLD Project Office, Natural Resources Canada. Alberta.

Goh, D.P.S., and T. Bunnell. 2013. "Recentering Southeast Asian Cities." *International Journal of Urban and Regional Research* 37, 3 : 825–33.

Goldblum, C., and T.-C. Wong. 2000. "Growth, Crisis and Spatial Change: A Study of Haphazard Urbanization in Jakarta, Indonesia." *Land Use Policy* 17: 29–37.

Gonsalves, J.B. 2006. *An Assessment of the Biofuels Industry in Thailand*. Geneva: UNCTAD.

Gossling, S., P. Peeters, J.-P. Ceron, G. Dubois, T. Patterson, R.B. Richardson. 2005. "The Eco-efficiency of Tourism." *Ecological Economics* 54: 417–34.

Goto, K. 2011. "Implications for Laos' Development of its Increasing Regional Integration and Chinese Influence." *Asian–Pacific Economic Literature* 25: 68–88.

Government of Laos (GoL). 2009. *Law on Investment Promotion No. 02 /NA, July*, 2009.

Greacen, C., and J. Footner. 2006. *Decentralizing Thai Power: Towards a Sustainable Energy System*. Bangkok, Greenpeace SE Asia.

Greacen, C.S., and C. Greacen. 2012. "Proposed Power Development Plan (PDP) 2012 and a Framework for Improving Accountability and Performance of Power Sector Planning." Bangkok: Palang Thai.

Gregory, P.J., J.S.I. Ingram, and M. Brklacich. 2005, "Climate Change and Food Security." *Philosophical Transactions of the Royal Society B*, 360, no. 1463, http://rstb.royalsocietypublishing.org/content/360/1463/2139.full.

Grumbine, E., and X. Jianchu. 2011. "Mekong Hydropower Development." *Science* 332: 178–9.

Grumbine, R., J. Dore, and J. Xu. 2012. "Mekong Hydropower: Drivers of Change and Governance Challenges." *Frontiers in Ecology and the Environment* 10: 91–98.

Guttal, S. 2006. "Development and Plunder in the Mekong Region." In *Revisiting Southeast Asian Regionalism*, 33–48. Bangkok: Focus on the Global South.

Hahn, M.B., Anne M.Riederer, and S.O. Foster. 2009. "The Livelihood Vulnerability Index: A Pragmatic Approach to Assessing Risks from Climate Variability and Change; A Case Study in Mozambique." *Global Environ. Change* 19, 1: doi:10.1016/j.gloenvcha. 2008.11.002.

Hall, D. 2011. "Land Grabs, Land Control, and Southeast Asian Crop Booms." *Journal of Peasant Studies* 38, 4: 837–57.

Hall, D., P. Hirsch, and T.M. Li. 2011a. *Powers of Exclusion: Land Dilemmas in Southeast Asia*. Singapore: NUS Press.

———. 2011b. "Post-agrarian Exclusions: Land Conversion." In *Powers of Exclusion: Land Dilemmas in Southeast Asia*, 118–44. Singapore: NUS Press.

Hamilton, L.C., and B.D. Keim. 2009. "Regional Variation in Perceptions about Climate Change." *International Journal of Climatology* 29, 15: 2348–52.

Hannah L., M. Ikegami, D.G. Hole, C. Seo, S.H.M. Butchart, A.T. Peterson, and P.R. Roehrdanz. 2013. "Global Climate Change Adaptation Priorities for Biodiversity and Food Security." *PLoS* ONE 8(8): e72590. doi:10.1371/journal.pone.0072590.

Hardoy, J.E., D. Mitlin, and D. Satterthwaite. 2001. *Environmental Problems in an Urbanizing World*. London: Earthscan.

Harris, N.L. et al. 2012. "Baseline Map of Carbon Emissions from Deforestation in Tropical Regions." *Science* 336, 6088: 1573–6.

Harvey, D. 1996. *Justice, Nature and the Geography of Difference*. Oxford: Blackwell Publishers.

Haughton, G. 1999. "Environmental Justice and the Sustainable City." *Journal of Planning Education and Research* 18, 3: 233–43.

Haughton, G., and G. McGranahan. 2006. "Editorial: Urban Ecologies." *Environment and Urbanization* 18, 1: 3–8.

Health Effects Institute (HEI), International Scientific Oversight Committee. 2010. *Outdoor Air Pollution and Health in the Developing Countries of Asia: A Comprehensive Review*. Special Report 18. Boston: HEI.

Heinimann, A., and P. Messerli. 2013. "Coping with a Land-Grab World: Lessons from Laos." *Global Change* 8: 1–15.

Helsinki University of Technology (TKK) and Southeast Asia START Regional Center (SEA START RC). 2009. "Water and Climate Change in the Lower Mekong Basin: Diagnosis and Recommendations for Adaptation." Bangkok: TKK and SEA START RC, Chulalongkorn University; Helsinki: TKK.

Heynen, N., M. Kaika, and E. Swyngedouw. 2006. "Urban Political Ecology: Politicizing the Production of Urban Natures." In *The Nature of Cities: Urban Political Ecology and the Politics of Urban Metabolism*, ed. N. Heynen, M. Kaika and E. Swyngedouw. London: Routledge.

Hirsch, P. n.d. "Underlying Causes of Deforestation in the Mekong Region." http://enviroscope.iges.or.jp/modules/envirolib/upload/1503/attach/3ws-20-hirsch.pdf.

Hirsch, P. 2010. "The Changing Political Dynamics of Dam Building on the Mekong." *Water Alternatives* 3, 2: 312 – 23.

Hoa, L.T.V., N.H. Nhan, E. Wolanski, T.T. Cong, and H. Shigeko 2007. "The Combined Impact on the Flooding in Vietnam's Mekong River Delta of Local Man-made Structures, Sea Level Rise and Dams Upstream in the River Catchment." *Estuarine, Coastal and Shelf Science* 71: 110–16.

Hoang, M.H. et al. 2013. "Benefit Distribution across Scales to Reduce Emissions from Deforestation and Forest Degradation (REDD+) in Vietnam." *Land Use Policy* 31 (March): 48–60.

Hoggart, K., ed. 2005. *The City's Hinterland: Dynamism And Divergence in Europe's Peri-urban Territories.* Hants: Ashgate.

Hortle, K.G., and S. Bush, 2003. "Consumption in the Lower Mekong Basin as a Measure of Fish Yield." http://www.fao.org/docrep/005/ad070e/ad070e0b.htm.

Hu, Tao, Jun Pang, and Lili Wang. 2011. A Decade in the WTO: Environmental Implications for China. In *A Decade in the WTO: Implications for Global Trade and Governance,* ed. R. Melendez-Ortiz, C. Bellman, and Shuaihua Cheng, 68–73. Geneva: ICTSD Programme on Global Economic Policy and Institutions.

Hu, Tao, Wu Yuping, Jun Pang, Guo Hongyan, and Song Peng. 2011. "Post-ante EIA on China's 10 Years WTO Accession." *Environment and sustainable development* 3: 21–2.

Hu, Tao, Wu Yuping, Shen Xiaoyue, Li Liping, Yu Hai, and Mao Xianqiang. 2007. "Environmental Deficit behind Trade Surplus." *China WTO Tribune* 8: 10–12.

Hue City Office for Statistics. 2011. *Statistical Yearbook 2010.* Hue: Office for Statistics.

Hue College of Science. 2011. "Sanitation Constraints Classification and Alternatives Evaluation for Asian Cities." Report of SaniCon-Asia Project. Hue: Hue College of Science.

Huguet, J.W. et al. 2011. *Thailand Migration Profile. Thailand Migration Report 2011.* Bangkok: IOM.

Hung, H.V., R. Shaw, and M. Kobayashi 2007. "Flood Risk Management for the RUA of Hanoi: Importance of Community Perception of Catastrophic Flood Risk in Disaster Risk Planning." *Disaster Prevention and Management* 16, 2: 245–58.

Hung, T., and Y. Yasuoka. 2000. "Remote Sensing and GIS to Study Sub-urbanization Dynamics: A Case Study of Northern Bangkok, Thailand." In *The Chao Phraya Delta: Historical Development, Dynamics and Challenges of Thailand's Rice Bowl.* http://std.cpc.ku,ac.th/delta/conf/Acrobat/Papers_Eng/Volume%202/Thanawat2.pdf.

Institute for Development Studies (IDS). 2007. "Governance Screening for Urban Climate Change Resilience Building and Adaptation Strategies in Asia: Assessment of Da Nang, Vietnam." Brighton: IDS, University of Sussex.

Institute for Global Environmental Strategies (IGES) and Southeast Asian Regional Center for Graduate Study and Research in Agriculture (SEARCA). 2012. "A Review of Issues and Challenges in Climate Change and Agriculture in Southeast Asia. Hayama: IGES.

Institute for Social and Environmental Transition (ISET). 2010. "The Shared Learning Dialogue: Building Stakeholder Capacity and Engagement for Resilience Action." ISET Climate Resilience in Concept and Practice Working Paper 1, Nov., ISET, Boulder, CO.

Inter Press Service (IPS). 2006. *Crossing Borders: Reportage from our Mekong*. Bangkok: IPS Asia Pacific.

Intergovernmental Panel on Climate Change (IPCC) 2001. *Special Report on Emission Scenarios*. Geneva: IPCC.

———. 2003. "Good Practice Guidance for Land Use, Land-use Change and Forestry." In *IPCC National Greenhouse Gas Inventories Programme*, ed. J. Penman et al. Hayama: IGES.

———. 2006. IPCC Guidelines for National Greenhouse Gas Inventories, ed. Eggleston, H.S. et al. IPCC National Greenhouse Gas Inventories Programme Institute for Global Environmental Strategies (IGES) Hayama, Kanagawa.

———. 2007a. "Climate: The Physical Science Basis." In *Contribution of Working Group I to the Fourth Assessment Report of the IPCC*. Cambridge: Cambridge University Press.

———. 2007b. *IPCC: Climate Change 2007. Synthesis Report*. Cambridge: Cambridge University Press.

. 2007c. Fourth Assessment Report (AR4). *Climate Change 2007: Synthesis Report AR4. 2007. Contribution of Working Groups I, II and III to the Fourth Assessment Report of the Intergovernmental Panel on Climate Change*. Geneva: IPCC.

———. 2007. *Climate Change 2007: Impacts, Adaptation and Vulnerability: Contribution of Working Group II to the Fourth Assessment Report of the Intergovernmental Panel on Climate Change*. Cambridge: Cambridge University Press.

International Center for Environmental Management (ICEM). 2003. "Economic Valuation: Its Use for Protected Area Management." In *Lessons Learned from Global Experience: Review of Protected Areas and Development in the Lower Mekong River Region*. Indooroopilly: ICEM.

———. 2010. *MRC Strategic Environmental Assessment (SEA) of Hydropower on the Mekong*. Hanoi: ICEM.

International Development Research Centre (IDRC). 2001. *Outcome Mapping: Building Learning and Reflection into Development Programs*. Ottawa: IDRC.

International Energy Agency (IEA). 2010. "Renewables in Southeast Asian Countries: Trends and Potentials." http://www.iea.org/countries/non-membercountries/vietnam.

———. 2013. "Feed-in Tariff for Distributed Solar Systems." http://www.iea.org/policiesandmeasures/pams/thailand/name.

International Monetary Fund (IMF). 2006. *Cambodia Poverty Reduction Strategy Paper, 2006*. http://www.imf.org/external/pubs/ft/scr/2006/cr06266.pdf.

———. 2008. *Lao Second Poverty Reduction Strategy Paper, 2008*. http://www.imf.org/external/pubs/ft/scr/2008/cr08341.pdf.

International Organization for Migration (IOM). 2006. *Review of Labor Migration Dynamics in Cambodia*. Phnom Penh: IOM.

International Telecommunications Union (ITU). 2010. *World Telecommunication/ICT Indicators Database*. http://www.itu.int/ITU-D/icteye/Indicators/Indicators.aspx# (accessed Jan. 30, 2011).

International Water Management Institute (IWMI-SEA). 2009. "Scoping Study on Natural Resources and Climate Change in Southeast Asia with a Focus on Agriculture." Colombo: IWMI-SEA.

Interwies, E. 2010. *The Economic and Social Value of the Guinea Current Ecosystem: A First Approximation*. Vienna: UNIDO.

Irwin, F., and J. Ranganathan. 2007. *Restoring Nature's Capital: An Action Agenda to Sustain Ecosystem Services*. Washington, D.C.: WRI.

Island Press. 2007. *Millennium Ecosystem Assessment: A Toolkit for Understanding and Action*. www.islandpress.org.

Janekarnkij, P., and D. Moran. 2005. "Economic Valuation of Ramsar Wetlands in Thailand: Chiang Saen Lake and Krabi River Estuary." Report presented to Implementation of the Ramsar Convention: Management and Protection of Wetland Areas, Bangkok.

Jha, S., D. Roland-Holst, and S. Sriboonchitta. 2010. *Regional Trade Opportunities for Asian Agriculture*. ADB Economics Working Paper Series No. 191, ADB, Manila.

Johnston, R., C.T. Haonh, G. Lacombe, A. Noble, V. Smakhtin, D. Suhardiman, K.S. Pheng, and C.P. Sze. 2009. "Scoping Study on Natural Resources and Climate Change in Southeast Asia with a Focus on Agriculture." Report Prepared for the Swedish International Development Cooperation Agency. Colombo: IWMI.

Johnston, R. et al. 2010. *Climate Change, Water and Agriculture in the Greater Mekong Subregion*. Colombo: IWMI Research Report 136.

Johnston, R., and M. Kummu. 2011. Water Resource Models in the Mekong Basin: A Review. *Water Resource Management*. doi:10.1007/s11269-011-9925-8.

Jun, L.T.F., and Xi Y. 2011. "The Energy Requirements and Carbon Dioxide Emissions of Tourism Industry of Western China: A Case of Chengdu City." *Renewable and Sustainable Energy Reviews* 15: 2887–94.

Kaisti, H. and M. Kakonen. 2012. "Actors, Interests and Forces Shaping the Energyscape of the Mekong Region." *Forum for Development Studies* 39, 2: 147–58.

Kammeier, H.D. 2003. *Rural Urban Sub-regional Linkages in the Mekong Region: A Holistic Approach to Development and Poverty Reduction*. Manila: ADB.

Kaosa-ard, M., ed. 2007. *Mekong Tourism: Blessings for All?* Chiang Mai: Social Research Institute, Chiang Mai University.

Kaosa-ard, M., and J. Dore, eds. 2003. *Social Challenges in the Mekong Region*. Bangkok: White Lotus.

Kapos, V., Ravilious, C., Leng, C., Bertzky, M., Osti, M., Clements, T., Dickson, B. 2010. *Carbon, Biodiversity and Ecosystem Services: Exploring Co-benefits. Cambodia*. UNEP. http://www.unep-wcmc.org/carbon-biodiversity-ecosystem-services-cambodia_570.html.

Kaur, A. 2010. "Labour Migration Trends and Policy Challenges in Southeast Asia." *Policy and Society*. doi:10.1016/j.polsoc.2010.09.001.

Kawai, M., and G. Wignaraja. 2010. "Asian FTAs: Trends, Prospects and Challenges." *Journal of Asian Economics*. doi:10.1016/j.asieco.2010.10.002.

Kennedy, C. et al. 2012. "Sustainable Urban Systems." *Journal of Industrial Ecology* 16, 6: 775–9.

Kenworthy, J.R. 2006. "The Eco-city: Ten Key Transport and Planning Dimensions for Sustainable City Development." *Environment and Urbanization* 18, 1: 67–85.

Khajuria A., and N.H. Ravindranath. 2012. "Climate Change Vulnerability Assessment: Approaches DPSIR Framework and Vulnerability Index." *Earth Sci Climat Change* 3: 109. doi:10.4172/2157-7617.1000109.

Khon Kaen Municipality. 2012. "Khon Kaen Municipality Population and Area Data." http://www.kkmuni.go.th/.

Kiyoshi, H. 2002. "Ethnic Tourism and Cultural Change in the Border Region of Yunnan Province: A Case Study of Xishuangbanna Dai Autonomous Prefecture." In *Inter-ethnic Relations in the Making of Mainland Southeast Asia and Southwestern China*, ed. Y. Hayashi and A. Wichienkeeo, 291–310. Kyoto: CSEAS, Kyoto University.

Koma, Y.S. 2011. "Building experiences with SRI Development and Dissemination in Cambodia (2000–2011)." Phnom Penh: CEDAC. http://www.cedac.org.kh/?page=detail&ctype=article&id=371&lg=en.

Kouphokham, K. 2009. "Recent Development in Lao PDR Energy Sector." Paper presented at Subregional Energy Forum "Special Meeting to Finalize the Road Map for Expanded GMS Energy Cooperation." Bangkok, Mar. 18–19.

Kozul-Wright, R., and P. Fortunato. 2012. "International Trade and Carbon Emissions." *European Journal of Development Research* 24, 4: 509–29.

Krittasudthacheewa, C. 2013. "Understanding and Modeling Climate Change Impact and Uncertainties: Current Effort in Climate Change Study and Possible Use of Robust Decision Making in the Mekong Region." Presentation, Workshop on Governing Critical Uncertainties: Climate Change and Decision-Making in Transboundary River Basins, Kantary Hills Hotel, Chiang Mai, Jan. 2013.

Kubes, J. 2013. "European Post-socialist Cities and their Near Hinterland in Intra-urban Geography Literature." *Bulletin of Geography. Socio-economic Series* 19: 19–43.

Kummu, M., M. Keskinen, and O. Varis, eds. 2008. *Modern Myths of the Mekong*. Helsinki: Helsinki University of Technology.

Kummu, M., and J. Sarkkula. 2008. "Impact of the Mekong River Flow Alteration on the Tonle Sap Flood Pulse." *Ambio* 37: 185–92.

Kyi, A. 2005. "Enhancing the Sustainable Development of Diverse Agriculture through CGPRT Crops in Myanmar: Current Status of CGPRT Crop Agriculture and Identification of its Development Constraints." UNESCAP CAPSA Working Paper No. 85.

Kyoto: Think Global, Act Local (KTGAL). 2009. *A Field Guide for Assessing and Monitoring Reduced Forest Degradation and Carbon Sequestration by Local Communities*, ed. Verplanke J.J. and E. Zahabu. Enschede: University of Twente.

Lacombe, G., M. McCartney, and G. Forkuor. 2012. "Drying Climate in Ghana over the Period 1960–2005: Evidence from the Resampling-based Mann-Kendall Test at Local and Regional Levels." *Hydrol. Sci. J.* 57, 8: 1594–1609.

Lamberts, D. 2006. "The Tonle Sap Lake as a Productive Ecosystem." *Water Resources Development* 22: 481–95.

———. 2008. "Little Impact, Much Damage: The Consequences of Mekong River Flow Alterations for the Tonle Sap Ecosystem." In *Modern Myths of the Mekong*, ed. Kummu, M., M. Keskinen, and O. Varis, 3–18. Helsinki: Helsinki University of Technology.

Lamichhane, K. 2010. "Sustainable Livelihood Approach in Assessment of Vulnerability to the Impacts of Climate Change: A Study of Chhekampar VDC, Gorkha District of Nepal." Project Work submitted to the Kathmandu University, Center For Development Studies, National College, Baluwatar.

Landell-Mills, N. and I.T. Porras. 2002. *Silver Bullet or Fools' Gold? A Global Review of Markets for Forest Environmental Services and Their Impact on the Poor*. London: IIED.

Lawlor, K., E. Weinthal, and L. Olander. 2010. "Institutions and Policies to Protect Rural Livelihoods in REDD+ Regimes." *Global Environmental Politics* 10, 4: 1–11.

Laws, E., and P. Semone. 2009. "The Mekong: Developing a New Tourism Region." In *River Tourism*, ed. B. Prideaux and M. Cooper, 55–73. Wallingford: CABI.

Leaf, M. 2002. "A Tale of Two Villages: Globalization and Peri-urban Change in China and Vietnam." *Cities* 19: 23–31.

LEAP/FAO. 2007. "Contract Farming in Lao PDR: Cases and Questions." In Fullbrook D., *Laos Extension for Agriculture Project (LEAP) for the Government-Donor Sub Working Group on Farmers and Agribusiness*, October 2007. http://www.fao.org/uploads/media/Library_Contract_Farming_in_Laos_Cases_Questions_2007_LEAP.pdf.

Lebel, L. et al. 2007. "Integrating Carbon Management into the Development Strategies of Urbanizing Regions in Asia: Implications of Urban Function, Form, and Role." *Journal of Industrial Ecology* 11, 2: 61–81.

Lebel, L., and S. Lorek 2008. "Enabling Sustainable Production–Consumption Systems." *Annual Review of Environment and Resources* 33, 1: 241–75.

Lebel, L., E. Nikitina, and B.T. Sinh, eds. 2008. "Climate Change and the Science and Practice of Managing Floods in Urbanizing Regions of Monsoon Asia." MAIRS Working Paper Series no. 4. Monsoon Asia Integrated Regional Study International Project Office and the Institute for Atmospheric Physics, Chinese Academy of Science, Beijing.

Lebel, L. and B. Sinh. 2009. "Risk Reduction or Redistribution? Flood Management in the Mekong Region." *Asian Journal of Environment and Disaster Management* 1, 1: 23–39.

Lebel, L. and R. Daniel. 2009. "The Governance of Ecosystem Services from Tropical Upland Watersheds." *Current Opinion in Environmental Sustainability* 1, 1: 61–8.

Lebel, L., A. Snidvongs, C.-T. A. Chen, and R. Daniel. 2009a. *Critical States: Environmental Challenges to Development in Monsoon Southeast Asia*. Petaling Jaya: SIRD.

Lebel, L. et al. 2009. "The Implications of Urbanization for Atmospheric Emissions at Multiple Levels." In *Critical States*, 112–32.

Lebel, L., R.T. Perez, T. Sukhapunnaphan, B.V. Hien, N. Vinh, and P. Garden. 2009. "Reducing Vulnerability of Urban Communities to Flooding." In *Critical States*, 381–99.

Lebel, L., T. Foran, P. Garden, and B.J. Manuta. 2009. "Adaptation to Climate Change and Social Justice: Challenges for Flood and Disaster Management in Thailand." In *Climate Change Adaptation in the Water Sector*, ed. F. Ludwig et al. 125–41. London: Earthscan.

Lebel, L., P. Lebel, and R. Daniel. 2010. "Water Insecurities and Climate Change Adaptation in Thailand." In *Climate Change Adaptation and Disaster Risk Reduction: An Asian Perspective*, ed. R. Shaw, J. Pulhin, and J. Pereira, 349–72. Bingley: Emerald.

Lebel, L., T. Grothmann, and B. Siebenhuner. 2010. "The Role of Social Learning in Adaptiveness: Insights from Water Management." *International Environmental Agreements* 10: 333–53.

Lebel, L., B.T. Sinh, N. Chinh, S. Boontaveeyuwat, and H. Kimkong. 2013. "Risk Communication and Adaptation Planning in Deltas and Coastal Settlements of the Mekong Region." *Successful Adaptation to Climate Change: Linking Science and Policy in a Rapidly Changing World*, ed. S. Moser and M. Boykoff, 253–69. New York: Routledge.

Leebouapao, L. and S. Voladeth. 2011. "Agricultural Development, Trade, and Regional Cooperation in an Integrating and Industrializing East Asia: The Case of Lao PDR." In *Agricultural Development, Trade and Regional Cooperation in Developing East Asia*, ed. P.S. Intal Jr, S. Oum, and M.J.O. Simorangkir, 269–306. Jakarta: Economic Research Institute for ASEAN and East Asia (ERIA).

Lefroy, R., L. Collet, and C. Grovermann. 2010. "Study on Potential Impacts of Climate Change on Land Use in the Lao PDR." International Center for Tropical Agriculture, Report prepared for: Lao–German Land Management and Registration Project (LMRP), Deutsche Gesellschaft für Technische Zusammenarbeit (GTZ) GmbH.

Lim, J.J., and Wi Y.C. 2005. "Digital Dreams and Divergent Regimes: The Impact of ICT on Pacific Asia." In *Pacific Asia 2022: Sketching Futures of a Region*, ed. S.S.C. Tay, 66–106. Tokyo: Japan Center for International Exchange.

Lin, S., and C. Grundy-Warr. 2012. "One Bridge, Two Towns and Three Countries: Anticipatory Geopolitics in the Greater Mekong Subregion." *Geopolitics* 17: 952–79.

Lo, D.S. 2011. "Climate Change, Hazards, Vulnerabilities and Risks, Overview for Asia, the Philippines and CdO." Powerpoint presented at Expert Seminar on "Regional Urban Networks: Urban Environmental and Risk Management in Southeast Asia." Cebu, Mar. 27–Apr. 6, 2011.

Loc, V.T.T., S.R. Bush, L.X. Sinh, and N. Khiem. 2010. "High and Low Value Fish Chains in the Mekong Delta: Challenges for Livelihoods and Governance." *Environment, Development and Sustainability* 12: 889–908.

Ma, J., A. Hoekstra, H. Wang, A. Chapagain, and D. Wang. 2006. "Virtual versus Real Water Transfers within China." *Phil.Trans.R.Soc. B* 361: 835–42.

Maibach, E., C. Roser-Renouf, and A.A. Leiserowitz 2008. "Communication and Marketing as Climate Change-Intervention Assets: A Public Health Perspective." *American Journal of Preventive Medicine* 35, 5: 488–500.

Manivong, V., R. Cramb, and J. Newby. 2012. "Rice and Remittances: The Impact of Labour Migration on Rice Intensification in Southern Laos." Paper presented at the 56th AARES (Australian Agriculture and Resource Economics Society) annual conference. Fremantle, Feb. 7–10.

Manarungsan, S., and S. Suwanjindar. 1992. "Contract Farming and Outgrower Schemes in Thailand." In *Contract Farming in Southeast Asia: Three Country Case Studies*, ed. D. Glover and Ghee L.T., 10–70. Kuala Lumpur: Institute Pengajian Tinggi.

Manoram, K. et al. 2011. "Cross-Border Contract Farming Arrangement: Variations and Implications in the Lao People's Democratic Republic." ADB Greater Mekong Subregion–Phnom Penh Plan for Development Management Research Report Series. 1, 2. http://mssrc.la.ubu.ac.th/2006/admin/Projects/18/cross-border-contract-farming.pdf.

Mantanarat, R. 2010. "Laos Enjoys 20% Increase in Visits." *TR Weekly*, 16 Dec. 2010. http://www.ttrweekly.com.

Maplecroft. 2011. "New Products and Analysis: Climate Change Vulernability Index (CCVI)." http://maplecroft.com/about/news/ccvi.html.

Marcotullio, P.J. 2003. "Globalization, Urban Form and Environmental Conditions in Asia–Pacific Cities." *Urban Studies* 40, 2: 219–47.

———. 2007. "Urban Water-related Environmental Transitions in Southeast Asia." *Sustainability Science* 2, 1: 27–54.

Marks, D. 2011. "Climate Change and Thailand: Impact and Response." *Contemporary Southeast Asia*, 33, 2: 229–58.

Martin, P. 2007. "The Economic Contribution of Migrant Workers to Thailand: Towards Policy Development." Bangkok: ILO.

Matthews, N. 2012. "Water Grabbing in the Mekong Basin: An Analysis of the Winners and Losers of Thailand's Hydropower Development in Lao PDR." *Water Alternatives* 5, 2: 392–411.

May, J. 2010. "International Financial Volatility and Commodity Exports: Evidence from the Thai Agricultural Sector." *American Journal of Agricultural Economics* 92: 763–75.

May Lay. 2012. "Farmers Want National Association to Protect Interests." *The Irrawaddy*, 18 September 2012. http://www.irrawaddy.org/archives/14280.

Mayrand, K., and M. Paquin. 2004. "Payments for Environmental Services: A Survey and Assessment of Current Schemes." Montreal: Unisfera Center International Center.

McDonald, K., P. Bosshard, and N. Brewer. 2009. "Exporting Dams: China's Hydropower Industry goes Global." *Journal of Environmental Management* 90: S294–S302.

McElwee, P. 2010. "The Social Dimensions of Adaptation to Climate Change in Vietnam." World Bank Discussion Paper No. 17. Washington, D.C.: World Bank.

McGee, T.G. 1991. "The Emergence of Desakota Regions in Asia: Expanding a Hypothesis." In *The Extended Metropolis: Settlement Transition in Asia*, ed. N. Ginsburg, B. Koppel and T.G. McGee, 3–25. Honolulu: University of Hawai'i Press.

McGranahan, G. 2007. "Urban Transitions and the Spatial Displacement of Environmental Burdens." In *Scaling Urban Environmental Challenges: From Local to Global and Back*, ed. P.J. Marcotullio and G. McGranahan, 18–4. Abingdon: Earthscan.

McKinsey & Company. 2009. "Pathways to a Low Carbon Economy, version 2 of the Global Greenhouse Gas Abatement Cost Curve." https://solutions.mckinsey.com/climatedesk/default.aspx.

Mekong Migration Network (MMN) and Asian Migrant Centre (AMC). 2013. *Migration in the Greater Mekong Subregion: Resource Book. In-depth Study: Border Economic Zones and Migration*. Chiang Mai: MMN and AMC.

Mekong River Commission (MRC). 2002. Fisheries. Mekong River Commission Basin Development Planning: Regional Sector Overview 2002. n.p.

———. 2004. *Distribution and Ecology of Some Important Riverine Fish Species of the Mekong River Basin*. MRC Technical Paper No. 10, May 2004. http://www.mrcmekong.org/assets/Publications/technical/tech-No10-distribution-n-ecology-of-important.pdf

———. 2005a. Annual Mekong Flood Report 2005: http://www.mrcmekong.org/assets/Publications/basin-reports/Annual-Mekong-Flood-Report-2005.pdf.

———. 2005b. *Annual Report, Apr. 2004–Mar. 2005*. Vientiane: The MRC Program for Fisheries Management and Development Corporation.

———. 2006a. *Annual Mekong Flood Report 2006*. http://www.mrcmekong.org/assets/Publications/basin-reports/Annual-Mekong-Flood-Report-2006.pdf.

———. 2006b. *Fish Migration Triggers in the Lower Mekong Basin and Other Tropical Freshwater Systems*. MRC Technical Paper No. 14, Dec. 2006. http://www.mrcmekong.org/assets/Publications/technical/tech-No14-fish-migration-triggers.pdf.

———. 2009a. *Adaptation to Climate Change in the Countries of the Lower Mekong Basin. MRC Management Information Booklet Series No.1*. http://www.mrcmekong.org/assets/Publications/report-management-develop/MRC-IM-No1-Adaptation-to-climate-change-in-LMB.pdf.

———. 2009b. *Annual Mekong Flood Report 2009*. http://www.mrcmekong.org/assets/Publications/basin-reports/Annual-Mekong-Flood-Report-2009.pdf.

———. 2010a. *Annual Flood report 2010*. http://www.mrcmekong.org/assets/Publications/basin-reports/Annual-Flood-Report-2010-2.pdf.

———. 2010b. *Results of Strategic Environmental Assessment of hydropower on the Mekong mainstream*. http://www.mrcmekong.org/MRC_news.

Mekonnen, G. 2012. *Urban Sprawl: Its Economic Impact on Hinterland Farmers*. Lambert Academic Publishing.

Menon. J. 2012. "Narrowing the Development Divide in ASEAN: The Role of Policy." ADB Working Paper Series on Regional Economic Integration No. 100. http:// http://aric.adb.org/pdf/workingpaper/WP100_Menon_Narrowing_the_Development_Divide.pdf.

Meadows, D., J. Randers, and D. Meadows. 2004. *Limits to Growth: The 30-year update*. White River Junction, VT: Chelsea Green.

Merkelsen, H. 2011. "Risk Communication and Citizen Engagement: What to Expect from Dialogue." *Journal of Risk Research* 14: 631–45.

Middleton, C. et al. 2009. "Old and New Hydropower Players in the Mekong Region: Agendas and Strategies." In *Contested Waterscapes in the Mekong Region*, 23 –54.

Middleton, C. 2012. "Transborder Environmental Justice in Regional Energy Trade in Mainland South-East Asia." *Austrian Journal of Southeast Asian Studies* 5, 2: 292–315.

Middleton, C., J. Garcia, and T. Foran. 2009. "Old and New Hydropower Players in the Mekong Region: Agendas and Strategies." In *Contested Waterscapes in the Mekong Region: Hydropower, Livelihoods and Governance*, ed. F. Molle, T. Foran, and M. Käkönen, 413. London: Earthscan.

Milattanapheng, C. 2010. "Country Presentation: Renewable Energy Strategy, EEP Mekong Regional Forum, Oct. 2010, Vientiane, LAO PDR." http://www.eepmekong.org/_downloads/_regional_forum/Presentations/2_Country_reports/1_Presentation_Lao%20PDR.pdf.

Millennium Ecosystem Assessment (MA). 2005a. *Ecosystems and Human Well-being: Synthesis*. Millennium Ecosystem Assessment. Washington, D.C.: Island Press.

———. 2005b. *Ecosystems and Human Well-being: A Framework for Assessment*. Millennnium Ecosystem Assessment. Washington, D.C.: Island Press.

Ministry of Agriculture and Forestry, Lao PDR. 2010. Strategy for Agricultural Development 2011-2020, Final Draft, Vientiane, Lao PDR.

Ministry of Commerce, PRC. 2011. *Statistics on Foreign Investment 2011*. http://english.mofcom.gov.cn/article/statistic/foreigninvestment/?2.

Ministry of Energy (Thailand). 2009. *National Energy Policy 2009*. http://www.eppo.go.th/doc/index.html#policy.

———. 2011. *Thailand 20-Year Energy Efficiency Development Plan (2011–2030).*, Bangkok: Ministry of Energy.

Ministry of Forestry. 2009. *Fourth National Report to the United Nations Convention on Biological Diversity*. Nay Pyi Taw: Government of the Union of Myanmar.

Ministry of Natural Resources (MONRE). 2011. *Development of Emission Factor of Viet Nam Grid Electricity*. Hanoi: MONRE.

Ministry of Planning 2012. *Migration in Cambodia: Report of the Cambodian Rural Urban Migration Project CRUMP*. Phnom Penh: Ministry of Planning, Royal Government of Cambodia.

Mitchell, A.W., Secoy, K., Mardas N., Trivedi, M., and Howard, R. 2008. *Forests NOW in the Fight Against Climate Change. Forest Foresight Report 1.v3.* Oxford: Global Canopy Programme.

Molle, F., T. Foran, and M. Kakonen, eds. 2009. *Contested Waterscapes in the Mekong Region: Hydropower, Livelihoods and Governance*. London: Earthscan.

Morton, J.F. 2007. "The Impact of Climate Change on Smallholder and Subsistence Agriculture." *Proceedings of the National Academy of Sciences of the United States of America* 104, 50: 19680–5.

Morton, T.A., A. Rabinovich, D. Marshall, and P. Bretschneider. 2011. "The Future that May (or May Not) Come: How Framing Changes Responses to Uncertainty in Climate Change Communications." *Global Environmental Change* 21: 103–9.

Moser, S.C. 2010. "Communicating Climate Change: History, Challenge, Process and Future Directions." *WIREs Climate Change* 1: 31–53. doi:10.1002/wcc.011.

Mukama, K., I. Mustalahti, and E. Zahabu. 2012. "Participatory Forest Carbon Assessment and REDD+: Learning from Tanzania." *International Journal of Forestry Research*. Article ID 126454, 14 pages. doi:10.1155/2012/126454.

Narain, V. 2009. "Growing City, Shrinking Hinterland: Land Aquisition, Transition and Conflict in Peri-urban Gurgaon, India." *Environment and Urbanization* 21, 2: 501–12.

———. 2010. "Periurban Water Security in a Context of Urbanization and Climate Change: A Review of Concept and Relationship." Periurban Water Security Discussion Paper Series, Paper No. 1, SaciWATERS, Sainikpuri.

Naritoom, C. 2000. "Contract farming in Central Plain: A Case Study of Asparagus Grower Groups in Nakhon Pathom Province." Paper presented at the International Conference on The Chao Phraya Delta: Historical Development, Dynamic and Challenges of Thailand's Rice Bowl, December 12–14, 2000, Kasetsart University, Bangkok.

National Academy of Sciences (NAS). 2004. *Translating Ecosystem Functions to the Value of Ecosystem Services: Case Studies.* http://www.nap.edu/openbook.php?record_id=11139&page=153.

———. 2005. *Valuing Ecosystem Services: Toward Better Environmental Decision-Making.* http://www.nap.edu/openbook.php?isbn=030909318X: National Academy of Sciences, National Academies Press, Washington, D.C.

National Bureau of Statistics (China). 2012. *2011 Statistical Bulletin of China's Outward Foreign Direct Investment.* Beijing: China Statistical Press.

———. 2004–2012. *China Statistical Yearbooks.* Beijing: China Statistical Press.

National Statistics Office (NSO). 2012. *Philippines in Figures.* http://www.census.gov.ph.

National Statistics Office of Thailand (NSOT). 2008. "The Intercensal Survey of Agriculture, 2008." Bangkok: NSOT.

Naughton-Treves, L. and C. Day, eds. 2012. *Lessons about Land Tenure, Forest Governance and REDD+. Case Studies from Africa, Asia and Latin America.* Madison: UW-Madison Land Tenure Center.

Navya, H., and M. Bhattarai. 2009. "Economics and Livelihoods of Small-scale Inland Fisheries in the Lower Mekong Basin: A Survey of Three Communities in Cambodia." *Water Policy* 11, 1: 31–51.

Neef, A. and D. Thomas. 2009. "Rewarding the Upland Poor for Saving the Commons? Evidence from Southeast Asia." *International Journal of the Commons* 3, 1: 1–15.

Nerlich, B., N. Koteyko, and B. Brown. 2010. "Theory and Language of Climate Change Communication." *WIREs Climate Change* 1: 97–110.

Nevins, J., and N.L. Peluso 2008. *Taking Southeast Asia to Market.* Ithaca: Cornell University Press.

Newman, P. 2006. "The Environmental Impact of Cities." *Environment and Urbanization* 18, 2: 275–95.

Newman, P., and I. Jennings 2008. *Cities as Sustainable Ecosystems: Principles and Practices.* Washington D.C.: Island Press.

Nguyen, N. 2011. "Floods of Dragon Year." *Baomoi.com.* http://www.baomoi.com/nam-thin-bao-lut/.

Nguyen, V.S., and Nguyen D.C. 2008. "Study on Local Community Institutions to Cope with the Flood Situation of the Mekong Region." SUMERNET Project No 6591 Report.

Nicholls, J., E. Lawlor, E. Neitzert, and T. Goodspeed. 2012. *A Guide to Social Return on Investment*. London: Office of the Third Sector, The Cabinet Office.

Nielsen, S.P., A. Sesartic, and M. Stucki. 2010. "The Greenhouse Gas Intensity of the Tourism Sector: The Case of Switzerland." *Environmental Science and Policy* 13: 131–40.

Nolintha, V. 2011. "Cities, SEZs, Connectivity in between Major Cities of Laos." In *Intra and Inter City Connectivity in the Mekong Region*, ed. M. Ishida. BRC Research Report No. 6. Bangkok Research Center, IDE-JETRO, Bangkok.

Norman-Lopez, A., and J. Innes. 2008. "Review of River Fisheries in Tropical Asia." In *Tropical River Fisheries Valuation: Background Papers to a Global Synthesis*, ed. A. Neiland, and C. Béne. Penang: World Fish.

Norton, B.G. 2012. "Valuing Ecosysterms." *Nature Education Knowledge* 3, 10: 2.

O'Brien K. et al. 2004. "Mapping Vulnerability to Multiple Stressors: Climate and Globalization in India." *Global Environmental Change* 14: 303–13.

Odufuwa, B.O., B.A. Odufuwa, and S.O. Fasina. 2012, "Climate Change and Livelihood: The Two Sides of the Coin." *Indonesian Journal of Geography* 44, 1: 47–61.

Oehlers, A. 2006. "A Critique of ADB Policies towards the Greater Mekong Sub-Region." *Journal of Contemporary Asia* 36, 4: 464–78.

Ohlemuller, R. et al. 2008. "The Coincidence of Climatic and Species Rarity: High Risk to Small-range Species from Climate Change." *Biology Letters*. http://www.biochange-lab.eu/wordpress/wp-content/uploads/ohlemuller-et-al-biolett20081.pdf.

Ooi, G.L. 2007. "Urbanization in Southeast Asia: Assessing Policy Process and Progress toward Sustainability." *Journal of Industrial Ecology* 11, 2: 31–42.

Opitz-Stapleton, S., G. Guilbert, K. MacClune, K. MacClune, L. Sabbag, and S. Tyler. 2010. "Only Death is Certain, Yet You Still Get Out of Bed in the Morning: Or Observations on the Use of Climate Information in Adaptation and Resilience Practice." In ISET Climate Resilience in Concept and Practice Working Paper 2, Nov. ISET.

Organisation for Economic Co-operation and Development (OECD). 2006. *Applying Strategic Environmental Assessment: Good Practice Guidance for Development Co-operation*. http://www.oecd.org/dataoecd/4/21/37353858.pdf.

———. 2009. *Integrating Climate Change Adaptation into Development Co-operation: Policy Guidance*. Paris: OECD.

———and FAO. 2011. "Sugar." In *OECD-FAO Agricultural Outlook 2011–2020*, Chap. 6. http://www.oecd.org/dataoecd/2/37/48184295.pdf.

———and International Transport Forum (OECD/ITF). 2009. "The Cost and Efficiency of Reducing Transport GHG Emissions. Preliminary Findings." Paris: OECD.

Pagiola, S., J. Bishop, and N. Landell-Mills, eds. 2002. *Selling forest environmental services: Market-based mechanisms for conservation and development*. London: Earthscan.

Pagiola, S., K. von Ritter, and J. Bishop. 2004. *How Much is an Ecosystem Worth? Assessing the Economic Value of Conservation*. Washington, D.C.: World Bank.

Panchamlong, S. 2010. "Contract Farming and Riverine Aquaculture: Realities of the 'Slave Contract' and the Risks that Farmers Must Bear." *Asian Labour Update*, No. 74 Agricultural Workers' Struggle in Asia. http://www.amrc.org.hk/node/1000.

Park, N.-W. 2005. "Underground Economy: Causes and Size." Presentation to Korean Institute of Public Finance.

Parnwell, M., and L. Wongsuphasawat. 1997. "Between the Global and the Local: Extended Metropolitanisation and Industrial Location Decision Making in Thailand." *Third World Planning Review* 19, 2: 119–38.

PEI-Lao PDR. 2011. "Assessment of Economic, Social and Environmental Costs and Benefits of Mitr Lao Sugar Plantation and Factory: Case Study in Savannakhet Province." Vientiane: IUCN Lao PDR and National Economic Research Institute (NERI), Ministry of Planning and Investment of Lao PDR. http://www.unpei.org/sites/default/files/dmdocuments/SVK%20Sugarcane%20Report_Final%20Apr%202011.pdf.

Peñalba, L.M. et al. 2012. "Climate Change Implications to Food Security and Livelihood of Small-Scale farmers." Center for Strategic Planning and Policy Studies, College of Public Affairs and Development, University of the Philippines Los Baños.

Perez, R.T., L.A. Amadore, and R.B. Feir. 1999. "Climate Change Impacts And Responses in The Philippines Coastal Sector." *Climate Research* 2: 97–107.

Pham K.N., and Tran V.H.S. n.d. "Recreational Value of the Coral Surrounding the Hon Mun Islands in Vietnam: A Travel Cost and Contingent Valuation." In *Economic Valuation and Policy Priorities for Sustainable Management of Coral Reefs*, ed. M. Ahmed, Chiew K.C., and H. Cesar. Penang: WorldFish Center.

Philippines Atmospheric, Geophysical and Astronomical Services Administration (PAGASA). 2011. *Climate Change in the Philippines*. Funded under the MDGF-1656 "Strenghthening the Philippines Institutional Capacity to Adapt to Climate Change." Manila: PAGASA.

Pholphirul, P., and P. Rukumnuaykit. 2010. "Economic Contribution of Migrant Workers to Thailand." *International Migration* 48: 174–202.

Pholsena, V., and R. Banomyong. 2006. *Laos: From Buffer State to Crossroads?* Chiang Mai: Mekong Press.

Phongpaichit, P., and P. Benyaapikul. 2013. "Political Economy Dimension of a Middle Income Trap: Challenges and Opportunities for Policy Reform: Thailand." Bangkok: Faculty of Economics, Chulalongkorn University.

Phonvisay, A., and S. Bush. 2001. "Baseline Study of Fish Trade from the Siphandone Fishery, Champassak Province." Living Aquatic Resources Research Centre (LARReC) Research Paper No. 4, LARReC, Vientiane.

Plungsricharoensuk, P. 2009. "Male Transnational Labour Returnees and Contested Meanings of 'khon pai klai'" [in Thai]. *Social Science Journal of the Faculty of Social Sciences*, Chiang Mai University 21: 59–100.

Pollution Control Department (PCD). 2002. *Domestic Wastewater and Treatment Systems*. Bangkok: PCD.

Poncet, S. 2006. "Economic Integration of Yunnan with the Greater Mekong Subregion." *Asian Economic Journal* 20: 303–17.

Pongyelar, S. 2007. "Thailand's Foreign Policy-making towards Myanmar since 1988: The Reflection from the Business Sector's Roles." *Kasetsart J. Soc. Sci.* 28: 127–40.

Potter, R.B., and T. Unwin 1989. *The Geography of Urban–Rural Interaction in Developing Countries*. London: Routledge.

Prachvuthy, M. et al. 2013. "The Impacts of Contract Farming on Households Who Grow Sugarcane or Rice in Countries of the Greater Mekong Sub-region." USER Working Paper WP-2013-12. Chiang Mai: USER, Chiang Mai University.

Provincial Agriculture and Forestry Office (PAFO). 2013. "Report on Agriculture Production and Forestry in Savannakhet in the Dry Season for the Fiscal Year 2011–2012 and the Annual Plan for Productive Seasons of 2012." Vientiane: PAFO.

Prowse, M. 2012. "Farming in Developing Countries: A Review." Research Department, French Agency for Development. Paris. February 2012. http://www.afd.fr/webdav/site/afd/shared/PUBLICATIONS/RECHERCHE/Scientifiques/A-savoir/12-VA-A-Savoir.pdf.

Quang N., and N. Sato. 2008. "The Role of Forest in People's Livelihood: A Case Study in North-eastern Vietnam." *J. Fac. Agr.* Kyushu Univ. 53, 1: 357–62.

Rab, M.A., N. Hap, M. Ahmed, L. Seng, and K. Viner. 2005. "Marketing Infastructure, Distribution Channels and Trade Pattern of Inland Fisheries Resources in Cambodia: An Exploratory Study." Penang: WorldFish.

Ramaswami, A. et al. 2012. "A Social-Ecological-Infrastructural Systems Framework for Interdisciplinary Study of Sustainable City Systems." *Journal of Industrial Ecology* 16, 6: 801–13.

Ramsar Convention Secretariat. 2010. *The Ramsar Strategic Plan 2009–2015: Goals, Strategies, and Expectations for the Ramsar Convention's Implementation for the Period 2009 to 2015*. Ramsar Handbooks for the Wise Use of Wetlands, 4th ed., vol. 21. Gland: Ramsar Convention Secretariat.

Ranganathan, J., F. Irwin, and C. Procopé Repinski. 2009. *Banking on Nature's Assets How Multilateral Development Banks Can Strengthen Development by Using Ecosystem Services*. Washington, D.C.: WRI.

Ranganathan, J. et al. 2008. *Ecosystem Services: A Guide for Decision Makers*. Washington, D.C.: WRI.

Ranger, N., and S.-L. Garbett-Shiels. 2011. "How can Decision-Makers in Developing Countries Incorporate Uuncertainty about Future Climate Risks into Existing Planning and Policy-making Processes?" Policy Paper. London: Centre for Climate Change Economics and Policy, Grantham Research Institute on Climate Change and the Environment.

Ravets, J., C. Fertner, and T.S. Nielson. 2013. "The Dynamics of Peri-Urbanization." In *Peri-urban Futures: Scenarios and Models for Land Use Change in Europe*, ed. K. Nilson et. al., 13–44. Berlin and Heidelberg: Springer-Verlag.

Redman, C.L., and N.S. Jones, 2004. "The Environment, Social and Health Dimensions of Urban Expansion." Paper presented at Population Environment Research Network, Cyberseminar "Urban Expansion: The Environmental and Health Dimension," Nov. 29–Dec. 15.

Ribot, J., and A.M. Larson. 2012. "Reducing REDD Risks: Affirmative Policy on an Uneven Playing Field." *International Journal of the Commons* 6, 2: 233–54.

Richards, M. and M. Jenkins. 2007. "Potential and Challenges of Payments for Ecosystem Services from Tropical Forests." Forest Policy Environment Programme (FPEP) Forestry Briefing 16, Overseas Development Institute, London.

Rigg, J. 2003. "Evolving Urban–Rural Relations and Livelihoods." In *Southeast Asia Transformed: A Geography of Change*, ed. C.L. Sien, 231–25. Singapore: ISEAS.

———. 2006. "Land, Farming, Livelihoods, and Poverty: Rethinking the Links in the Rural South." *World Development* 34, 1: 180–202.

Rimmer, P.J., and H. Dick 2009. *The City in Southeast Asia: Patterns, Processes and Policy*. Singapore: NUS Press.

Ringler, C., and X. Cai. 2006. "Valuing Fisheries and Wetlands Using Integrated Economic-Hydrologic Modeling—Mekong River." SPECIAL ISSUE: Economic-Engineering Analysis of Water Resource Systems. *Journal of Water Resources Management* 132, 6: 480–87.

Robbins, P. 2004. *Where to Now? Political Ecology: A Critical Introduction*. Oxford: Blackwell Publishing.

Rohrmann, B. 2008. "Risk Perception, Risk Attitude, Risk Communication, Risk Management: A Conceptual Appraisal." In Global Co-operation in Emergency and Disaster Management, ed. International Emergency Management Society. Conference booklet.

Rondinelli, D.A. 1985. *Applied Methods of Regional Analysis: The Spatial Dimension of Development Policy*. Boulder: Westview.

Royal Govt of Cambodia (RGoC). 2010. "Policy Paper on the Promotion of Paddy Production and Rice Export, Royal Government of Cambodia, Council of Ministers." http://www.card.gov.kh/tl_files/CARDIG/Agriculture_policy/Rice%20policy%20in%20Engilish%20version.pdf.

———. 2011. *Sub-decree on Contract Farming Royal Government of Cambodia*, No. 36 ANKR BK. Royal Government of Cambodia, Council of Ministers. http://www.kuratapepper.com/Contract_Farming_Law.pdf.

Russell, C. 2008. "Infrastructure and Regional Market Development: Is the Greater Mekong Sub-Region viable?" In *Transborder Issues in the Greater Mekong Sub-region*, ed. S. Thaweesit, P. Vail, and R. Sciortino, 419–50. Ubon Ratchathani: Mekong Sub-region Social Research Center, Ubon Ratchathani University.

Rutherford, J., K. Lazarus, and S. Kelley. 2008. *Rethinking Investments in Natural Resources: China's Emerging Role in the Mekong Region*. Phnom Penh: Heinrich Boll Stiftung Cambodia.

Sajor, E.E., and R. Ongsakul 2007. "Mixed Land Use and Equity in Water Governance in Peri-Urban Bangkok." *International Journal of Urban & Regional Research* 31, 4: 782–801.

Sari, B.R. 2010. "The Impact of Organic Rice Contract Farming on Farmers' Livelihood and Land Tenure in Cambodia: A Case Study in Kampong Speu Province." M.A. Thesis, Chulalongkorn University, Bangkok.

Sarkkula, J., M. Keskinen, J. Koponen, M. Kummu, J. Richey, and O. Varis. 2009. Hydropower in the Mekong Region: What are the Impacts on Fisheries? In

Contested Waterscapes in the Mekong Region: Hydropower, Livelihoods and Governance, ed. F. Molle, T. Foran, and M. Käkönen, 227–51. London: Earthscan.

Sathirathai, S., and E. Barbier. 2001. "Valuing Mangrove Conservation in Southern Thailand." *Contempory Economic Policy* 19, 2: 109–22.

Satterthwaite, D. 2006. "Outside the Large Cities: The Demographic Importance of Small Urban Centres and Large Villages in Africa, Asia and Latin America." London: IIED.

Satterthwaite, D., and C. Tacoli 2003. "The Urban Part of Rural Development: The Role of Small and Intermediate Urban Centres in Rural and Regional Development and Poverty Reduction." IIED Working Paper Series, No. 9. London: IIED.

Savage, V.R. 2006. "Ecology Matters: Sustainable Development in Southeast Asia." *Integrated Research System for Sustainability Science* 1, 37–63.

Schneider, F. 2002. *Hiding in the Shadows: The Growth of the Underground Economy*. Washington, D.C.: IMF.

Schönweger, O., A. Heinimann, M Epprecht, J. Lu, and P. Thalongsengchanh. 2012. "Concessions and Leases in the Lao PDR: Taking Stock of Land Investments." Bern: Centre for Development and Environment.

Schönweger, O., and A. Üllenberg. 2009. *Foreign Direct Investment (FDI) in land in the Lao PDR*. Eschborn: GTZ.

Schulz, N.B. 2010. "Delving into the Carbon Footprints of Singapore: Comparing Direct and Indirect Greenhouse Gas Emissions of a Small and Open Economic System." *Energy Policy* 38, 9: 4848–55.

Seabrook, S. 2007. *Cities*. Norwich: Pluto Press.

Seenprachawong, U. 2003. "Economic Valuation of Coral Reefs at Phi Phi Islands, Thailand." *International Journal of Global Environmental Issues* 3, 1: 104–14.

Segrave, M. 2009. "Order at the Border: The Repatriation of Victims of Trafficking." *Women's Studies International Forum* 32: 251–60.

Selvakkumaran, S. and B. Limmeechokchai. 2013. "Energy Security and Co-benefits of Energy Efficiency Improvement in Three Asian Countries." *Renewable and Sustainable Energy Reviews* 20 (April): 491–503.

Setboonsarng, S. 2008. "Global Partnership in Poverty Reduction: Contract Farming and Regional Cooperation." ADB Institute Discussion Paper No. 89, ADBI, Manila. http://www.adbi.org/files/dp89.global.partnership.poverty.reduction.pdf.

Setboonsarng, S., P. Leung, and J. Cai. 2006. "Contract Farming and Poverty Reduction: The Case of Organic Rice Contract Farming in Thailand." ADB Institute Discussion Paper No. 49.

Shackley S., and R. Deanwood. 2003. "Constructing Social Futures for Climate-Change Impacts and Response Studies: Building Qualitative and Quantitative Scenarios with the Participation of Stakeholders." *Climate Research* 24: 71–90.

Shah, K.U, H.B. Dulal, C. Johnson, and A. Baptiste 2013. "Understanding Livelihood Vulnerability to Climate Change: Applying the Livelihood Vulnerability Index in Trinidad and Tobago." *Geoforum* 47: 125–37.

Shamshub, H. 2010. "Enhancing Sustainable Tourism in Thailand: A Policy Perspective." In *Sustainable Production Consumption Systems: Knowledge, Engagement and Practice*, ed. L. Lebel, S. Lorek, and R. Daniel, 211–35. Dordrecht: Springer.

Sheng, L. 2003. "China–ASEAN Free Trade Area: Origins, Developments and Strategic Motivations." Institute of Southeast Asian Studies Working Paper: International Politics & Security Issues Series No. 1, ISEAS, Singapore.

Shepherd, G. 2004. *The Ecosystem Approach: Five Steps to Implementation* Cambridge: IUCN.

Shi, W. 2008. "Rubber Boom in Luang Namtha: A Transnational Perspective." Vientiane: GTZ.

Simon, D. 2008. "Urban Environments: Issues on the Peri-Urban Fringe." *Annual Review of Environment and Resources* 33: 167–85.

Simpson, A. 2007. "The Environment–energy Security Nexus: Critical Analysis of an Energy 'Love Triangle' in Southeast Asia." *Third World Quarterly* 28: 539–54.

Singh, S. 2006a. "Trade in Natural Resources in Stung Treng Province: An assessment of Trade in the MWBP demonstration site in Cambodia." TRAFFIC and MWBP.

Singh, S. 2006b. "Contract farming System in Thailand." *Economic and Political Weekly* 40, 53: 5578–86.

Sirisambhand, N. 1985. "The Forest Colonization Process: Case Studies of Two Communities in North-east and South-east Thailand." In *Agricultural Expansion and Pioneer Settlements in the Humid Tropics*, ed. W. Manshard and W.B. Morgan. Kuala Lumpur: United Nations University.

Skutsch, M.M., M.K. McCall, Bhaskar Karky, E. Zahabu, and G. Peters-Guarin. 2009. *Community Measurement of Carbon Stock Change for REDD. Case Studies on Measuring and Assessing Forest Degradation*. Forest Resources Assessment Programme Working Paper No. 156, FAO, Rome.

Smith T.F., D.C. Thomsen, S. Gould, K. Schmitt, and B. Schlege. 2013. "Cumulative Pressures on Sustainable Livelihoods: Coastal Adaptation in the Mekong Delta." *Sustainability* 5: 228–241. doi: 10.3390/su5010228.

Sneddon, C., and C. Fox. 2012. "Inland Capture Fisheries and Large River Systems: A Political Economy of Mekong Fisheries." *Journal of Agrarian Change* 12: 279–99.

Socialist Republic of Vietnam. 2011a. *Decision: Approval of the National Master Plan for Power Development for the 2011–2020 Period with the Vision to 2030 21 July 2011. No. 1208/QD-TTg*. Hanoi: Office of the Prime Minister.

———. 2011b. *Decision No. 2139/QD-TTg: National Climate Change Strategy 5 December 2011*. Hanoi: Office of the Prime Minister.

Sorapipatana, C., and S. Yoosin. 2011. Life Cycle Cost of Ethanol Production from Cassava in Thailand. *Renewable and Sustainable Energy Reviews* 15, 2: 1343.

Sova, C., A. Chaudhury, A. Helfgott, and C. Corner-Dolloff. 2012. "Community-Based Adaptation Costing: An Integrated Framework for the Participatory Costing of Community-based Adaptation to Climate change in Agriculture." Working Paper No. 16. Cali: CGIAR Research Program on Climate Change, Agriculture and Food Security (CCAFS).

Sovanna, T. 2010. "The Current Status of Renewable Energy, Energy Efficiency Development in Cambodia." Paper presented at EEP Mekong Regional Forum, Oct. 2010, Vientiane, LAO PDR.

Springate-Baginski, O., and E. Wollenberg, eds. 2010. *REDD, Forest Governance and Rural Livelihoods: The Emerging Agenda*. Bogor: CIFOR.

Sriboonchitta, S., and A. Wiboonpoongse. 2008. "Overview of Contract Farming in Thailand: Lessons learned." ADBI Institute Discussion Paper No. 112, ADB Institute, Tokyo.

Stockholm Environment Institute (SEI). 2009. "Insight to Policy SUMERNET Phase 2: A Proposal to Sida from the Stockholm Environment Institute." Bangkok: SEI.

Stork, N.E. et al. 2007. "Tropical Rainforest Canopies and Climate Change." *Austral Ecology* 32: 105–12.

Sturgeon, J.C. 2013. "Cross-border Rubber Cultivation between China and Laos: Regionalization by Akha and Tai Rubber Farmers." *Singapore Journal of Tropical Geography* 34: 70–85.

Sugar Poll. 2012. "Sugarcane Growers and Millers Vote in August to Form Sugar Research Australia (SRA)." http://sugarpoll.com.au/.

Sullivan, C. 2002. "Calculating a Water Poverty Index." *World Development* 30: 1195–210.

Sunderlin, W.D., A.M. Larson, and P. Cronkleton. 2009. "Forest Tenure Rights and REDD+: From Inertia to Policy Solutions." *Realising REDD+: National Strategy and Policy Options*, ed. A. Angelsen, 139–50. Bogor Barat: CIFOR.

Sussangkarn, C. 2006. "CLMV and East Asian integration." *TDRI Quarterly Review* 21: 3–10.

Sverdrup-Jensen, S. 2002. " Fisheries in the Lower Mekong Basin: Status and Perspectives", *MRC Technical Paper No. 6*, Mekong River Commission, Phnom Penh.

Tacoli, C. 2003. "The Links between Urban and Rural Development." *Environment and Urbanization* 15, 1: 3–12.

Tallis, H., P. Kareiva, M. Marvier, and A. Chang. 2008. *An Ecosystem Services Framework to Support Both Practical Conservation and Economic Development*. Proceedings of the National Academy of Sciences of the United States of America (PNAS) 105, 28: 9457–64.

Tapaneeyangkul, P. 2011. "Thailand Government Policy & Initiatives for Sugar Production in the Next 5 Years." Office of the Cane and Sugar Board. http://www.ocsb.go.th/upload/news/fileupload/2094-3397.ppt.

Tapp, N. 2010. "Yunnan: Ethnicity and Economies-Markets and Mobility." *Asia Pacific Journal of Anthropology* 11: 97–110.

TEEB. 2010. *Mainstreaming the Economics of Nature: A Synthesis of the Approach, Conclusions and Recommendations of TEEB*. 2010. The Economics of Ecosystems and Biodiversity (TEEB). http://www.teebweb.org/publication/mainstreaming-the-economics-of-nature-a-synthesis-of-the-approach-conclusions-and-recommendations-of-teeb.

Than, M. 2006. "Intraregional, and Cross-border Economic Cooperation." In *The Mekong Arranged and Rearranged*, ed. M. Diokno and N. Chinh, 141–70. Chiang Mai: Mekong Press.

Thang Nam Do, and J. Bennett. 2005. "An Economic Valuation of Wetlands in Vietnam's Mekong Delta: A Case Study of Direct Use Values in Camau Province." Working Papers No. 05–8. Asia Pacific School of Economics and Government, The Australian National University (ANU), Canberra. http://apseg.anu.edu.au.

———. 2007. "Estimating Wetland Biodiversity Values: A Choice Modelling Application in Vietnam's Mekong River Delta." Economics and Environment Network Working Paper EEN0704, ANU, Canberra.

Thaung, N.N. 2011. "Integration of Myanmar Domestic Agricultural Marketing into ASEAN." NyeinThaungYezin Agricultural University, Pyinmana. http://www.na-businesspress.com/JMPP/ThaungNN_Web12_5_.pdf.

Theeravit, K. 2003. "Relationships Within and Between the Mekong Region in the Context of Globalisation." In *Social Challenges for the Mekong Region*, ed. M. Kaosa-ard and J. Dore, 2nd ed. Bangkok: White Lotus.

Thein, M., and M. Myint. 2008. BIMSTEC-Japan Cooperation in Energy Sector: Myanmar Perspective. CSIRO Discussion Paper No. 39, CSIRO, Canberra.

Thongyou, M., and R. Savangnokg. 2011. "Negotiating Livelihoods in an Urban Hinterland: A Study of Fishing Households at Thung Sang Lake, Khon Kaen Province." In *International Proceedings of Economics Development and Research: Humanities, Society and Culture*, 80–83. Singapore: IACSIT Press.

Thorne, C., G. Annandale, K. Jensen, E. Jensen, A. Green, and J. Koponen. 2011. "Review of Sediment Transport, Morphology, and Nutrient Balance." Report to the Mekong River Commission Secretariat prepared as part of the Xayaburi MRCS Prior Consultation Project Review Report, Nottingham University.

Torres, H., H. Alves, and M. Oliveira. 2007. "Sao Paulo Peri-urban Dynamics: Some Social Causes and Environmental Consequences." *Environment and Urbanization* 19: 207–23.

Torrey, B. 2004. "Urbanization: An Environmental Force to be Reckoned With." *Population Reference Bureau*. http://www.prb.org/articles/2004.

Tourism Authority of Thailand. 2011. *ASEAN Tourism Forum 2011*. http://www.tatnews.org/asean-tourism-forum-2011.

Tran Dang Hong. 2012. "Floods of Dragon Year." http://www.trandang.net/Default.aspx?g=posts&t=1284.

Truong, T.-H., and B. King. 2009. "An Evaluation of Satisfaction Levels among Chinese Tourists in Vietnam." *International Journal of Tourism Research* 11: 521–35.

Trumper, K. et al. 2009. *The Natural Fix? The Role of Ecosystems in Climate Mitigation. A UNEP Rapid Response Assessment*. Cambridge: UNEP and WCMC.

Tsuji, G.Y., G. Uehara, and S. Balas, eds. 1994. *Decision Support System for Agrotechnology Transfer (DSSAT) Version 3*. Honolulu: University of Hawai'i.

Turner, R.K. et.al. 2000. "Ecological–Economic Analysis of Wetlands: Scientific Integration for Management and Policy." *Ecological Economics* 35: 7–23.

Turner, R.K., and G.C. Daily. 2008. "The Ecosystem Services Framework and Natural Capital Conservation." *Environmental Resource Economics* 39, 1: 25–35.

Tyler, S., and L. Fajber. 2009. "Land and Water Resource Management in Asia: Challenges for Climate Adaptation." Winnipeg: International Institute for Sustainable Development.

U Ye Min Aung. 2011. "Private Sector's Perspectives on Myanmar Rice Industry: Addressing Potentials and Opportunities, Challenges and Risks." Myanmar Rice Industry Association (MRIA) Third Development Partnership Forum, Nay Pyi Taw. 27 June 2011. http://www.unescap.org/pdd/calendar/CSN-3DPF/3DP-private-sector-perspectives.pdf.

Üllenberg, A. 2009. *Foreign Direct Investment (FDI) in Land in Cambodia*. Eschborn: GTZ.

United Nations Conference on Trade and Development (UNCTAD). 2013. *UNCTAD Statistical Database*. http://unctadstat.unctad.org/ReportFolders/reportFolders.aspx.

United Nations Department of Economic and Social Affairs (UNDESA). 2012. *World Urbanization Prospects: The 2011 Revision*, CD-ROM.

United Nations Development Programme (UNDP). 2006. *Human Development Report 2006*. New York: UNDP.

———. 2007. *Human Development Report*. New York: UNDP.

———. 2009. *Making the Economic Case: A Primer on the Economic Arguments for Mainstreaming Poverty–Environment Linkages into Development Planning*. UNDP/UNEP Poverty–Environment Initiative. New York: UNDP.

———. 2010a. *The State of the Environment in Myanmar*. Myanmar: http://www.mm.undp.org/UNDP_Publication_PDF/Env%20Sheet.pdf.

———. 2010b. "Improving the Resilience of the Agriculture Sector in Lao PDR to Climate Change Impacts." Project Document. Vientiane: UNDP.

——— and Government of Lao PDR (GoL). 2008. *Millenium Development Goals Progress Report: Lao PDR*. Vientiane: GoL and UNDP.

United Nations Economic and Social Commission for Asia and the Pacific (UNESCAP).

———. 2004. *Trade and Investment Policies for the Development of the Information and Communication Technology Sector of the Greater Mekong Subregion*. Studies in Trade and Investment No. 52. Bangkok: UNESCAP.

———. 2009. *Innovative Socio-economic Policy for Improving Environmental Performance: Payments for Ecosystem Services*. Bangkok: ESCAP.

———. 2011. *Urbanization: Statistical Yearbook for Asia and the Pacific*. Bangkok: ESCAP.

United Nations Educational, Scientific and Cultural Organization (UNESCO). 2013. *Biosphere Reserves: Learning Sites for Sustainable Development*. Ecological Sciences for Sustainable Development. http://www.unesco.org/new/en/natural-sciences/environment/ecological-sciences/biosphere-reserves.

United Nations Environment Programme (UNEP). 2004. *State of Waste Management in South East Asia*. Osaka: UNEP.

———. 2006. *Mekong River, Global International Waters Assessment Mekong River*. Kalmar: GIWA Regional Assessment 55.

———. 2007. *Guidelines for Conducting Economic Valuation of Coastal Ecosystem Goods and Services*. UNEP/GEF/SCS Technical Publication No. 8. Osaka: UNEP.

———. 2009. *Ecofacts: Climate Change in Bangkok*. Osaka: UNEP.

———. 2012. *Promoting Upstream–downstream Linkages Through Integrated Ecosystem Management in the Greater Mekong Subregion*. Osaka: UNEP.

——— and Thailand Environment Institute (TEI). 2007. *Greater Mekong Environmental Outlook*. Bangkok: UNEP and TEI.

——— et al. 2008. *Green Jobs: Towards Decent Work in a Sustainable, Low-Carbon World.* Washington, D.C.: Worldwatch Institute.

United Nations Habitat (UNHABITAT). 2003. *The Challenge of the Slums: Global Report on Human Settlements*. Nairobi: UNHABITAT.

———. 2009. *Global Report on Human Settlements: Planning Sustainable Cities*. London: Earthscan.

United Nations Framework on Climate Change Convention (UNFCCC). CP 16, Cancun Agreements. 2010. Decision 1/CP.16. The Cancun Agreements: Outcome of the work of the Ad Hoc Working Group on Long-term Cooperative Action under the Convention. FCCC/CP/2010/7/Add.1.

United Nations Programme on Reducing Emissions from Deforestation and Forest Degradation (UNREDD). 2010. Cambodia REDD+ Roadmap UN-REDD Programme 5th Policy Board Meeting, Nov. 4–5, Washington, D.C.

United Nations University (UNU), Institute of Advanced Studies (IAS). 2003. *Urban Ecosystems Analysis: Identifying Tools and Methods*. Toyko: IAS.

United States Central Intelligence Agency (CIA). *CIA World Factbook*. https://www.cia.gov/library/publications/the-world-factbook/.

Urothody, A.A., and H.O. Larsen 2010. "Measuring Climate Change Vulnerability: A Comparison of Two Indexes." *Banko Janakari* 20, 1. doi:10.3126/banko.v20i1.3503.

Uyen, N.N., and H. Schnitzera. 2009. "Sustainable Solutions for Solid Waste Management in Southeast Asian Countries." *Waste Management* 29, 6: 1982–95.

Valbo-Jorgensen, J., D. Coates, and K. Hortle. 2009. "Fish Diversity in the Mekong River Basin." In *The Mekong: Biophysical Environment of an International River Basin*, ed. Ian C. Campbell. Amsterdam: Elsevier.

Van den Berg, L.M., M.S. van Vijik,, and P. van Hoi 2003. "The Transformation of Rural Life in Hanoi." *Environment and Urbanization* 15, 1: 35–51.

Van Zalinge et al. 2003. "The Mekong River System." Paper presented at the Second International Symposium on the Management of Large Rivers for Fisheries, Phnom Penh, Feb. 11–14. http://www.ibcperu.org/doc/isis/2954.pdf.

Verbiest, J.-P. A. 2013. "Regional Cooperation and Integration in the Mekong Region." *Asian Economic Policy Review* 8, 1: 148–64.

Vermeulen, S.J. et al. 2013. "Addressing Uncertainty in Adaptation Planning for Agriculture." *Proceedings of the National Academy of Sciences of the United States of America*: 1–6. http://www.ncbi.nlm.nih.gov/pubmed/23674681.

Vietnam Union of Science and Technology Association (VUSTA). 2007. *Assessment of Vietnam Power Development Plan*. Hanoi: VUSTA.

Vind, I. and N. Fold. 2010. "City Networks and Commodity Chains: Identifying Global Flows and Local Connections in Ho Chi Minh City." *Global Networks* 10, 1: 54–74.

Vo Thi, T.L., B. Simon, S. Le Xuan, N. Hap, and Nguyen T.K. 2009. "Value Chains for Sustainable Mekong Fisheries: The Case of *Pangasius hypopthalmus* and *Henicorhynchus/Labiobarbus* in Vietnam and Cambodia." Bangkok: SUMERNET.

Walker, B., L. Gunderson, A. Kinzig, C. Folke, S. Carpenter, and L. Schultz. 2006. "A Handful of Heuristics and Some Propositions for Understanding Resilience in Social-Ecological Systems." *Ecology and Society* 11, 1: 13.

Walker, G. 2012. *Environmental Justice: Concepts, Evidence and Politics.* London: Routledge.

Walsh, J., and F. Amponstira. 2013. "Infrastructure Development and the Repositioning of Power in Three Mekong Region Capital Cities." *International Journal of Urban and Regional Research* 37, 3: 879–93.

Wassmann, R., NX. Hien, C.T. Hoanh, and T.P. Tuong. 2004. "Sea Water Rise Affecting the Vietnamese Mekong Delta: Water Elevation in the Flood Season and Implications for Rice Production." *Climate Change* 66: 89–107.

Watcharejyothin, M., and R.M. Shrestha. 2009. "Regional Energy Resource Development and Energy Security under CO2 Emission Constraint in the Greater Mekong Sub-region Countries." *Energy Policy* 37: 4428–41.

Water Resources and Environment Administration (WREA). 2009. *National Adaptation Programme of Action to Climate Change.* Vientiane: WREA.

Weber, E.U. 2010. "What Shapes Perceptions of Climate Change?" Wiley Interdisciplinary Reviews: *Climate Change,* 1, 3: 332–42.

Webster, D. 2005. "Urbanization: New drivers, new outcomes." In *Thailand: Beyond the Crisis,* ed. P. Warr, 285–314. London: Routledge.

Wenfa, Xiao, Dai Guangcu, and Zhang Sheng. 2010. *China's Strategy and Financing for Forestry Sustainable Development.* Beijing: UN Forum on Forests Ad Hoc Expert Group on Forest Financing Country Case Study.

Wilby, R.L. and C.W. Dawson. 2004. Using SDSM Version 3.1 – A Decision Support Tool for the Assessment of Regional Climate Change Impacts. (Software User Manual).

Wilby, R.L., C.W. Dawson, and E.M. Barrow. 2002. "SDSM: A Decision Support Tool for the Assessment of Regional Climate Change Impacts." *Environmental and Modelling Software* 17: 145–57.

Wong, E., N. Mistilis, and L. Dwyer. 2010. "Understanding ASEAN Tourism Collaboration: The Preconditions and Policy Framework Formulation." *International Journal of Tourism Research* 12: 291–302.

World Bank. 1997. *Environmental Assessment Sourcebook,* No 19. Washington, D.C.: World Bank.

———. 2001–2010. *The Little Green Data Book.* Washington, D.C.: World Bank.

———. 2010. "Economics of Adaptation to Climate Change: Vietnam." Washington, D.C.: World Bank Group.

———. 2012a. *World Development Indicators 2012.* Washington D.C.: World Bank.

———. 2012b. "Improving Service Delivery." In *Thailand: Public Finance Management Report.* Bangkok: World Bank.

World Business Council for Sustainable Development (WBCSD) and World Research Institute (WRI). 2004. The Greenhouse Gas Protocol: A Corporate Accounting and Reporting Standard. Rev. ed. Washington, D.C.: WBCSD and WRI.

World Commission on Environment and Development (WCED). 1987. *Our Common Future: The World Commission on Environment and Development WCED.* Oxford: Oxford University Press.

World Economic Forum (WEF). 2012. *The ASEAN travel and tourism competitiveness report 2012.* Geneva: WEF.

World Food Program (WFP). 2009. *Climate Change and Hunger: Responding to the Challenge*. Rome: WFP.

World Health Organization (WHO), and UNICEF 2010. *Progress on Sanitation and Drinking-water: 2010 Update*. Geneva: WHO and UNICEF.

World Investment Report. 2012. http://unctad.org/en/Pages/DIAE/World%20Investment%20Report/WIR2012_WebFlyer.aspx.

World Tourism Organization (WTO) and United Nations Environment Programme (UNEP) 2008. *Climate Change and Tourism: Responding to Global Challenges*. Madrid: WTO and Paris: UNEP.

World Travel and Tourism Council (WTTC). 2012a. *Travel and Tourism: Economic Impact 2012 Thailand*. London: WTTC.

———. 2012b. *Travel and Tourism: Economic Impact 2012 Vietnam*. London: WTTC.

World Wildlife Fund (WWF) Philippines. 2002. A Survey of Climate Change Impacts in the Philippines. Unpublished report.

———. 2008. *First Contact in the Greater Mekong: Greater Mekong a Biological Treasure Trove: More than 1000 New Species Discovered in a Decade*. http://wwf.panda.org.

———. 2009. The Greater Mekong and Climate Change: Biodiversity, Ecosystem Services and Development at Risk. WWF Greater Mekong Programme.

———. 2010. "Mekong Dams Threaten Rare Giant Fish." http://wwf.panda.org/wwf_news/?194313.

———. 2013. "Ecosystems in the Greater Mekong: Past Trends, Current Status, Possible Futures."

Wu, L.L, J. Lu, C.F. Tong, and C.Q. Liu. 2003. "Valuation of Ecosystem Services in the Yangtze Basin." *Resource and Environmental Economics* 12, 5: 411–16.

Wunder, S., B.D. The, and E. Ibarra. 2005. "Payment is Good, Control is Better: Why Payments for Forest Environmental Services in Vietnam Have So Far Remained Incipient." Bogor Barat: Center for International Forestry Research (CIFOR).

Wunder, S. 2005. "Payment for Environmental Services: Some Nuts and Bolts." CIFOR Occasional Paper No. 42, CIFOR, Bogor Barat.

———. 2007. "The Efficiency of Payments for Environmental Services in Tropical Conservation." *Conservation Biology* 21, 1: 48–58.

Xi, Jiao. 2009. "Valuation of Ecosystem Services in Xishuangbanna Biodiversity Conservation Corridors Initiative Pilot Site, China." Greater Mekong Subregion Core Environment Program Final Draft.

Xie, Y., M. Yu, Y. Bei, and X. Xing. 2006. "Ecological Analysis of an Emerging Urban Landscape Pattern. Desakota: A case study in Suzhou, China." *Landscape Ecology* 21: 1297–309.

Xing, Y. 2010. "China's Exports in Information Communication Technology and Its Impact on Asian Countries." *Econ. Change Restruct*. doi:10.1007/s10644-010-9093-5.

Yamakushi, H., and B. Promphakping. 2007. "Regional Development and Well-being in the Mekong Region." Paper presented at International Conference on Happiness and Public Policy, UN Conference Centre, Bangkok, 18–19 July 2007.

Yap K.S. 2010. "Good Urban Governance in Southeast Asia." *Environment and Urbanization Asia* 1, 2: 131–47.

Yap K.S., and L. Lebel 2009. "Incomplete Urbanization." In *Critical States: Environmental Challenges to Development in Monsoon Southeast Asia,* ed. L. Lebel, A. Snidvongs, C.-T. A. Chen, and R. Daniel, 35–50. Petaling Jaya: SIRD.

Yapa, P.W.S. and K. Jacobs. 2010. "Accounting in Transition in the Transitional Economy: The Case of Cambodia." Sixth Accounting History International Conference (6AHIC), Wellington. http://www.victoria.ac.nz/sacl/about/events/past-conferences/6ahic/publications/6AHIC-113_FINAL_Paper.pdf.

Yu, B., T. Zhu, C. Breisinger, and N.M. Hai. 2010. "Impacts of Climate Change on Agriculture and Policy Options for Adaptation: The Case of Vietnam." IFPRI Discussion Paper 01015, IFPRI, Washington, D.C.

Yu, X. 2003. "Regional Cooperation and Energy Development in the Greater Mekong Sub-region." *Energy Policy* 31: 1221–4.

Yuen, B. 2009. "Revisiting Urban Planning in East Asia, South-east Asia and the Pacific." Regional study prepared for Planning Sustainable Cities. Global Report on Human Settlements 2009. UNHABITAT, http://www.unhabitat.org/grhs/2009.

Yuen, B., and L. Kong 2009. "Climate Change and Urban Planning in Southeast Asia." *SAPIENS* 2, 3. http://sapiens.revues.org/881.

Yusuf, A.A., and H.A. Francisco. 2009. *Climate Change Vulnerability Mapping for Southeast Asia.* Singapore: Economy and Environment Program for Southeast Asia (EEPSEA).

———. 2010. *Hotspots! Mapping Climate Change Vulnerability in Southeast Asia.* Economy and Environment Program for Southeast Asia (EEPSEA). http://www.eepsea.net/pub/book/Coffee_Book_Final_29Sep10.pdf.

Zarin, D.J. 2012. "Carbon from Tropical Deforestation." *Science* 336 (6088): 1518–19.

Ziegler, A.D., J.M. Fox, and J.C. Xu. 2009. "The Rubber Juggernaut." *Science* 324: 1024–5.

Ziegler, A.D. et al. 2013. "Dams and Disease Triggers on the Lower Mekong River." *PLoS Negl Trop Dis* 7(6): e2166: doi:10.1371/journal.pntd.0002166.

Xinhuanet. 2011. "ASEAN-China Free Trade Area." http://news.xinhuanet.com/english2010/china/2011-11/18/c_131254269.htm.

Contributors

AMIT, MAYO GRACE C., Assistant Professor, College of Public Affairs and Development, University of the Philippines Los Baños

AUNG WIN HTUT, Asian Development Research Institute, Myanmar

BACH TAN SINH, Deputy-Director, Research Centre of Science Policy, National Institute of S&T, Policy and Strategy Studies, Hanoi, Vietnam

BOUNTHAVY SOSAMPHANH, Department of Geography, Faculty of Social Sciences, Lao National University, Vientiane, Lao PDR

BUI DUC TINH, Hue College of Economics, Hue University, Vietnam

BUSH, ANGELA L., Faculty of Public Health, Khon Kaen University, Thailand

CAO CHUNMIAO, School of Environment, Renmin University of China, Beijing, China

CHAYANIS KRITTASUDTHACHEEWA, Stockholm Environment Institute–Asia, Bangkok, Thailand

CHETPHONG BUTTHEP, National Research Council of Thailand, Thailand

CHU THAI HOANH, International Water Management Institute, Vientiane, Lao PDR

CLARKE, YVETTE, Social Researcher, Research Center for Disaster Risk Reduction and Climate Change, Asian Management and Development Institute, Hanoi, Vietnam

CORNER-DOLLOFF, CAITLIN, Decision and Policy Analysis Research Area, International Center for Tropical Agriculture, Colombia

DANIEL, RAJESH, Stockholm Environment Institute–Asia, Bangkok, Thailand

Do Xuan Lan, Department of Science, Technology, and Environment, Ministry of Agriculture and Rural Development, Vietnam

Douangkham Singhanouvong, Living Aquatic Resources Research Centre, Lao PDR

Dusita Krawanchid, Stockholm Environment Institute–Asia, Bangkok, Thailand

Eagleton, Graham, Faculty of Public Health, Khon Kaen University, Thailand

Ekawit Meteejaroenwong, School of Environment, Resources and Development, Asian Institute of Technology, Thailand

Elazegui, Dulce D., University Researcher, College of Public Affairs and Development, University of the Philippines Los Baños

Elbert, Eleanor, International Union for Conservation of Nature, Lao PDR

Faderogao, Francis John F., Research Associate, College of Public Affairs and Development, University of the Philippines Los Baños

Featherston, Pippa J., Faculty of Public Health, Khon Kaen University, Thailand

Glemet, Raphael, International Union for Conservation of Nature, Lao PDR

Goteti, Srujana, School of Environment, Resources and Development, Asian Institute of Technology, Thailand

Hap Navy, Inland Fisheries Research and Development Institute, Lao PDR

Hu Tao, World Wildlife Fund, Washington, D.C., United States

Khanh Linh, Hue City Center for International Cooperation, Hue City, Vietnam

Kumar, Sivannapan, School of Environment, Resources and Development, Asian Institute of Technology, Thailand.

Kyoko Kusakabe, Department of Environmental Science, College of Science, Hue University

Lailai Li, Country Director for World Resources Institute, China

LANSIGAN, FELINO P., Professor, College of Arts and Sciences and Co-chair, Interdisciplinary Program on Climate Change, University of the Philippines Los Baños

LEBEL, LOUIS, Unit for Social and Environmental Research, Chiang Mai University, Thailand

MANIEMAI THONGYOU, Centre for Research on Plurality in the Mekong Region, Faculty of Humanities and Social Sciences, Khon Kaen University, Thailand

MEN PRACHVUTHY, Department of Community Development, Faculty of Development Studies, Royal University of Phnom Penh, Cambodia

MIDDLETON, CARL, Lecturer, Master of Arts in International Development Study program, Faculty of Political Science, Chulalongkorn University, Thailand

MOLL-ROCEK, JULIAN, Decision and Policy Analysis Research Area, International Center for Tropical Agriculture

MONCHAI PHONGSIRI, Centre for Research on Plurality in the Mekong Region, Faculty of Humanities and Social Sciences, Khon Kaen University, Thailand

MUANGPONG JUNTOPAS, Stockholm Environment Institute–Asia, Bangkok, Thailand

NGO CONG CHINH, Director of Research Center for Disaster Risk Reduction and Climate Change, Asian Management and Development Institute

NGUYEN DUY CAN, College of Rural Development, Cantho University, Vietnam

NGUYEN HUNG MANH, Researcher, Research Center for Disaster Risk Reduction and Climate Change, Asian Management and Development Institute, Hanoi, Vietnam

NGUYEN TRI KHIEM, Senior Researcher/Lecturer, An Giang University, Vietnam

NGUYEN XUAN GIAP, Forest Science Institute of Vietnam, Vietnam

ORN-UMA POLPANICH, Stockholm Environment Institute–Asia, Bangkok

OUTHAI SOUKKHY, Director, Northern Agriculture and Forestry College, Luang Prabang, Lao PDR

Peñalba, Linda M., Associate Professor, College of Public Affairs and Development, University of the Philippines Los Baños

Penporn Janekarnkij, Faculty of Economics, Kasetsart University, Bangkok

Phong Tran, Hue College of Economics, Hue University, Vietnam

Phung Van Khoa, Forestry University of Vietnam, Vietnam

Pornchai Uttaruk, Faculty of Science, Mahasarakham University, Thailand

Pradhan, Pravakar, School of Environment, Resources and Development, Asian Institute of Technology, Thailand

Sakaradhorn Boontaveeyuwat, Lecturer, Department of Civil Engineering, Faculty of Engineering, Kasetsart University, Si Racha Campus, Thailand

Samek, Jay H., Global Observatory for Ecosystem Services, Forestry Department, Michigan State University

Saykham Voladet, Director of Policy Research Division, National Economic Research Institute, Lao PDR

Seak Sophat, Head, Department of Natural Resource Management and Development, Royal University of Phnom Penh, Cambodia

Shrestha, Pujan, School of Environment, Resources and Development, Asian Institute of Technology, Thailand

Sithong Thongmanivong, Faculty of Forestry, National University of Laos, Lao PDR

Skole, David L., Global Observatory for Ecosystem Services, Forestry Department, Michigan State University, United States

Sok Khom, Cambodian National Mekong Committee, Cambodia

Sopon Naruchaikusol, Stockholm Environment Institute–Asia, Bangkok, Thailand

Tatirose Vijitpan, Stockholm Environment Institute–Asia, Bangkok, Thailand

Teerawong Laosuwan, Faculty of Science, Mahasarakham University, Thailand

THANAPAUGE CHAMARATANA, Institute of Skill Development Regional VI (Khon Kaen), Ministry of Labour, Thailand

TRAN HUU TUAN, Department of Environmental Science, College of Science, Hue University

TRAN ANH TUAN, Hue College of Economics, Hue University, Vietnam

TRINNAWAT SUWANPRIK, Chiang Mai Municipality, Thailand

USA KINHOM, Faculty of Science, Mahasarakham University, Thailand

VO HONG TU, College of Rural Development, Cantho University, Vietnam

WINSTON SET AUNG, Asia Development Organization, Myanmar

YANYONG INMUONG, Faculty of Public Health, Khon Kaen University, Thailand

Index

Page numbers in italics indicate figures, tables, or boxes.